World Conference on Carbon (Carbon 2024)

Carbon for the Future

Shenzhen, China
14-19 July 2024

ISBN: 979-8-3313-1204-6

Printed from e-media with permission by:

Curran Associates, Inc.
57 Morehouse Lane
Red Hook, NY 12571

Printed with permission by Curran Associates, Inc. (2025)

Carbon 2024
c/o Hui-Ming Cheng

cheng@imr.ac.cn

Additional copies of this publication are available from:

Curran Associates, Inc.
57 Morehouse Lane
Red Hook, NY 12571 USA
Phone: 845-758-0400
Fax: 845-758-2633
Email: curran@proceedings.com
Web: www.proceedings.com

Carbon 2024 Shenzhen China
The World Conference on Carbon

Carbon for the Future

The World Conference on Carbon 2024

FORUM BROCHURE

Shenzhen · China

14-19 July, 2024

Thank You to Sponsors and Supporters!

Title Banquet

BTR 贝特瑞

Gold Sponsorship

NGIC Graphene Innovation Center

清华大学深圳国际研究生院 材料与器件检测技术中心 Testing Technology Center of Materials and Devices

Silver Sponsorship

深圳石墨烯创新中心有限公司 SHENZHEN GRAPHENE INNOVATION CENTER CO., LTD.

SIAMC

Regular booth

Dissol

VERDER scientific

MDPI Academic Open Access Publishing since 1996

方大集团 FANGDA GROUP

新威 NEWARE Since 1998

CANRD

清華大學出版社 Tsinghua University Press

東揚精測系統 TOYO Corporation China

中欣新碳（厦门）材料

清能博创 TsingBosch

Carbontech CONFERENCE & EXPOSITION

Boyee博亿

CONTENT

Foreword

The World Conference on Carbon – Carbon 2024
Carbon for the Future “碳”索未来

July 14th to 19th   Shenzhen, China

It is our great honor to announce that **the 2024 World Conference on Carbon (Carbon 2024)** organized by Shenzhen University of Advanced Technology, Tsinghua Shenzhen International Graduate School and Institute of Metal Research, CAS is **held at Tsinghua Shenzhen International Graduate School, University Town of Shenzhen, China, from July 14th to 19th, 2024.**

This congress is the most important one in carbon science and technology. By tradition it takes place in the America, Europe, or Asia, and this year is held, for the third time, in China with the support of the Shenzhen Science Technology and Innovation Bureau. Carbon 2024 has received more than 400 abstracts from over 30 countries. 5 plenary lectures, 60 keynote lectures, 160 oral presentations, and around 120 posters, of excellent scientific quality, will be presented. The next conference for carbon scientists is Carbon 2025 in France, where we hope to converge once again.

It is our great pleasure having all of you to attend Carbon 2024 in Shenzhen!

Prof. Hui-Ming Cheng

Shenzhen Institute of Advanced Technology, CAS, China
Institute of Metal Research, CAS, China
Shenzhen Geim Graphene Center, China

Prof. Feiyu Kang

Tsinghua Shenzhen International Graduate School, China
Shenzhen Geim Graphene Center, China

Conference Committee

Supporting Organization

Shenzhen Science Technology and Innovation Bureau

Organizers

Shenzhen University of Advanced Technology
Tsinghua Shenzhen International Graduate School
Institute of Metal Research, Chinese Academy of Sciences

Co-Organizers

Institute of Technology for Carbon Neutrality, SIAT, CAS
Shenzhen Geim Graphene Center
Guangdong Graphene Innovation Center
Advanced Battery Materials Industry Cluster
Elsevier B.V.
Editorial Office of New Carbon Materials
Shenzhen Power Supply Technology Society

Host

Tsingsun Power Institute of Shenzhen

Co-Chairman

Prof. Hui-Ming Cheng Shenzhen Institutes of Advanced Technology, CAS, Institute of Metal Research, CAS, Shenzhen Geim Graphene Center, China
Prof. Feiyu Kang Tsinghua Shenzhen International Graduate School, Shenzhen Geim Graphene Center, China

Organizing Committee

International Advisory Board

1.Peter. A. Thrower, The Pennsylvania State University, USA
2.A. John Hart, Massachusetts Institute of Technology (MIT), USA
3.Yury Gogotsi, Drexel University, USA
4.Rodney Andrews, University of Kentucky Center for Applied Energy Research, USA
5.Mauricio Terrones, The Pennsylvania State University, USA
6.Robert H. Hurt, Brown University, USA
7.Nikhil Koratkar, Rensselaer Polytechnic Institute, USA
8.Michael Strano, Massachusetts Institute of Technology, USA
9.Paola Barbara, Georgetown University, USA
10.Andre Geim, The University of Manchester, UK
11.Gareth B. Neighbour, The Open University, UK
12.Gao-Qing Lu, University of Surrey, UK
13.Andrea C. Ferrari, University of Cambridge, UK
14.Geoff Fowler, Imperial College London, UK
15.Nathalie Job, University of Liège, Belgium
16.Conchi O. Ania, Centre national de la recherche scientifique (CNRS), France
17.Alberto Bianco, CNRS Alsace Delegation, France
18.Chris Ewels, Centre national de la recherche scientifique (CNRS), France
19.Marc Monthioux, Center for Materials Preparation and Structural Studies (CEMES), France
20.Arndt Peter Schinkel, Orion Engineered Carbons, Germany
21.Shoushan Fan, Tsinghua University, China
22.Zhongfan Liu, Peking University, China
23.Yuliang Li, Institute of Chemistry, CAS, China
24.Jin Zhang, Peking University, China
25.Hejun Li, Northwestern Polytechnical University, China
26.Licheng Ling, East China University of Science and Technology, China
27.Konstantin Novoselov, National University of Singapore, Singapore
28.Sumio IIJIMA, Meijo University, Japan
29.Riichiro Saito, Tohoku University, Japan
30.Hiroaki Hatori, National Institute of Advanced Industrial Science and Technology (AIST), Japan
31.Yutaka Ohno, Nagoya University, Japan
32.Yasushi Soneda, National Institute of Advanced Industrial Science and Technology (AIST), Japan
33.Yoong Ahm Kim, Chonnam National University, Korea
34.Rodney S. Ruoff, Ulsan National Institute of Science and Technology, Korea
35.Chong Rae Park, Seoul National University, Korea
36.Jae-seung Noh, Kumoh Institute of Technology, Korea
37.Yuan Chen, The University of Sydney, Australia
38.Shizhang Qiao University of Adelaide, Australia
39.Emilia Morallon, University of Alicante, Spain

Organizing Committee

International Advisory Board

40.Juan Tascon, Spanish Scientific Research Council, Spain
41.Maria Jesus Lazaro Elorri, Dipartimento di Scienze e Tecnologie Chimiche, Spain
42.Ado Jorio, Federal University of Minas Gerais, Brazil
43.Elzbieta Frackowiak, Poznan University of Technology, Poland
44.Aneta Frączek-Szczypta, AGH University of Krakow, Poland
45.Zulkhair Mansurov, Al-Farabi Kazakh National University, Kazakhstan
46.Rene Rangel-Mendez, Instituto Potosino de Investigación Científica Y Tecnológica, Mexico
47.M.Ferhat YARDIM, Istanbul Technical University, Türkiye
48.Lalit M. Manocha, Inter-University Accelerator Centre, India

Local Organizing Committee

1.Jieshan Qiu, Beijing University of Chemical Technology, China
2.Mao-Sheng Cao, Beijing Institute of Technology, China
3.Jiaxing Huang, Westlake University, China
4.Yongsheng Chen, Nankai University, China
5.Fei Wei, Tsinghua University, China
6.Qingwen Li, Suzhou Institute of Nano-Tech and Nano-Bionics (SINANO), CAS, China
7.Chunxiang Lv, Institute of Coal Chemistry, CAS, China
8.Yan Li, Peking University, China

Secretariat

1.Baohua Li, Tsinghua Shenzhen International Graduate School, China
2.Bilu Liu, Tsinghua Shenzhen International Graduate School, China
3.Yongbing Tang, Shenzhen Institute of Advanced Technology, CAS, China
4.Yan-Bing He, Tsinghua Shenzhen International Graduate School, China
5.Quan-Hong Yang, Tianjin University, China
6.Ruitao Lv, Tsinghua University, China
7.Feng Ding, Shenzhen University of Advanced Technology, China
8.Feng Li, Institute of Metal Research, CAS, China
9.Chang Liu, Institute of Metal Research, CAS, China
10.Da-Wei Wang, Shenzhen University of Advanced Technology, China
11.Ling Qiu, Tsinghua Shenzhen International Graduate School, China
12.Ke Shen, Hunan University, China
13.Yanzhen Liu, Institute of Coal Chemistry, Chinese Academy of Sciences, China
14.Chen Zhang, Tianjin University, China
15.Zhexu Zhang, Tsinghua Shenzhen International Graduate School, China

Organizing Committee

Topic sessions and Chairs

Session 1: Nano carbons: graphene, graphdiyne, CNTs and fullerenes
Prof. Bilu Liu Tsinghua Shenzhen International Graduate School, China
Prof. Chang Liu Institute of Metal Research, CAS, China

Session 2: Graphite, diamond, porous carbon, coal, coke, etc.
Prof. Baohua Li Tsinghua Shenzhen International Graduate School, China
Prof. Ke Shen Hunan University, China

Session 3: Carbon fiber and their composites
Prof. Yan Song Institute of Coal Chemistry, CAS, China
Prof. Ling Qiu Tsinghua Shenzhen International Graduate School, China

Session 4: Carbon for energy storage and conversion
Prof. Quan-Hong Yang Tianjin University, China
Prof. Feng Li Institute of Metal Research, CAS, China
Prof. Hirotomo Nishihara Tohoku University, Japan
Prof. Yan-Bing He Tsinghua Shenzhen International Graduate School, China

Session 5: Carbon for environment
Prof. Ruitao Lv Tsinghua University, China
Prof. Yang Su Tsinghua Shenzhen International Graduate School, China

Session 6: Carbon for biology, health, and medicine
Prof. Dingcai Wu Sun Yat-sen University, China
Prof. Xiaomin Xu Tsinghua Shenzhen International Graduate School, China

Session 7: Carbon for electronics and optoelectronics
Prof. Dongming Sun Institute of Metal Research, CAS, China
Prof. Zhiyong Zhang Peiking University, China
Prof. Guodan Wei Tsinghua Shenzhen International Graduate School, China

Session 8: Carbon 2024 Emerging Young Scientist Symposium
Prof. Da-Wei Wang Shenzhen University of Advanced Technology, China
Prof. Ke Shen Hunan University, China
Prof. Yin Tao Tianjin University, China

Conference Time and Site

Theme	Time	Site
Registration	Sunday July 14th 14:00 - 18:00	First Floor Lobby
Reception Dinner	Sunday July 14th 18:30 - 19:30	2nd Floor Happiness Canteen
Opening and Welcome	Monday July 15th 08:30 - 09:00	Conference Hall Building C
Plenary Lecture 1 Prof. Konstantin Novoselov	Monday July 15th 09:00 - 10:00	
Session 1	Monday July 15th 10:30 - 12:00 14:00 - 16:00	Room 209 Building A
Session 2	Monday July 15th 10:30 - 12:00 14:00 - 16:00	Room 210 Building A
Session 4	Monday July 15th 10:30 - 12:00 14:00 - 16:00	Conference Hall Building C
Session 5	Monday July 15th 10:30 - 12:00 14:00 - 16:00	Room 211 Building A
Poster Session and Coffee Break	Monday July 15th 16:00 - 17:30	First Floor Lobby
Invited Dinner	Monday July 15th 18:00 - 19:00	Sheraton third floor Junior Ballroom
Plenary Lecture 2 Prof. Jin Zhang	Tuesday July 16th 09:00 - 10:00	Conference Hall Building C
Session 1	Tuesday July 16th 10:30 - 12:00 14:00 - 16:00	Room 209 Building A
Session 2	Tuesday July 16th 10:30 - 12:00 14:00 - 16:00	Room 210 Building A
Session 4	Tuesday July 16th 10:30 - 12:00 14:00 - 16:00	Conference Hall Building C
Session 5	Tuesday July 16th 10:30 - 12:00 14:00 - 16:00	Room 211 Building A
Session 8	Tuesday July 16th 10:30 - 12:00 14:00 - 16:00	Room 212 Building A
Poster Session and Coffee Break	Tuesday July 16th 16:00 - 17:30	First Floor Lobby
Carbon Editor Workshop	Tuesday July 16th 16:00 - 18:00	Room 209 Building A

Conference Time and Site

Theme	Time	Site	
Plenary Lecture 3 Prof. Weiming LU	Wednesday July 17th 09:00 - 10:00	Conference Hall Building C	International Building (Section I), Tsinghua SIGS, University Town of Shenzhen
Session 1	Wednesday July 17th 10:30 - 12:00	Room 209 Building A	
Session 2	Wednesday July 17th 10:30 - 12:00	Room 210 Building A	
Session 4	Wednesday July 17th 10:30 - 12:00	Conference Hall Building C	
Session 6	Wednesday July 17th 10:30 - 12:00	Room 213 Building A	
Social Event	Wednesday July 17th 13:30 - 19:00		
Plenary Lecture 4 Prof. Alberco Bianco	Thursday July 18th 09:00 - 10:00	Conference Hall Building C	International Building (Section I), Tsinghua SIGS, University Town of Shenzhen
Session 1	Thursday July 18th 10:30 - 12:00 14:00 - 17:30	Room 209 Building A	
Session 2	Thursday July 18th 10:30 - 12:00 14:00 - 17:30	Room 210 Building A	
Session 3	Thursday July 18th 10:30 - 12:00 14:00 - 17:30	Room 211 Building A	
Session 4	Thursday July 18th 10:30 - 12:00 14:00 - 17:30	Conference Hall Building C	
Session 6	Thursday July 18th 10:30 - 12:00 14:00 - 17:30	Room 213 Building A	
Session 7	Thursday July 18th 10:30 - 12:00 14:00 - 17:30	Room 212 Building A	
Excellent Poster and Best Emerging Young Scientist Oral Presentation Awarding Ceremony Banquet for all registered participants	Thursday July 18th 18:30 - 20:00	Sheraton Shenzhen Nanshan Grand Ballroom	
Plenary Lecture 5 Prof. Seong-Ho Yoon	Friday July 19th 09:00 - 10:00	Conference Hall Building C, International Building (Section I), Tsinghua SIGS, University Town of Shenzhen	
Round Table Dialogue Session	Friday July 19th 10:30 - 11:30		
Upcoming Carbon Conference Announcement	Friday July 19th 11:30 - 11:40		
Closing Remarks	Friday July 19th 11:40 - 11:50		

Report Time: PL Plenary Lecture-45min, KL Keynote Lecture-30min, OP Oral Presentation-20min

Conference Venue Layout

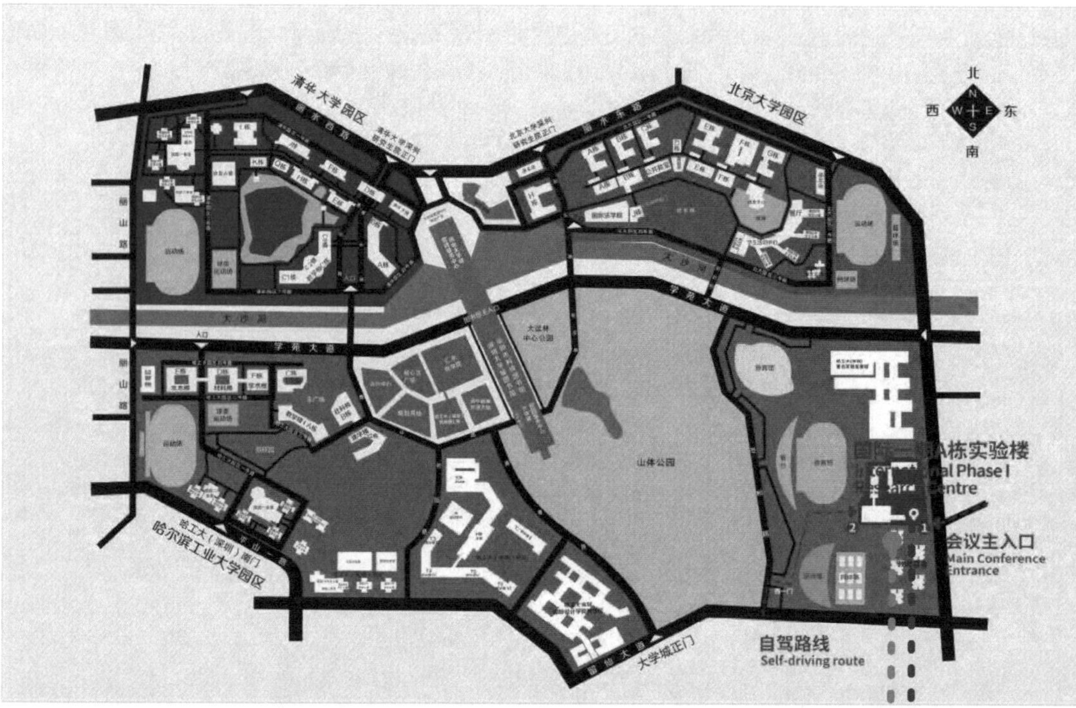

深圳市南山区西丽大学城清华大学深圳国际研究生院国际一期会议室
Conference Room, International Building (Section I), Tsinghua Shenzhen International Graduate School, University Town of Shenzhen, Nanshan District, Shenzhen

主会场 Main Venue	**国际一期A栋报告厅** International Phase I Research Centre
分会一 Session 1	**教室A209** Classroom A209
分会二 Session 2	**教室A210** Classroom A210
分会三 Session 3	**教室A211** Classroom A211
分会四 Session 4	**国际一期A栋报告厅** International Phase I Research Centre
分会五 Session 5	**教室A211** Classroom A211
分会六 Session 6	**教室A213** Classroom A213
分会七 Session 7	**教室A212** Classroom A212
分会八 Session 8	**教室A212** Classroom A212

深圳大学城幸福餐厅
University Towno Shenzhen Happiness Canteen

Conference Venue Layout

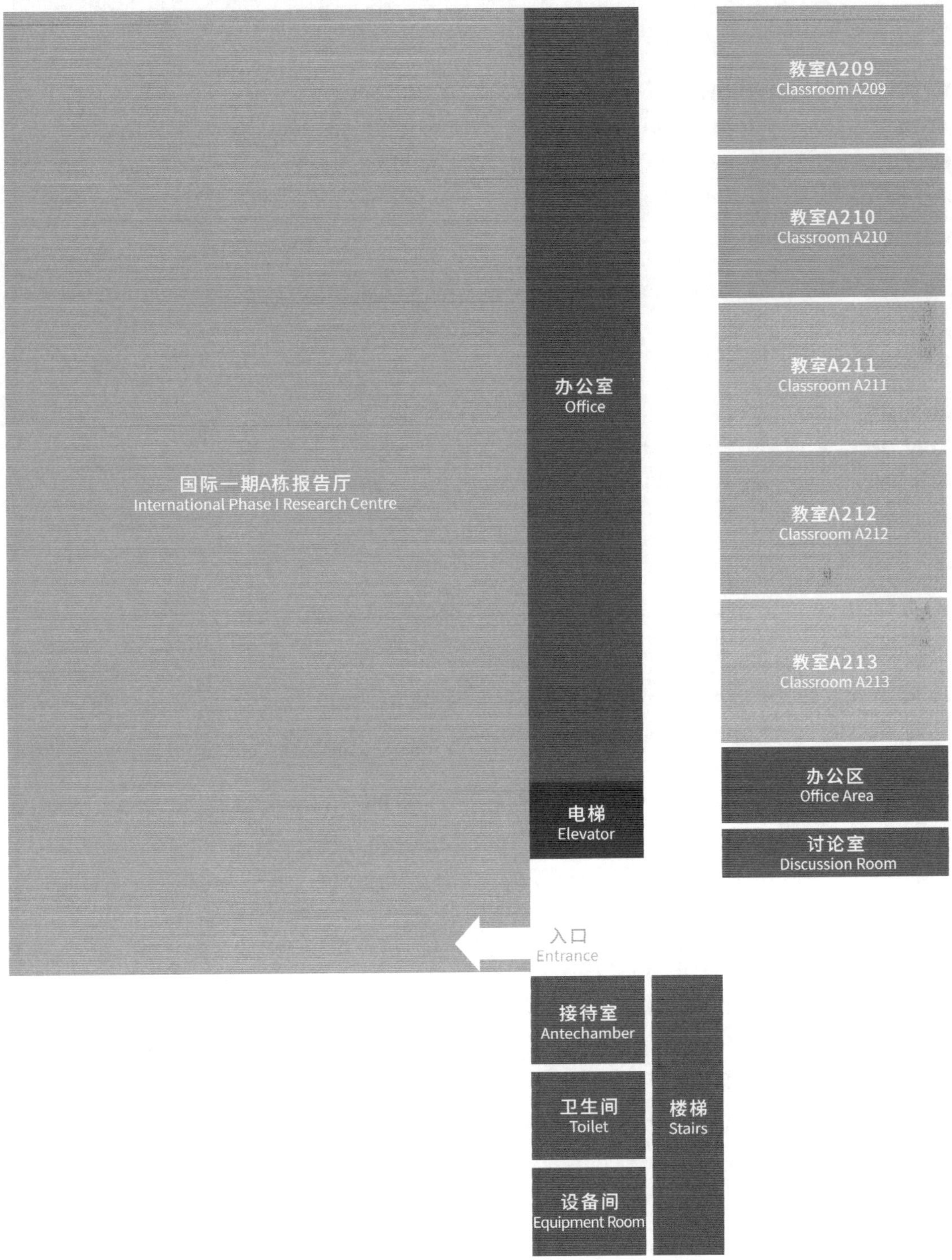

Service Information

Registration

1. Time: 14:00-18:00, 14 July, 2024/08:00-18:00, 15July, 2024.
2. Address: 1st Floor of International Building (Section I), Tsinghua Shenzhen International Graduate School, University Town of Shenzhen, Nanshan District, Shenzhen Attendees can get conference materials, including attending, badges. conference handbook and dining coupons, etc. at the registration desk.

Conference Venue Wi-Fi

Account: tsinghua-guest
Password: Mobile SMS authentication
Due to the bandwidth limit, the Wifi speed may be slow when it is fully loaded.

Time and Location for Meal

July 15-19 Lunch 12:00-13:30
Venue: University Town Shenzhen happiness canteen 1F, 2F, 4F (Buffet)
Distance: about 50 meters on foot
Note: The lunch venue is divided into the first, second and fourth floors. To avoid overcrowding, please pay attention to the dining area on the floor indicated on your voucher.

July 18th 18:30-20:30 (Sign in at 18:00)
Address: Sheraton 3F Grand Ballroom
Transportation: A shuttle bus will be arranged by the conference organizers from Tsinghua University's International Research and Development Institute (IRDI) in Shenzhen to the Sheraton Bolin Tianrui Hotel, a 2.5-kilometre and 10-minute drive from the hotel.

Service Information

Vehicle Arrangement

Date	Time	Specific itinerary arrangement
July 14th	19:30	International Building (Section I), Tsinghua SIGS ➔ Sheraton Shenzhen Nanshan/Vienna International Hotel (Shenzhen University Town)
July 15th	8:00	Sheraton Shenzhen Nanshan/Vienna International Hotel (Shenzhen University Town) ➔ International Building (Section I), Tsinghua SIGS
	18:00	International Building (Section I), Tsinghua SIGS ➔ Sheraton Shenzhen Nanshan/Vienna International Hotel (Shenzhen University Town)
July 16th	8:30	Sheraton Shenzhen Nanshan ➔ International Building (Section I), Tsinghua SIGS/Vienna International Hotel (Shenzhen University Town)
	18:00	International Building (Section I),Tsinghua SIGS ➔ Sheraton Shenzhen Nanshan/Vienna International Hotel (Shenzhen University Town)
July 17th	8:30	Sheraton Shenzhen Nanshan ➔ International Building (Section I), Tsinghua SIGS/Vienna International Hotel (Shenzhen University Town)
	13:00	International Building (Section I),Tsinghua SIGS ➔ Sheraton Shenzhen Nanshan/Vienna International Hotel (Shenzhen University Town)
July 18th	8:30	Sheraton Shenzhen Nanshan ➔ International Building (Section I), Tsinghua SIGS/Vienna International Hotel (Shenzhen University Town)
	18:00	International Building (Section I), Tsinghua SIGS ➔ Sheraton Shenzhen Nanshan
	20:00	Sheraton Shenzhen Nanshan ➔ Vienna International Hotel (Shenzhen University Town)
July 19th	8:30	Sheraton Shenzhen Nanshan ➔ International Building (Section I), Tsinghua SIGS/Vienna International Hotel (Shenzhen University Town)
	13:00	International Building (Section I), Tsinghua SIGS ➔ Shenzhen Bao'an International Airport/Shenzhen North Railway Station

Service Information

Meeting Notice

1. Please take care of all the materials delivered by organizing committee,
including attendee badges, conference handbook, etc. Replacement is not possible if lost.
2. During the forum, please wear the included Forum Pass (badge) with you.
3. During the forum, please turn your mobile phone to vibration mode during the forum.
4. Any questions during the conference please contact the Meeting Affair Service of the organizing committee.

Location for Exhibition and Poster

Site for Exhibition:
Lobby, International Building (Section I), Tsinghua Shenzhen International Graduate School,
University Town of Shenzhen, Nanshan District, Shenzhen

Symposium Secretariat

Zhexu Zhang ,Tsinghua Shenzhen International Graduate School, China
Tel:18038109461 (wechat)
Email:graphene@sz.tsinghua.edu.cn

Conference services

Lin Lixia
Tel: 13510491207 (wechat)
Hongmin Niu
Tel:15099911887 (wechat)
Guangxian Mai
Tel: 13699779087 (wechat)

Conference Agenda

Plenary Lectures

Conference Hall Building C, International Building (Section I), Tsinghua SIGS

Day 1: Opening Ceremony & Plenary Lecture

Time: 08:30-10:30, Monday Morning, 15th July, 2024

Time	Activities
	Chairman: Prof. Hui-Ming Cheng
08:30-09:00	**Opening Ceremony**
09:00-10:00	**PL: Konstantin Novoselov,** National University of Singapore, Singapore **Progress in graphene research**
10:00-10:30	**Coffee Break**

Plenary Lectures

Conference Hall Building C, International Building (Section I), Tsinghua SIGS

Day 2: Plenary Lecture

Time: 09:00-10:30, Tuesday Morning, 16th July, 2024

Time	Activities
	Chairman: Prof. Dingcai Wu
09:00-10:00	**PL: Jin Zhang,** Peking University, China **Growth of single-walled carbon nanotubes with controlled structure**
10:00-10:30	**Coffee Break**

Plenary Lectures

Conference Hall Building C, International Building (Section I), Tsinghua SIGS

Day 3: Plenary Lecture

Time: 09:00-10:30, Wednesday Morning, 17th July, 2024

Time	Activities
Chairman: Prof. Yan Song	
09:00-10:00	**PL: Weiming LU,** Collins Aerospace, an RTX Business, USA **Carbon fiber and Its applications in composites**
10:00-10:30	**Coffee Break**

Plenary Lectures

Conference Hall Building C, International Building (Section I), Tsinghua SIGS

Day 4: Plenary Lecture

Time: 09:00-10:30, Thursday Morning, 18th July, 2024

Time	Activities
Chairman: Prof. Ke Shen	
09:00-10:00	**PL: Alberto Bianco,** Centre National de la Recherche Scientifique (CNRS) Alsace Delegation, France **Are graphitic carbon materials a concern for health?**
10:00-10:30	**Coffee Break**

Plenary Lectures

Conference Hall Building C, International Building (Section I), Tsinghua SIGS

Day 5: Plenary Lecture & Closing Ceremony

Time: 09:00-12:00, Friday Morning, 19th July, 2024

Time	Activities
Chairman: Prof. Feiyu Kang	
09:00-10:00	**PL: Seong-Ho Yoon,** Kyushu University, Japan **Understanding mesophase pitch from a lyotropic liquid crystalline perspective**
10:00-10:30	**Coffee Break**
10:30-11:30	**Round Table Dialogue Session: Carbon for the Future** **Coordinator: Feiyu Kang,** Tsinghua SIGS, China **Invited Guest:** **Mauricio Terrones,** The Pennsylvania State University, USA **Marc Monthioux,** CEMES, CNRS, Université de Toulouse, France **Yuan Chen,** The University of Sydney, Australia **Ken Haenen,** Hasselt University, Belgium **Da-Wei Wang,** Shenzhen University of Advanced Technology, China
11:30-11:40	**Upcoming Carbon Conference Announcement** **Nathalie Job,** Université De Liège, Belgium **Carbon 2026 in France (Vedio)**
11:40-11:50	**Upcoming Carbon Conference Announcement** **Mauricio Terrone,** The Pennsylvania State University, USA **Carbon 2026 in Charleston, SC, USA (PPT)**
11:50-12:00	**Closing Remarks**

Session 1: Nano Carbons: Graphene, Graphdiyne, CNTs and Fullerenes
Room 209 Building A, International Building (Section I), Tsinghua SIGS

Day 1: Keynote Lectures & Oral Presentations & Poster

Time: 10:30-17:30, Monday, 15th July, 2024

Time	Activities
	Chairman: Prof. Bilu Liu
10:30-11:00	**KL: Fabien Vialla,** Université Claude Bernard Lyon 1, CNRS, France **Extreme in-plane strain to engineer folds or sp3 bonds in graphene and few-layer graphite**
11:00-11:20	**OP: You Zeng,** Institute of Metal Research, Chinese Academy of Sciences, China **Graphene-based hierarchical structure for significant enhancement in thermal conductivity of carbon fiber/epoxy composites with high mechanical reinforcement**
11:30-11:50	**OP: Jian Chen,** Sichuan University of Science and Engineering, China **A new discovery on the surface activity of carbon black**
11:50-12:10	**OP: Li Lin,** Peking University, China **Large-area transfer of two-dimensional materials free of cracks, contamination and wrinkles**
12:00-13:30	**Lunch**
	Chairman: Prof. Chang Liu
13:30-14:00	**KL: Dan Li,** The University of Melbourne, Australia **Graphene for emerging iontronics**
14:00-14:30	**KL: Xufeng Zhou,** Ningbo Institute of Materials Technology and Engineering, Chinese Academy of Sciences, National Graphene Innovation Center (NGIC), China **Technical progresses and prospects of graphene @ lithium batteries**
14:30-14:50	**OP: Feng Zhang,** Institute of Metal Research, CAS, China **Efficient controlled growth of single-wall carbon nanotubes and their derivatives**

Session 1: Nano Carbons: Graphene, Graphdiyne, CNTs and Fullerenes
Room 209 Building A, International Building (Section I), Tsinghua SIGS

Day 1: Keynote Lectures & Oral Presentations & Poster

Time: 10:30-17:30, Monday, 15th July, 2024

Time	Activities
	Chairman: Prof. Chang Liu
14:50-15:10	**OP: Kazuyuki Takai,** Hosei University, Japan **Tuning structure and electron transport properties of graphene by chemical modification using ion-beam irradiation**
15:10-15:30	**OP: Zheng Wang,** Institute of Coal Chemistry, CAS, China **β-resin cross-linking graphitization for highly thermally conductive graphene film**
15:30-15:50	**OP: Lili Zhang,** Institute of Metal Research, Chinese Academy of Sciences, China **Atomic Origin of the Axis-Symmetrical Breaking of a Growing Single-Wall Carbon Nanotube**
16:00-17:30	**Poster Session and Coffee Break**

Session 1: Nano Carbons: Graphene, Graphdiyne, CNTs and Fullerenes
Room 209 Building A, International Building (Section I), Tsinghua SIGS

Day 2: Keynote Lectures & Oral Presentations & Poster
Time: 10:30-17:30, Tuesday, 16th July, 2024

Time	Activities
	Chairman: Prof. Pengzhan Sun
10:30-11:00	**KL:Wencai Ren,** Institute of Metal Research, Chinese Academy of Sciences, China **Graphene oxide: Synthesis, assembly and film applications**
11:00-11:30	**OP: Xin Zhao,** Shenzhen Yick Xin Tech LTD. Co., China **Applications of vertical graphene thin-flim materials for bio-electronics, sensors and electrochemical energy devices**
11:30-11:50	**OP: Zhong-Hai Ji,** Institute of Metal Research CAS, China **'Pre-dissolving carbon' in metal catalysts for growing high-crystallinity vertically aligned carbon nanotube arrays**
12:00-13:30	**Lunch**
	Chairman: Prof. Junying Wang
13:30-13:50	**OP: Xin Li,** Institute of Metal Research CAS, China **Metallic-enriched SWCNTs grown from CoxRe1-x by quasi-static CVD method**
13:50-14:10	**OP: Pengzhan Sun,** University of Macau, China **Molecular and ionic transport through graphene membranes**
14:10-14:30	**OP: Zhaofan Zhou,** Northwestern Polytechnical University, China **Construction and differential growth mechanism of uniform and controllable CNTS and CNWS by zif-67 precursor**

Session 1: Nano Carbons: Graphene, Graphdiyne, CNTs and Fullerenes
Room 209 Building A, International Building (Section I), Tsinghua SIGS

Day 2: Keynote Lectures & Oral Presentations & Poster
Time: 10:30-17:30, Tuesday, 16th July, 2024

Time	Activities
Chairman: Prof. Junying Wang	
14:30-14:50	**OP: Qing Zhang,** Tsinghua Shenzhen International Graduate School, China **Fabrication of high-quality graphene films and mechanism of graphitization promoted by lattice defects**
14:50-15:10	**OP: Songfeng Pei,** Institute of Metal Research, Chinese Academy of Sciences, China **The role of water on the electrochemical synthesis of graphene oxide**
16:00-17:30	**Poster Session and Coffee Break**

Session 1: Nano Carbons: Graphene, Graphdiyne, CNTs and Fullerenes
Room 209 Building A, International Building (Section I), Tsinghua SIGS

Day 3: Keynote Lectures & Oral Presentations

Time: 10:30-12:00, Wednesday Morning, 17th July, 2024

Time	Activities
	Chairman: Dr. Javier Quílez-Bermejo
10:30-11:00	**KL: Rong Xiang,** Zhejiang University, China **One-dimensional van der waals heterostructures**
11:00-11:30	**KL: Yuan Chen,** The University of Sydney, Australia **Mapping and Alleviating Fe-N-C Catalyst Degradation**
11:30-11:50	**OP: Zhao-Qing Gao,** Institute of Metal Research, Chinese Academy of Sciences, China **Interfacial enhancing of carbon nanotube fiber with Cu shell for future lightweight cables**
12:00-13:30	**Lunch**
14:00-17:30	**Social Event**

Session 1: Nano Carbons: Graphene, Graphdiyne, CNTs and Fullerenes
Room 209 Building A, International Building (Section I), Tsinghua SIGS

Day 4: Keynote Lectures & Oral Presentations

Time: 10:30-17:30, Thursday, 18th July, 2024

Time	Activities
	Chairman: Prof. Chang Liu
10:30-11:00	**KL: Javier Quílez-Bermejo,** Paul Scherrer Institut (PSI), Switzerland **Electron density transfer from encapsulated Fe3C to the outermost n-doped carbon layer for superior ORR**
11:00-11:20	**OP: Marc Monthioux,** CEMES, CNRS, Université de Toulouse, France **One-side hydrogenation of single-layer graphene: etching or only hydrogen chemisorption?**
11:20-11:40	**OP: Beibei Wang,** Xi' an Aeronautical Institute, China **Grafting CNTS on carbon fabrics with enhanced mechanical and thermal properties for tribological applications of carbon fabrics/ phenolic composites**
12:00-13:30	**Lunch**
	Chairman: Prof. Marc Monthioux
13:30-14:00	**KL: Junying Wang,** Institute of Coal Chemistry, CAS, China **Graphene-based catalytic or multi-functional materials via interface engineering**
14:00-14:20	**OP: Lei Shi, Sun Yat-sen University, China** **Confined synthesis of graphene nanoribbons inside single-walled carbon nanotubes**
14:20-14:40	**OP: Bo Tian,** Nanyang Technological University, Singapore **Wafer-Scale Single-Crystal 2D Materials Growth on Insulating Substrates**
15:00-15:30	**Coffee Break**
18:30-20:00	**Excellent Poster and Best Emerging Young Scientist Oral Presentation Awarding Ceremony** **Banquet for all registered participants**

Session 2: Graphite, Diamond, Porous carbon, Coal, Coke, etc.
Room 210 Building A, International Building (Section I), Tsinghua SIGS

Day 1: Keynote Lectures & Oral Presentations & Poster

Time: 10:30-17:30, Monday, 15th July, 2024

Time	Activities
Chairman: Prof. Baohua Li	
10:30-11:00	**KL: Marc Monthioux,** CEMES, CNRS, Université de Toulouse, France **A protocol for accurately describing carbon materials**
11:00-11:30	**KL: Zhoutong He,** Shanghai Institute of Applied Physics, Chinese Academy of Sciences, China **Assessing the compatibility and irradiation behaviors of nuclear graphite for molten salt reactors**
11:30-11:50	**OP: Xun Zhang,** University of Manchester, UK **In situ mechanical testing of IG-110 graphite for high-temperature gas-cooled reactors**
11:50-12:10	**OP: Guma Yeli,** Tsinghua University, China **Study on oxidation rates and gas product for matrix fuel element under high temperature steam oxidation**
12:00-13:30	**Lunch**
Chairman: Prof. Marc Monthioux	
13:30-14:00	**KL: Ke Shen,** Hunan University, China **The irradiation defect and irradiation induced dimensional change of graphite crystal**
14:00-14:30	**KL: Toshihira Irisawa,** Gifu University, Japan **Development of circularly designed CFRPs and sustainable carbon fibers**
14:30-14:50	**OP: Minghao Li,** Kyushu University, Japan **129Xe NMR study of adsorption interaction between porous carbons and Xe molecules**

Session 2: Graphite, Diamond, Porous carbon, Coal, Coke, etc.
Room 210 Building A, International Building (Section I), Tsinghua SIGS

Day 1: Keynote Lectures & Oral Presentations & Poster
Time: 10:30-17:30, Monday, 15th July, 2024

Time	Activities
	Chairman: Prof. Marc Monthioux
14:50-15:10	**OP: Qifan Zhong,** Central South University, National Engineering Research Center of Low-carbon Nonferrous Metallurgy, China **Microstructural representations and reaction mechanisms during multiple high-temperature processes of complex amorphous carbon materials**
15:10-15:30	**OP: Kui Wang,** Hefei university of technology, CAS, China **Elucidating the intrinsic core-shell structure of carbon nanospheres from glucose hydrothermal carbonization**
15:30-15:50	**OP: Ding Nan,** Inner Mongolia University, China **Graphene oxide reinforced epoxy nanocomposites coatings with superior corrosion resistant and mechanical performance**
15:50-16:10	**OP: Zhenxing Zhu,** Tsinghua University, China **Molecular evolutionary growth and assembly of ultrapure semiconducting carbon nanotubes**
16:10-17:30	**Poster Session and Coffee Break**

Session 2: Graphite, Diamond, Porous carbon, Coal, Coke, etc.
Room 210 Building A, International Building (Section I), Tsinghua SIGS

Day 2: Keynote Lectures & Oral Presentations & Poster

Time: 10:30-17:30, Tuesday, 16th July, 2024

Time	Activities
Chairman: Prof. Oliver Williams	
10:30-11:00	**KL: Mingguang Yao,** Jilin University, China **New structures from high pressure densed carbon phases**
11:00-11:30	**KL: Ken Haenen,** Hasselt University, Belgium **Doping of diamond: from nanocrystalline to monocrystalline CVD diamond films**
11:30-11:50	**OP: Lantao Liu,** Petrochemical Research Institute, Petrochina Company Llimited, China **Preparation of graphitized mesocarbon microbeads as anode material for lithium-ion battery by conductive graphene induction**
11:50-12:10	**OP: Mamta Devi,** Indian Institute of Technology Mandi, India **Investigating the relationship between nitrogen content and the structural and electronic properties in nitrogen-carbon (n-c) materials**
12:00-13:30	**Lunch**
Chairman: Prof. Mingguang Yao	
13:30-14:00	**KL: Yan Song,** Institute of Coal Chemistry, Chinese Academy of Sciences, China **Investigation of pyrolysis behavior of pitch and preparation of pitch-based functional carbon materials**
14:00-14:30	**KL: Oliver Williams,** Cardiff University / Cardiff Diamond Foundry, UK **Making a miniature sun: Ultra high temperature carbon devices**
14:30-14:50	**OP: Jian Xiao,** China University of Mining and Technology, China **Preparation of coal tar pitch-based multidimensional carbon materials and their light metal anode performance**

Session 2: Graphite, Diamond, Porous carbon, Coal, Coke, etc.
Room 210 Building A, International Building (Section I), Tsinghua SIGS

Day 2: Keynote Lectures & Oral Presentations & Poster

Time: 10:30-17:30, Tuesday, 16th July, 2024

Time	Activities
Chairman: Prof. Mingguang Yao	
14:50-15:10	**OP: Chuanjun Tu,** Hunan University, China **Favorable component regulation of coke for high-performance graphite blocks**
15:10-15:30	**OP: Rui Tang,** Hunan University, China **Capacitance of porous carbon materials unraveled by graphene mesosponge**
15:30-15:50	**OP: Wataru Takarada,** Shinshu University, Japan **Structure control of carbon films by addition of graphene oxide in the precursor film**
16:00-17:30	**Poster Session and Coffee Break**

Session 2: Graphite, Diamond, Porous carbon, Coal, Coke, etc.
Room 210 Building A, International Building (Section I), Tsinghua SIGS

Day 3: Keynote Lectures & Oral Presentations

Time: 10:30-12:00, Wednesday Morning, 17th July, 2024

Time	Activities
Chairman: Prof. Ken Haenen	
10:30-11:00	**KL: Quangui Guo,** Institute of Coal Chemistry, CAS, China **Microstructure and Properties of bulk high-density isotropic graphite derived from MCMB**
11:00-11:20	**OP: Xiaoxiao Qu,** Henan Polytechnic University, China **Controllable preparation and application of coal-based functional carbon materials in energy storage**
11:20-11:40	**OP: Xiazhe Zhai,** Kyushu University, Japan **Influence of oxygen-containing surface functional groups on fluorination of porous carbons**
11:40-12:00	**OP: Gaofeng Shao,** Nanjing University of Information Science and Technology, China **Multiscale structural engineering of carbon aerogels for broadband electromagnetic wave absorption**
12:00-13:30	**Lunch**
14:00-17:30	**Social Event**

Session 2: Graphite, Diamond, Porous carbon, Coal, Coke, etc.
Room 210 Building A, International Building (Section I), Tsinghua SIGS

Day 4: Keynote Lectures & Oral Presentations

Time: 10:30-17:30, Thursday, 18th July, 2024

Time	Activities
	Chairman: Prof. Z.A. Mansurov
10:30-11:00	**KL: John Texter,** Eastern Michigan University, USA **Concentrated nanocarbon dispersions in water stabilized by polymerized ionic liquids**
11:00-11:20	**OP: Qi Han,** Dalian University of Technology, China **Hard carbons derived from different condensation degree COPNA resin and their sodium ions storage performances**
11:20-11:40	**OP: Yuhang Xie,** Dalian University of Technology, China **Heavy organic matters based hard carbon anodes via mechanochemical method and their sodium storage performance**
11:40-12:00	**OP: Kening Xing,** Dalian University of Technology, China **Peg template hypercrosslinked coal liquefaction asphalt based closed-pore hard carbon anode for sodium ions batteries**
12:00-13:30	**Lunch**
	Chairman: Prof. Yan Song
13:30-14:00	**KL: Alex Theodosiou,** University of Manchester, UK **Investigating graphite oxidation via real-time offgas analysis**
14:00-14:20	**OP: Z.A. Mansurov,** Institute of Combustion Problems, Kazakhstan **Efficient combustion of fixed coal layer in the advanced chamber of the low-power boiler**
14:20-14:40	**OP: Zheng-Long Xu,** The Hong Kong Polytechnic University, China **Solvated-metal-ion cointercalation chemistry in graphite**

Session 2: Graphite, Diamond, Porous carbon, Coal, Coke, etc.

Room 210 Building A, International Building (Section I), Tsinghua SIGS

Day 4: Keynote Lectures & Oral Presentations

Time: 10:30-17:30, Thursday, 18th July, 2024

Time	Activities
Chairman: Prof. Yan Song	
14:40-15:00	**OP: Xining Zang,** Tsinghua University, China **Laser extreme manufacturing of nanocarbon materials from low rank carbonaceous materials (coal, tar, pitch, and beyond)**
15:00-15:30	**Coffee Break**
18:30-20:00	**Excellent Poster and Best Emerging Young Scientist Oral Presentation Awarding Ceremony** **Banquet for all registered participants**

Session 3: Carbon Fiber and Their Composites

Room 211 Building A, International Building (Section I), Tsinghua SIGS

Day 4: Keynote Lectures & Oral Presentations

Time: 10:30-17:30, Thursday, 18th July, 2024

Time	Activities
	Chairman: Prof. Chong Ye
10:30-11:00	**KL: Koji Nakabayashi,** Kyushu University, Japan **Research on the production of carbon materials using stranded resources**
11:00-11:30	**OP: Sora Lee,** Korea Institute of Science and Technology, Korea **Calcium compound-assisted catalytic graphitization to improve crystallinity of pan-based carbon fibers**
11:30-11:50	**OP: Sharma Prabhjot,** Lovely Professional University, India **Fatigue behavior in CFRP composites: Unravelling the complex relationship of cyclic loading and structural performance**
12:00-13:30	**Lunch**
	Chairman: Prof. Ling Qiu
13:30-14:00	**KL: Chao Gao,** Zhejiang University, China **Graphene fiber and its multifunctional composites**
14:00-14:30	**KL: Chong Ye,** Hunan University, China **Advancements in high thermal conductive carbon fibers for enhanced phase change energy storage**
14:30-14:50	**OP: Bhavika Chouhan,** Indian Institute of Technology Mandi, India **Inducing inherent oxidation resistance in carbon fibers by incorporation of oxidation inhibitors in the precursor**
14:50-15:10	**OP: Motoki Tanaka,** Tokyo Institute of Technology, Japan **Changes of structure and compressive strength of carbon fibers with fatigue**
15:10-15:30	**Coffee Break**

Session 3: Carbon Fiber and Their Composites

Room 211 Building A, International Building (Section I), Tsinghua SIGS

Day 4: Keynote Lectures & Oral Presentations

Time: 10:30-17:30, Thursday, 18th July, 2024

Time	Activities
	Chairman: Prof. Chao Gao
15:30-16:00	**KL Qingwen Li,** Suzhou Institute of Nano-Tech and Nano-Bionics, CAS, China **Scalable manufacturing technology for carbon nanotube fibers**
16:00-16:20	**OP: Shu-Huei Hsieh,** National Formosa University **Impact of various carbon fiber types on the properties of Pb-C composites**
16:20-16:40	**OP: Jianxiao Yang,** Hunan University, China **Developing strategy of low-cost mesophase pitch-based carbon fibers with high performance**
16:40-17:00	**OP: Jiahui Lei,** Zhejiang University, China **Upcycling waste wind turbine blades into flash graphene**
17:00-17:20	**OP: Yan Yang,** Institute of Coal Chemistry, CAS, China **Synthesis and structural evolution of coal-based mesophase pitch by hydrogenation-vacuum thermal poly condensation**
18:30-20:00	**Excellent Poster and Best Emerging Young Scientist Oral Presentation Awarding Ceremony** **Banquet for all registered participants**

Session 4: Carbon for Energy Storage and Conversion

Conference Hall Building C, International Building (Section I), Tsinghua SIGS

Day 1: Keynote Lectures & Oral Presentations & Poster

Time: 10:30-17:30, Monday, 15th July, 2024

Time	Activities
	Chairman: Prof. Feng Li
10:30-11:00	**KL: Hirotomo Nishihara,** Tohoku University, Japan **Seamless 3D Graphene Framework with Enriched Topological Defects for Battery-Related Applications**
11:00-11:15	**OP: Kun Ni,** University of Science and Technology of China, China **Design and Modulation of Novel Carbon Materials**
11:15-11:30	**OP: Masaaki Yoshikawa,** Kyoto University, Japan **Porous Carbons for Positive Electrode of Zn-Carbon Rechargeable Battery**
11:30-11:45	**OP: Biao Zhang,** Hong Kong Polytechnic University, China **Structural and Interfacial Engineering of Carbon Materials for Advanced Potassium-ion Batteries**
11:45-12:00	**OP: Zhenyu Guo,** Imperial College London, UK **Hard Carbon Promises a Superior Anode in Sodium-Ion Batteries**
12:00-12:15	**OP: Zonglin Yi,** Institute of Coal Chemistry, CAS, China **Prediction of Electrochemical Reaction Potentials in Energy Storage Devices Regarding Carbon Materials**
12:15-13:30	**Lunch**
	Chairman: Prof. Hirotomo Nishihara
13:30-14:00	**KL: Jianguo Ren,** BTR New Material Group Co., Ltd., China **The Application and Development Direction of Carbon Materials in Lithium-ion Batteries**
14:00-14:15	**OP: Hae Ri Lee,** Korea Institute of Science and Technology, Korea **Bifunctional Role of Nitrogen-Doped Carbon Quantum Dots for CoS2/ Carbon Composites in Performance Enhancement of Li-ion Batteries**

Session 4: Carbon for Energy Storage and Conversion

Conference Hall Building C, International Building (Section I), Tsinghua SIGS

Day 1: Keynote Lectures & Oral Presentations & Poster

Time: 10:30-17:30, Monday, 15th July, 2024

Time	Activities
	Chairman: Prof. Hirotomo Nishihara
14:15-14:30	**OP: Jian Wang,** Taiyuan University of Technology, China **Advanced Multielement (O, N, S) Modified Strategies for Pitch-Based Carbon Anodes toward Both Lithium and Sodium Storage**
14:30-14:45	**OP: Vladimir Pavlenko,** Institute of Combustion Problems, Kazakhstan **Study on Effect of Carbon Nanomaterials as Conductive Additives in EDLCs**
14:45-15:00	**OP: Yidan Cao,** Tsinghua Shenzhen International Graduate School, China **Ultralight and Dual-Conductive Carbon-Based Anode Current Collector for Anode-Free Lithium Metal Battery**
15:00-15:15	**Coffee Break**
	Chairman: Prof. Kun Ni
15:15-15:45	**KL: Qiang Zhang,** Tsinghua University, China **The Promotion of Emerging Carbon Materials for Next-Generation Batteries through Lithium Bond Chemistry**
15:45-16:00	**OP: Gwanwon Lee,** Korea Institute of Science and Technology, Korea **Selective Synthesis of Soft and Hard Carbons from a Single Precursor Through Tailor-Made Stabilization for Anode in Sodium-ion Batteries**
16:00-16:15	**OP: Fang-Yuan Su,** Institute of Coal Chemistry, CAS, China **Structure Regulation and Electrochemical Performances of Hard Carbon Anode Materials for Sodium-ion Batteries**
16:15-17:30	**Poster Session and Coffee Break**

Session 4: Carbon for Energy Storage and Conversion

Conference Hall Building C, International Building (Section I), Tsinghua SIGS

Day 2: Keynote Lectures & Oral Presentations & Poster

Time: 10:30-17:30, Tuesday, 16th July, 2024

Time	Activities
Chairman: Prof. Jarrn-Horng Lin	
10:30-11:00	**KL: Nianjun Yang,** Hasselt University, Belgium **Diamond Materials for Electrochemical Energy Applications**
11:00-11:15	**OP: Jun Zhang,** NaCun (Tianjin) Technology Co., Ltd., China **Sieving carbon anodes power the high-energy sodium-ion batteries**
11:15-11:30	**OP: Meruyert Nazhipkyzy,** Institute of Combustion Problems, Kazakhstan **Cucumber Peel Derived Porous Carbons for Electrochemical Electrodes**
11:30-11:45	**OP: Junzhong Wang,** Anhui University, China **Graphene Materials: Green Synthesis and Robust Energy Conversion**
11:45-12:00	**OP: Gabriel Alemany-Molina,** University of Alicante, Spain **Pt Nanoparticles for Improving the Performance and Durability of FeNx Based Materials towards ORR in ADMFCs**
12:00-13:30	**Lunch**
Chairman: Prof. Nianjun Yang	
13:30-14:00	**KL: Jarrn-Horng Lin,** National University of Tainan **Biomass Wastes Derived Hierarchical Porous Carbons Used for High-Performance Electric Double-Layer Capacitors**
14:00-14:15	**OP: Huanxin Li,** University of Oxford, UK **New Carbon Allotropes: Graphene-P-Phenyl-Graphene**

Session 4: Carbon for Energy Storage and Conversion

Conference Hall Building C, International Building (Section I), Tsinghua SIGS

Day 2: Keynote Lectures & Oral Presentations & Poster

Time: 10:30-17:30, Tuesday, 16th July, 2024

Time	Activities
Chairman: Prof. Nianjun Yang	
14:15-14:30	**OP: Xinran Yang,** Osaka University, Japan **Facile Synthesis of S-Doped Co2P Nanoparticles on A P, S Co-Doped Carbon Substrate by Using Ion Exchange Resin for Hydrogen Evolution Reaction**
14:30-14:45	**OP: Ang Li,** Beijing University of Chemical Technology, **ChinaMicrostructures and energy conversion performance of carbon materials**
14:45-15:00	**OP: Xiaowen Wu,** China University of Geosciences (Beijing), China **Aligned Channel Gelatin@Nanographite Aerogel Supported Form-Stable Phase Change Materials for Solar-Thermal Energy Conversion and Storage**
15:00-15:15	**Coffee Break**
Chairman: Prof. Fangyuan Su	
15:15-15:45	**KL: Fei Xu,** Northwestern Polytechnical University, China **Design of Polymeric Derived Porous Carbon Spheres with in-situ Nanoconfinement for Energy Storage**
15:45-16:00	**OP: Yangming Lin,** Fujian Institute of Research on The Structure of Matter, CAS, China **Mechanism Understanding of Metal-Free Carbon-Based Electrocatalysis with Model Catalysts**
16:00-16:15	**OP: Gabriel Alemany-Molina,** University of Alicante, Spain **Synthesis of Fe-Nx Based Electrocatalysts from Different Biomass Residues by A Single Heat Treatment and Avoiding the Washing Step**
16:15-17:30	**Poster Session and Coffee Break**

Session 4: Carbon for Energy Storage and Conversion

Conference Hall Building C, International Building (Section I), Tsinghua SIGS

Day 3: Keynote Lectures & Oral Presentations

Time: 10:30-12:00, Wednesday Morning, 17th July, 2024

Time	Activities
Chairman: Prof. Huanxin Li	
10:30-11:00	**KL: Jilei Liu,** Hunan University, China **Differential Design Strategies for High-Capacity Hard Carbon Anodes in Lithium, Sodium, Potassium-ion Batteries**
11:00-11:15	**OP: Zhan Liang,** East China University of Science and Technology, China **Structural Design of Carbonaceous Porous Materials for Lithium-Sulfur Battery and Their Electrochemical Behavior**
11:15-11:30	**OP: Paula Riquelme-García,** University of Alicante, Spain **Exploring the Effects Behind the Outstanding Catalytic Performance of PdAg Catalysts Supported on Almond Shell-Derived Activated Carbon towards the Dehydrogenation of Formic Acid**
11:30-11:45	**OP: Fang Yi,** Sun Yat-Sen University, China **Carbonized Polymer Enhanced Electrochemical Performance of T_3C_2Tx MXene**
11:45-12:00	**OP: Jiaojing Shao,** Guizhou University, China **Design and preparation of multifunctional two-dimensional layered film barrier for lithium sulfur batteries**
12:00-12:15	**OP: Hironori Ogata,** Hosei University, Japan **Electrocatalytic Properties of Carbon Nanotube-Based Metal-Free Nanocomposites**
12:00-13:30	**Lunch**
14:00-17:30	**Social Event**

Session 4: Carbon for Energy Storage and Conversion
Conference Hall Building C, International Building (Section I), Tsinghua SIGS
Day 4: Keynote Lectures & Oral Presentations
Time: 10:30-17:30, Thursday, 18th July, 2024

Time	Activities
Chairman: Prof. Biao Zhang	
10:30-11:00	**KL: Shichao Wu,** Tianjin University, China **Carbon Marries Silicon for High-Performance Anodes**
11:00-11:15	**OP: Jyoti Shikhar,** Indian Institute of Technology Mandi, India **Fabrication of Non-Graphitizing Carbon Electrodes for Sodium-Ion Batteries**
11:15-11:30	**OP: Zipeng Jiang,** Qinghai University, China **Highly Uniform Distribution of Metallic Lithium and Carbon in Nanoscale for Highly Stable Carbon-Based Lithium Metal Anode**
11:30-11:45	**OP: Nathalie Job,** Université De Liège, Belgium **CVD-Coated Carbon Xerogels with Various Nodule Size for High Performance Na-Ion Battery Negative Electrode**
11:45-12:00	**OP: Yilin Wang,** Institute of Coal Chemistry, CAS, China **Releasing Free Radicals in Precursor Triggers the Formation of Closed Pores in Hard Carbon for Sodium-Ion Batteries**
12:00-12:15	**OP: Shaohong Liu,** Sun-Yat Sen University, China **Controllable Preparation of Novel Polymer-based Porous Carbon Materials for Electrochemical Energy Storage**
12:00-13:30	**Lunch**
Chairman: Prof. Jiaojing Shao	
13:30-14:00	**KL: Nikolay G. Prikhodko,** Institute of Combustion Problems, Kazakhstan **Increasing The Electrochemical Properties of Supercapacitors Due to Graphene Deposited in A Flame on A Nickel Current Collector and Active Graphene-Like Carbon Material from Plant Biomass**
14:00-14:15	**OP: Zoya V. Bobyleva,** Lomonosov Moscow State University, Russia **Pretreatment Strategies for Saccharide-Based Hard Carbon Anodes with Advanced Performance in Sodium-ion Batteries**

Session 4: Carbon for Energy Storage and Conversion

Conference Hall Building C, International Building (Section I), Tsinghua SIGS

Day 4: Keynote Lectures & Oral Presentations

Time: 10:30-17:30, Thursday, 18th July, 2024

Time	Activities
Chairman: Prof. Jiaojing Shao	
14:15-14:30	**OP: Yaxin Chen,** China University of Mining and Technology, China **Coal-Based Carbon Materials for Ion-Batteries**
14:30-14:45	**OP: Qi Yang,** Beijing University of Chemical Technology, China **Functional Carbon Materials and the Application in Zn/Na-Ion Batteries**
14:45-15:00	**OP: Yonghong Ye,** Institute of Coal Chemistry, CAS, China **Hybrid Carbon Spheres Derived from Polystyrene and Pitch for Optimized Sodium-Ion Batteries**
15:00-15:15	**OP: Zhiguo Zhang,** Tsing Bosch, China **Characterization of Graphite Anode by Ultrasonic Imaging Technology on The Influence of Electrolyte Unwetting**
15:15-15:30	**Coffee Break**
Chairman: Prof. Ming Liu	
15:30-16:00	**KL: Liang Peng,** City University of Hong Kong, China **Carbon-based mesoporous nanomaterials for energy storage and environment applications**
16:00-16:15	**OP: Yeru Liang,** South China Agricultural University, China **Nanoconfined Carbonization Enabling High-Density Porous Carbon for Jointly Superior Gravimetric and Volumetric Zinc-ion Storage**
16:15-16:30	**OP: Xiaomin Yang,** Jilin University, China **Fabrication of Biomass-Based Functional Carbon Materials for Energy Storage**

Session 4: Carbon for Energy Storage and Conversion

Conference Hall Building C, International Building (Section I), Tsinghua SIGS

Day 4: Keynote Lectures & Oral Presentations

Time: 10:30-17:30, Thursday, 18th July, 2024

Time	Activities
Chairman: Prof. Ming Liu	
16:30-16:45	**OP: Shiyong Zhao,** Institute of Metal Research, CAS, China **Multiple Dimensional Carbon Supported Single Atom Catalysts for Electrochemical Energy Application**
16:45-17:00	**OP: Xiaoqi Yu,** Beijing University of Chemical Technology, China **Rationally Designing Closed-Pores Hard Carbon Structures to Boost the Low-Voltage Plateau Capacity of Sodium-Ion Batteries**
17:00-17:15	**OP: Shuzhang Niu,** Shenzhen Technology University, China **Functional Design and Performance Study of Carbon-Based Composites Used for Lithium Metal Batteries**
17:15-17:30	**OP: Ting-Ting Yao,** Institute of Coal Chemistry, CAS, China **A Gas Diffusion Layer Based on Synergistic Effect of Whisker-Carbon Nanotubes and Carbon Black for Proton Exchange Membrane Fuel Cells**
18:30-20:00	**Excellent Poster and Best Emerging Young Scientist Oral Presentation Awarding Ceremony** **Banquet for all registered participants**

Session 5: Carbon for Environment

Room 211 Building A, International Building (Section I), Tsinghua SIGS

Day 1: Keynote Lectures & Oral Presentations & Poster

Time: 10:30-17:30, Monday, 15th July, 2024

Time	Activities
Chairman: Prof. Jieshan Qiu	
10:30-11:00	**KL: Mauricio Terrones,** The Pennsylvania State University, USA **New Insights into Carbide-Dichalcogenide Heterostructures: Synthesis, Superconductivity, Electrocatalysis and more**
11:00-11:30	**KL: Anhui Lu,** Dalian University of Technology, China **Porous Carbons for Gas Separation**
11:30-11:50	**OP: Yang Cheng,** Tsinghua Shenzhen International Graduate School, China **Hydrodynamic tearing of bacteria on carbon-coated nanotips for sustainable water disinfection**
11:50-12:10	**OP: Jiahui Ding,** Dalian University of Technology, China **Hard carbon anodes derived from crosslinked waste PS and PVC and their sodium ions storage performances**
12:00-13:30	**Lunch**
Chairman: Prof. Juan Matos Lale	
13:30-14:00	**KL: Jieshan Qiu,** Beijing University of Chemical Technology, China **Functional carbon materials for advanced electro-technology**
14:00-14:30	**KL: Juan Matos Lale,** Universidad Autónoma De Chile, Spanish **Which are the photocatalytic active sites in carbocatalysts? Examples in various photocatalytic reactions**
14:30-14:50	**OP: Yu Lei,** Tsinghua Shenzhen International Graduate School, China **Single molecule capturing and detection on graphene analogues**

Session 5: Carbon for Environment

Room 211 Building A, International Building (Section I), Tsinghua SIGS

Day 1: Keynote Lectures & Oral Presentations & Poster

Time: 10:30-17:30, Monday, 15th July, 2024

Time	Activities
	Chairman: Prof. Juan Matos Lale
14:50-15:10	**OP: Bei Zhang,** Zhejiang University, China **Biomass-derived 3d porous carbon materials for advanced environmental applications**
15:10-15:30	**OP: Alzhan Baimenov,** Satbayev University, Kazakhstan **Mercury removal from water using mxene-modified carbon aerogels from rice husk lignin**
15:30-15:50	**OP: Paula Riquelme-García,** University of Alicante, Spanish **Potential bifunctional carbon-supported pd-based catalysts for the production and storage of H2 attained by formic acid-bicarbonate couple**
16:00-17:30	**Poster Session and Coffee Break**

Session 5: Carbon for Environment

Room 211 Building A, International Building (Section I), Tsinghua SIGS

Day 2: Keynote Lectures & Oral Presentations & Poster

Time: 10:30-17:30, Tuesday, 16th July, 2024

Time	Activities
	Chairman: Prof. Mauricio Terrones
10:30-11:00	**KL: Conchi O. Ania,** CEMHTI, CNRS (UPR 3079), Université d' Orléans, France **Photocatalytic regeneration of saturated activated carbons used in water treatment: a pilot scale study**
11:00-11:20	**OP: Lei Qin,** Taiyuan University of Technology, China **Molecularly imprinted adsorbent based on β-cof/graphene aerogel for phenol adsorption and removal**
11:20-11:40	**OP: Dingxin Xu,** Tsinghua Shenzhen International Graduate School, China **two-dimensional nanosheets with 1-nm-thick adlayers for efficient and near-instant removal of PFAS**
11:40-12:00	**OP: Zhiyan He,** Wuhan Textile University, China **Synergistic effect of indium single atoms and pt-in alloy nanoparticles on carbon support for selective oxidation of glycerol**
12:00-13:30	**Lunch**
	Chairman: Prof. Conchi O. Ania
13:30-14:00	**KL: Jia Zhu,** Nanjing University, China **Carbon materials for interfacial solar evaporation: Fundamental & applications**
14:00-14:30	**KL: Rufan Zhang,** Tsinghua University, China **Controlled synthesis and properties of ultralong carbon nanotubes**
14:30-14:50	**OP: Z.A. Mansurov,** Institute of Combustion Problems, Kazakhstan **Gas-generator compositions based on activated carbon from a gas mask for destruction of rocks**

Session 5: Carbon for Environment

Room 211 Building A, International Building (Section I), Tsinghua SIGS

Day 2: Keynote Lectures & Oral Presentations & Poster

Time: 10:30-17:30, Tuesday, 16th July, 2024

Time	Activities
Chairman: Prof. Conchi O. Ania	
14:50-15:10	**OP: C. Srinivasakannana,** Khalifa University, United Arab Emirates **Thermocatalytic decomposition of methane to hydrogen and nano filamentous carbon over Ni-Ce bimetallic catalysts on porous carbon**
15:10-15:30	**OP: Jiayi Xie,** Tsinghua University, China **Life-cycle carbon emission and decarbonization potentials of carbon fiber and carbon nanotube in China**
15:30-15:50	**OP: Die Gong,** Sichuan University of Science and Engineering, China **Self-supporting spiral carbon forest clusters prepared by ethanol flame method and their performance in capacitive deionization**
16:00-17:30	**Poster Session and Coffee Break**

Session 6: Carbon for Biology, Health, and Medicine

Room 213 Building A, International Building (Section I), Tsinghua SIGS

Day 3: Keynote Lectures & Oral Presentations

Time: 10:30-12:00, Wednesday Morning, 17th July, 2024

Time	Activities
Chairman: Prof. Nyan-Hwa Tai	
10:30-11:00	**KL: Yanglong Hou,** Sun Yat-sen University, China **Fabrication and performance regulation of magnetic nanostructures**
11:00-11:30	**KL: Yuen Wu,** University of Science and Technology of China, China **Environmental air regulation by electrochemical device**
11:30-11:50	**OP: Guido Perrin,** CEMHTI, CNRS (UPR 3079), Université d' Orléans, France **Photoluminescent properties of metal-containing carbon dot nanostructures: Potential applications in MRI imaging**
12:00-13:30	**Lunch**
14:00-17:30	**Social Event**

Session 6: Carbon for Biology, Health, and Medicine

Room 213 Building A, International Building (Section I), Tsinghua SIGS

Day 4: Keynote Lectures & Oral Presentations

Time: 10:30-17:30, Thursday, 18th July, 2024

Time	Activities
	Chairman: Prof. Yu' en Wu
10:30-11:00	**KL: Bai Yang,** Jilin University, China **Carbonized polymer dots: Syntheses, properties and applications**
11:00-11:30	**KL: Nyan-Hwa Tai,** Tsing Hua University in Taiwan **Non-invasive salivary glucose biosensor based on conductive polymer hydrogels/carbon nanotubes**
11:30-11:50	**OP: Rongkang Huang,** The Sixth Affiliated Hospital, Sun Yat-sen University, China **Super-structured porous hydrogel composites and carbons for soft tissue repair**
12:00-13:30	**Lunch**
	Chairman: Prof. Yu' en Wu
13:30-14:00	**KL: Wei Ma,** Xi' an Jiaotong University, China **An organic electrochemical transistor for multi-modal sensing, memory and processing**
14:00-14:20	**OP: Qianming Gong,** Tsinghua university, China **Tailoring pore structure of hollow carbon spheres for blood purification**
14:20-14:40	**OP: Hao Yan,** Tsinghua Shenzhen International Graduate School, China **Planted graphene quantum dots for long-term visualization of nanoparticles in vivo**
15:00-15:30	**Coffee Break**

Session 6: Carbon for Biology, Health, and Medicine
Room 213 Building A, International Building (Section I), Tsinghua SIGS

Day 4: Keynote Lectures & Oral Presentations

Time: 10:30-17:30, Thursday, 18th July, 2024

Time	Activities
	Chairman: Prof. Xiaomin Xu
15:30-15:50	**KL: Dingsheng Wang,** Tsinghua University, China **Atomic-scale precision synthesis**
15:50-16:10	**KL: Xuemei Sun,** Fudan University, China **Carbon nanotube based fiber bioelectronic materials and devices**
16:10-16:30	**OP: Bingna Zheng,** The Eighth Affiliated Hospital Sun Yat-sen University, China **Functional porous polymers and carbons towards efficient tissue repair**
16:30-16:50	**OP: Yu Lei,** Tsinghua Shenzhen International Graduate School, China **Probe the interactions between the graphene analogues and bio-molecules**
16:50-17:10	**OP: Lin Chen, Taiyuan University of Technology, China** **Targeting carbon dot complex for integration of diagnosis and treatment**
17:10-17:30	**OP: Xiuyun Chuan ,** Peking University, China **The mechanical performance of the composite of the graphene and biofilm**
18:30-20:00	**Excellent Poster and Best Emerging Young Scientist Oral Presentation Awarding Ceremony** **Banquet for all registered participants**

Session 7: Carbon for Electronics and Optoelectronics
Room 212 Building A, International Building (Section I), Tsinghua SIGS

Day 4: Keynote Lectures & Oral Presentations

Time: 10:30-17:30, Thursday, 18th July, 2024

Time	Activities
	Chairman: Prof. Dong-Ming Sun
10:30-11:00	**KL: Jianwen Zhao,** Suzhou Institute of Nano-Tech and Nano-Bionics, CAS, China **Flexible carbon nanotube thin film transistors for optoelectronic neuromorphic devices and CMOS circuits**
11:00-11:20	**OP: Ashish Jaswal,** Indian Institute of Technology Mandi, India **Nitrogen-containing carbon-based flexible devices for electrochemical detection of Malathion pesticide**
11:20-11:40	**OP: Qingbin Zheng,** The Chinese University of Hong Kong, China **Carbon networks for soft electronics**
11:20-12:00	**OP: Chi Liu,** Institute of Metal Research, Chinese Academy of Sciences, China **Novel Hot-Carrier Transistors based on Low-dimensional Materials**
12:00-13:30	**Lunch**
	Chairman: Prof. Zhiyong Zhang
13:30-14:00	**KL: Youfan Hu ,** Peking University, China **Carbon nanotube-based high-performance bioelectronics**
14:00-14:20	**OP: Jia Si,** Peking University, China **Carbon Nanotube based Steep Slope Transistors and Tensor Processing Unit**
14:20-14:40	**OP: Haotian Hao,** Taiyuan University of Technology, China **Enhancing performance of carbon quantum dot-based memristors through boron doping for artificial synapse application**
14:40-15:00	**OP: Xinghua Liu,** Taiyuan University of Technology, China **Highly stable carbon dots enabled by structural regulation for white laser lighting diode with high color rendering index**
15:00-15:30	**Coffee Break**

Session 7: Carbon for Electronics and Optoelectronics

Room 212 Building A, International Building (Section I), Tsinghua SIGS

Day 4: Keynote Lectures & Oral Presentations

Time: 10:30-17:30, Thursday, 18th July, 2024

Time	Activities
	Chairman: Prof. Zhiyong Zhang
15:30-15:50	**KL: Yutaka Ohno,** Nagoya University, Japan **Low-frequency noise in carbon nanotube thin-film transistors and its impact on analog integrated circuits**
15:50-16:10	**OP: Chengcheng Wu,** Tsinghua-Berkeley Shenzhen Institute, Tsinghua Shenzhen International Graduate School, China **Enhanced Efficiency and Durability of Blue Phosphorescent OLEDs Employing Iridium Carbene Pincer Emitters**
16:10-16:30	**OP: Jiazhi Meng,** Tsinghua Shenzhen International Graduate School, China **Deuterium-substituted organic molecules for high-performance perovskite solar cell**
16:30-16:50	**OP: Wei He,** Tsinghua Shenzhen International Graduate School, China **High color rending index OLED for display application**
16:50-17:10	**OP: Kai-Ning Tong,** Tsinghua Shenzhen International Graduate School, China **Exceedingly Stable Luminescent Electron Transporting Material with doped Hole Blocking Layer for High-Performance Deep-Red OLEDs with LT95 over 7500 h at 1000 cd m−2**
18:30-20:00	**Excellent Poster and Best Emerging Young Scientist Oral Presentation Awarding Ceremony** **Banquet for all registered participants**

Session 8: Carbon 2024 Emerging Young Scientist Symposium
Room 212 Building A, International Building (Section I), Tsinghua SIGS

Day 1: Oral Presentations

Time: 10:30-17:30, Monday, 15th July, 2024

Time	Activities
	Chairman: Prof. Ke Shen
10:30-10:45	**OP: Xiaoyang Du,** The University of Melbourne, Australia **Revisiting self-discharge of nanoporous electrodes: insights from the multilayered graphene membrane-based electrodes**
10:45-11:00	**OP: Xiaotian Li,** Beijing University of Chemical Technology, China **Deciphering the triple relationship: Hard Carbon (HC) precursor-closed pore-plateau capacity for sodium-ion batteries**
11:00-11:15	**OP: Fei Li,** Tsinghua University, China **Highly efficient and selective extraction of gold by reduced graphene oxide**
11:15-11:30	**OP: Wang Zhou,** Hunan University, China **Rational design of high-capacity hard carbon anodes for Li- and Na/K-ion batteries**
11:30-11:45	**OP: Ke Wang,** Wuhan University of Science and Technology, China **Preparation and thermodynamic properties of mesophase pitch based carbon fiber mat reinforced 3D-C/C composites**
11:45-12:00	**OP: Yafang Zhang,** Dalian University of Technology, China **Engineering Atomically Dispersed Reactive Centers over Carbon Supports: Evolution Mechanism and Electrocatalysis Application**
12:00-12:15	**OP: Tan Yi,** Sun Yat-sen University, China **N/S Co-doped carbon nanosheets for efficient electrochemical extraction of uranium from seawater**
12:00-13:30	**Lunch**
	Chairman: Prof. Dawei Wang
13:30-13:50	**OP: Wenbin Li,** Dalian University of Technology, China **Design of functionalized carbon-based materials and electrocatalytic conversion of CO_2**

Session 8: Carbon 2024 Emerging Young Scientist Symposium
Room 212 Building A, International Building (Section I), Tsinghua SIGS

Day 1: Oral Presentations

Time: 10:30-17:30, Monday, 15th July, 2024

Time	Activities
Chairman: Prof. Dawei Wang	
13:45-14:00	**OP: Renlu Yuan,** Beijing University of Chemical Technology, China **"Disruption" vs. "Repairing" of carbon materials by ball-milling**
14:00-14:15	**OP: Yuting Luo,** University of Toronto, Canada **Electrocatalytic electrodes for high-current-density electrochemical synthesis and devices**
14:15-14:30	**OP: Ruogu Xu,** Institute of Metal Research, Chinese Academy of Sciences, China **Structural regulation of polymer electrolyte for high performance solid-state Li battery**
14:30-14:45	**OP: Zhihong Luo,** Guizhou University, China **Two-dimensional montmorillonite-based heterostructure for high-rate and long-life lithium-sulfur batteries**
14:45-15:00	**OP: Chuang Qiu,** Beijing University of Chemical Technology, China **One-step construction of closed pores enabling high plateau-capacity hard carbon anodes for sodium-ion battery: Closed-pore formation and energy storage mechanisms**
15:00-15:15	**OP: Qiaowei Lin,** Shenzhen Institute of Advanced Technology, China **Carbon surface: A key to improved anode coulombic efficiency in sodium-ion batteries**
15:15-15:30	**OP: Xue Li,** Xi'an University of Science and Technology, China **Preparation, structure regulation and application of bituminous-based hard carbon materials**
15:30-15:45	**OP: Ning Sun,** Beijing University of Chemical Technology, China **Efficient resource utilization of discarded polypropylene masks for high-value carbon anode materials**

Session 8: Carbon 2024 Emerging Young Scientist Symposium

Room 212 Building A, International Building (Section I), Tsinghua SIGS

Day 1: Oral Presentations

Time: 10:30-17:30, Monday, 15th July, 2024

Time	Activities
	Chairman: Prof. Dawei Wang
15:45-16:00	**OP: Pratteek Das,** Dalian Institute of Chemical Physics, Chinese Academy of Sciences, China **Graphene and MXene for High performance micro-supercapacitors**
16:00-16:15	**OP: Jiao Tan,** Hunan University, China **Improved self-sintering mechanical performance of carbon blocks by regulating composition of green petroleum coke**
16:15-17:30	**Poster Session and Coffee Break**

Session 8: Carbon 2024 Emerging Young Scientist Symposium
Room 212 Building A, International Building (Section I), Tsinghua SIGS

Day 2: Oral Presentations

Time: 10:30-17:30, Tuesday, 16th July, 2024

Time	Activities
	Chairman: Prof. Ying Tao
10:30-10:45	**OP: Qiqi Li,** Wuhan University of Science and Technology, China **Noble metal nanoparticles @ graphite edges for efficient pH-universal hydrogen evolution reaction**
10:45-11:00	**OP: Chuannan Geng,** Tianjin University, China **Catalytic conversion of Li2S for high-performance lithium-sulfur batteries**
11:00-11:15	**OP: Xiaoqi Yu,** Beijing University of Chemical Technology, China **Rationally designing closed-pores hard carbon structures to boost the low-voltage plateau capacity of sodium-ion batteries**
11:15-11:30	**OP: Biao Li,** Hunan University, China **Epitaxial growth of carbonaceous mesophase on coke surface**
11:30-11:45	**OP: Ziyun Zhao,** Tianjin University, China **High-purity rayon-based carbon fiber felt prepared by halogen gas purification for superior thermal insulation and oxidation resistance**
11:45-12:00	**OP: He Huang,** Wuhan University of Science and Technology, China **Preparation and thermal conductivity of 2.5D Carbon/Carbon composites**
12:00-12:15	**OP: Dongtian Miao,** Sun Yat-sen University, China **Carbon nanozyme-loaded star polymer hydrogel with long-lasting retention and shear-thinning properties facilitates durable long-term tumor therapy**
12:00-13:30	**Lunch**
	Chairman: Prof. Ying Tao
13:30-13:50	**OP: Yue Liu,** Hunan University, China **Controlling surface chemistry of activated carbon fibers for high product sulfate selectivity of H2S at room temperature catalytic oxidation**

Session 8: Carbon 2024 Emerging Young Scientist Symposium

Room 212 Building A, International Building (Section I), Tsinghua SIGS

Day 2: Oral Presentations

Time: 10:30-17:30, Tuesday, 16th July, 2024

Time	Activities
Chairman: Prof. Ying Tao	
13:45-14:00	**OP: Dong Liu,** Insititute of Coal Chemistry, Chinese Academiy of Science, China **New insight into boron catalytic graphitization of graphene lateral heat spreader**
14:00-14:15	**OP: Hongwei Zhong,** Wuhan University of Science and Technology, China **Hierarchical Graphite/SiCw matrix for phase change composites toward high-efficiency thermal harvesting**
14:15-14:30	**OP: Lidan Tan,** Wuhan University of Science and Technology, China **Spontaneous electrochemical reconstruction of NiNb2O6@C for high-rate lithium-ion batteries**
14:30-14:45	**OP: Zhiyuan Han,** Tsinghua-Berkeley Shenzhen Institute (TBSI), Tsinghua University, China **Carbon-based atom catalysts for high-performance lithium-sulfur batteries**
14:45-15:00	**OP: Yibo Zhang,** Tianjin University, China **Interfacial mechanism of carbon anode for sodium-ion batteries**
15:00-15:15	**OP: Zhuo Deng,** Wuhan University of Science and Technology, China **Exceptional thermal interfacial properties of highly resilient three-dimensional carbon foam/CNTs matrices**
15:15-15:30	**OP: Yuanbin Wang,** The Sixth Affiliated Hospital of Sun Yat-sen University **Exudate triggered electrical dressing for complicated wound management**
15:30-15:45	**OP: Wei Zhao,** Wuhan University of Science and Technology, China **Preparation of nitrogen-doped carbon points and their fluorescence characteristics by one-step electrolysis of graphite based on ionic liquid**

Session 8: Carbon 2024 Emerging Young Scientist Symposium
Room 212 Building A, International Building (Section I), Tsinghua SIGS

Day 2: Oral Presentations

Time: 10:30-17:30, Tuesday, 16th July, 2024

Time	Activities
Chairman: Prof. Ying Tao	
15:45-16:00	**OP: Xincheng Yao,** Hunan University, China **Delayed pyrolysis of binder pitch in fine-grained isostatic carbon blocks**
16:00-16:15	**OP: Pengwei Ma,** Sun Yat-sen University, China **Carbon nanosphere molecular brushes for treating inflammatory bowel disease**
16:15-17:30	**Poster Session and Coffee Break**

Day 2: Carbon Editor Workshop

Room 209 Building A, International Building (Section I), Tsinghua SIGS

Time: 16:00-18:00, Tuesday Afternoon, 16th July, 2024

Time	Activities
	Chairman: Prof. Mauricio Terrones
	Mauricio Terrones, The Pennsylvania State University, USA **Introduction: How to publish in Carbon: perspective and challenges**
	Discussion on the future of Carbon material science **Theme: Importance of providing a specific format for publication of key topics**
	Announcement of the recipient of the Carbon Journal Prize 2024 for outstanding PhD thesis

Plenary Lectures

Growth of Single-Walled Carbon Nanotubes with Controlled Structure

Jin Zhang*

Peking University, Beijing 100871, P. R. China.
Email: jinzhang@pku.edu.cn

Synthesis determines the future. In order to promote carbon nanotubes to the industrial production and real application, the controlled growth of SWNTs is imperative. Therefore, our research focuses on the structural controlled growth of SWNTs, aiming at realizing the direct growth of wafer-scale, high-density, structurally controllable horizontal arrays of SWNTs. This report mainly includes: the controlled growth of SWNTs with specific chirality through the design of solid catalysts and the regulation of nucleation thermodynamics and growth kinetics; growth of ultra-high-density SWNT arrays using Trojan catalysts; the development of a new catalyst loading method, ion implantation, and a vertical spray CVD system to achieve the direct growth of wafer-scale uniform, high-density SWNT arrays; and the establishment of a new characterization technique combining AFM with polarization optics to realize high-throughput characterization of wafer-scale arrays. Most recently, we developed an artificial intelligence (AI)-driven platform that integrates transformer-based language models tailored for carbon materials, robotic CVD, and data-driven machine learning models. Over a period of only 43 days, it exhibited marked advances in both catalyst innovation and controllable growth of SWNT arrays, accelerating the research of carbon nanotube synthesis.

Are Graphitic Carbon Materials a Concern for Health?

Alberto Bianco

CNRS, Immunology, Immunopathology and Therapeutic Chemistry, UPR 3572,
University of Strasbourg, ISIS, 67000 Strasbourg, France
E-mail: a.bianco@ibmc-cnrs.unistra.fr

In the last quarter of century, the new allotropic forms of carbon, i.e. carbon nanotubes and graphene, have been at the forefront of attention due to their potential toxic effects in leaving systems, including mammalian species and the environment [1]. While carbon nanotubes are still under strong debate if they should be definitely banned for use, graphene has been suspected to have the same behaviour, as sharing the same graphitic nature of the tubes. But not all graphitic carbon materials are alike in terms of impact on health and the environment [2].

In this lecture, I' ll cover our twenty-year research on the development of safe carbon nanotubes and graphene for biomedical applications. The health risks associated to the similarity between carbon nanotubes and asbestos fibres can be removed by appropriate chemical modifications, that can also trigger their degradability [3]. The assessment of the potential impact of graphene-based materials on health and the environment is also a key step in the development of current and future technologies based on these materials. In this context, I will highlight the potential risks of using graphene-based materials and the solutions to overcome them, in view of their registration as new and "safe" chemicals [4-6].

References

[1] Kim, M. et al. Goerzen, D.; Jena, P.V.; Zeng, E.; Pasquali, M.; Meidl, R. A.; Heller, D. A. Nat. Rev. Mater. 2024, 9, 63–81. DOI:10.1038/s41578-023-00611-8

[2] https://euon.echa.europa.eu/nanopinion/-/blogs/what-does-graphene-really-look-like-and-why-is-it-not-carbon-nanotubes-

[3] Ma, B.; Martìn, C.; Kurapati, R.; Bianco, A. Chem. Soc. Rev. 2020, 49, 6224-6247. DOI:10.1039/c9cs00822e

[4] Fadeel, B.; Bussy, C.; Guijarro, S. M.; Vázquez, E.; Flahaut, E.; Mouchet, F.; Evariste, L.; Gauthier, L.; Koivisto, A. J.; Vogel, U.; Martín, C.; Delogu, L. G.; Bürki, T.; Wick, P.; Beloin-Saint-Pierre, D.; Hirschier, R.; Pelin, M.; Candotto Carniel, F.; Tretiach, M.; Cesca, F.; Benfenati, F.; Scaini, D.; Ballerini, L.; Kostarelos, K.; Prato, M.; Bianco A.

ACS Nano 2018, 12, 10582-10620. DOI:10.1021/acsnano.8b04758

[5] Lin, H.; Buerki-Thurnherr, T.; Kaur, J.; Wick, P.; Pelin, M.; Tubaro, A.; Candotto Carniel, F.; Tretiach, M.; Flahaut, E.; Iglesias, D.; Vázquez, E.; Cellot, G.; Ballerini, L.; Castagnola, V.; Benfenati, F.; Armirotti, A.; Sallustrau, A.; Keck, M.; Taran, F.; Bussy, C.; Vranic, S.; Kostarelos, K.; Connolly, M.; Navas, J. M.; Mouchet, F.; Gauthier, L.; Baker, J.; Suarez-Merino, B.; Kanerva, T.; Prato, M.; Fadeel, B.; Bianco, A. ACS Nano 2024. DOI:10.1021/acsnano.3c09699

[6]https://euon.echa.europa.eu/documents/2435000/3268573/echa_2021_286_graphene_study.pdf/16cc871b-ca06-caf4-fdb0-3c14113b3e2a?t=1670235623023

Understanding mesophase pitch from a lyotropic liquid crystalline perspective

Taisei Tomaru [1], Hiroki Shimanoe [2], Young-Pyo Jeon [3], Koji Nakabayashi [1,2], Jin Miyawaki [1,2], Seong-Ho Yoon [1,2]

[1]Institute for Materials Chemistry and Engineering, Kyushu University, Kasuga, Fukuoka, 816-8580, Japan,

[2]Interdisciplinary Graduate School of Engineering Sciences, Kyushu University, Kasuga, Fukuoka, 816-8580, Japan,

[3]C1 Gas & Carbon Convergent Research Center, Korea Research Institute of Chemical Technology, Gajeong-ro Yuseong-gu, Daejeon, 34114, Korea

Mesophase pitch has been applicated as an unique precursor for pitch-based high-performance carbon fiber and artificial graphite anodes for Li-ion batteries, but it is necessary to separate and identify the characteristics of mesogen for more diverse applications as liquid crystal. In this presentation, some phenomenological results of the lyotropic liquid crystal of the mesophase pitch (MP) were explained, which revealed that it behaves as a typical lyotropic liquid crystal. The mesogenic and solvent components of SMP were separated using solvent extraction method. The threshold concentration (TC) that needs for SMP to achieve 100 vol% anisotropy was defined as the minimum amount of mesogenic component. The TCs of various SMPs were compared and the effect of the solvent component on them was discussed. Concerning the mesogenic component, the correlations between the size of layered molecular stacking units and anisotropic contents of SMP were quantitatively shown. A novel MP manufacturing through the thermal mixing of mesogenic and solvent components that were prepared separately with different conditions was tried. Finally, we discuss how this approach could be used to enhance SMP yield by modifying the contents of anisotropic mesogenic and solvent components.

KEYWORDS: Mesophase pitch, lyotropic liquid crystal

Keynote Lectures

Extreme in-plane strain to engineer folds or sp^3 bonds in graphene and few-layer graphite

F. Vialla[1], A.F. Juarez Saborio[1], R. Galafassi[1], V. Rajaji[1] and A. San Miguel[1]

[1]Université Claude Bernard LYON 1, CNRS, ILM, Villeurbanne FRANCE

Graphene and few-layer graphite correspond to intrinsically two-dimensional (2D) structures showing the stiffest mechanical properties. However, when a significantly high in-plain stress is applied, strain needs to be relaxed in the third dimension. It usually leads to stochastic and uncontrolled phenomena [1] such as ripples, wrinkles, stacking faults or sheet delamination. Here we introduce extreme in-plane strain engineering through the use of compressible substrates supporting the 2D nanocarbon and experimentally demonstrate novel behaviors.

With uniaxial strain transmitted to single layer graphene from a polymer previously stretched and micro-patterned, we nucleate folds which are deterministic in position, tunable in height, and reversible, thus defining the building block for a future implementation of graphene origami [2]. With biaxial strain transmitted to few-layer graphite from a SiO2/Si substrate compressed under high-pressure in an anvil cell, we catalyze a sp2–to-sp3 interlayer phase transition in water environment, yielding a 2D diamond-like structure obtained at room temperature and relatively low high-pressure,and thus defining a novel synthesis route for 2D nanomaterials [3].

References

[1] C. Bousige, et al. Biaxial Strain Transfer in Supported Graphene. Nano letters 17(1), 21–27 (2017)

[2] A. F. Juarez Saborio, et al. Deterministic Graphene Folds through Ultrafast Laser Nanotexturing. Just submitted

[3] V. Rajaji, et al. Biaxial Strain Effects in 2D Diamond Formation from Graphene Stacks. Available on arXiv.

Graphene for Emerging Iontronics

Dan Li

Department of Chemical Engineering, The University of Melbourne, Melbourne, VIC 3010, Australia
Email: dan.li1@unimelb.edu.au

Iontronics is a field that explores the transport of ions to create new devices and systems. Despite its promising potential to revolutionize next-generation technologies in energy storage and conversion, ion separation, biomedicine, and neuromorphic computing, iontronics is confronted with a plethora of both scientific and technical challenges that remain unresolved. In this presentation, I will explore these challenges and share insights from our recent investigations into ion transport through multilayered graphene membranes. This discussion will demonstrate how graphene-based multilayered membranes offer a distinctive material platform to unlock the vast possibilities of iontronics for future technological advancements.

References

[1]. Cheng, C., Jiang, G., Simon, G. P., Liu, J. L. & Li, D. Nature Nano., 13, 685–690 (2018).

[2]. Jing, X., Zhan, H., Wang, X., Xu, Z., Xiong, Z., Zhang, K., Simon, G. P., Liu, Z. J., & Li, D., Nature Nano., 15, 683–689. (2020).

[3]. Zhan, H., Xiong, Z., Cheng, C., Liang, Q., Liu, J. Z. & Li, D., Adv. Mater., 32, 1904562. (2020)

[4]. Cao,Y., Xiong, Z., Xia, Y., Franks, G. V., Zu, L., Wang, X., Hora, Y., Mudie, S., He, Z., Qu, L, Xing, Y., & Li, D., Adv. Funct. Mater., DOI: 10.1002/adfm.202201535 (2022).

Graphene-based catalytic or multi-functional materials via interface engineering

Junying Wang*[,1] , Junzhong Wang[2]

[1]Institute of Coal Chemistry, Chinese Academy of Sciences, Taiyuan 030001, China

Email: wangjy@sxicc.ac.cn

[2]Anhui Graphene Carbon Fiber Engineering Research Center, Anhui University, Hefei 230601, China

It is important for controlling surface and interfaces to reach high-performance multi-functional or catalytic materials. Firstly, we found conductive solid-electrolyte interphase (SEI) is useful to prepare graphene efficiently through electrochemical exfoliation of graphite or microcrystalline graphite minerals and even graphene-based catalytic materials [1-2]. I will show that iron-based SEI layer onto graphene surface can be formed through the redox decomposition of iron-carbon species during the electrochemical process. High-performance Fe-N-C/graphene electrocatalysts for oxygen or carbon dioxide reduction reaction were obtained by tuning SEI composition with surface modification [1]. Next, I will present that we have fabricated graphene integrating carbon fiber and hierarchical porous carbon as a robust flexible "carbon-concrete" multi-functional film through carbon-fiber bundle spreading and hybridizing with graphene flakes [3]. Also, graphene can assist conversing pitch to sheet-like microporous carbon/graphene composites with a high specific surface area of 2930 m2/g and a high carbon yield of 72%. The porous carbon cathode impregnated with 71.2% of elemental sulfur achieved high Coulombic Efficiency for a Li-S battery [4]. Graphene microsheets reinforced polyurethane exhibits overall improvement of multi-functional performances including mechanical properties of tensile strength, modulus and shape memory, thermal and electrical conductivity, thermal stability and flame retardance [5].

References

[1] J. Wang, J. Wang*, etal. Energy Stor. Mater., 2018, 12,1-7.

[2] J. Wang, J. Wang*, etal. J. Mater. Chem. A, 2015, 3, 3144-3150.

[3] H. Zhang, J. Wang*, etal. Carbon, 2018,126, 500-506.

[4] M. Cheng, J. Wang*, etal. Green Chem., 2018, 20, 4675-4683

[5] Q. Chen, J. Wang*, etal. Chem. Eng. J., 2024, 482, 148668.

One-dimensional van der Waals heterostructures

Rong Xiang [1]

Zhejiang University, Hangzhou 310027, China

We present the experimental synthesis of one-dimensional (1D) van der Waals heterostructures, a class of materials where different atomic layers are coaxially stacked.[1] A typical structure is a 5-nanometer–diameter heteronanotube consisting of an inner SWCNT, a middle three-layer BN nanotube (BNNT), and an outer MoS2 nanotube (MoS2NT). Electron diffraction verifies that all shells in this structure are single crystals. This work suggests that all the materials in the current 2D library could be rolled into their 1D counterparts and a plethora of function-designable 1D heterostructures may be realized. We also present our follow-up study on the structural details and formation mechanism of 1D heterostructures.[2] Edge structures, nucleation sites, crystal epitaxial relationships are clearly revealed using a nondestructive transmission electron microscopy.[3] Later, we summarize the current understanding on the properties of 1D heterostructures [4,5,6]. Lastly, we will present our recent progress and perspective on more sophisticated 1D heterostructures. [7,8]

References

[1] Science (2020), 367, 537.

[2] PNAS (2021), 118, e2107295118

[3] Science Advances (2019), 5-5, eaat9459.

[4] Small Science, (2021), 1, 2000039.

[5] ACS Nano (2021), 15-3, 5600.

[6] Small, (2021), 2102585.

[7] Nature Review Elect. Eng., (2024), 1, 149.

[8] unpublished.

Cu-containing carbon materials as catalysts for electrochemical CO2 reduction

Javier Quílez-Bermejo[1], Juan Herranz[1]

[1]Electrochemistry Laboratory, Paul Scherrer Institut (PSI), 5232 Villigen, Switzerland

Copper (Cu) stands out as the sole metal element known for the efficient electrochemical conversion of CO2 into hydrocarbons, with high catalytic performance and faradaic efficiencies [1]. This characteristic has generated considerable interest among researchers in the development of Cu-containing electrocatalysts. Nevertheless, recent works have revealed that seemingly similar Cu active sites show massive variations in selectivity and catalytic performance. These inconsistencies could stem from the catalytic support, often based on diverse types of carbon materials (e.g., with different pore size distributions, graphitization extents and/or surface functionalities).

Specifically, the features of the carbon support can exert a significant influence on the electronic properties of Cu active sites, thereby modifying the catalytic performance for CO2 reduction of Cu-containing carbon-based electrocatalysts. Therefore, in this contribution we have evaluated the effect of factors such as the Cu:C ratio, Cu nanoparticle size and carbon support on the CO2-reduction activity and selectivity of a series of Cu-containing carbon materials. These span from traditional nanoparticulate samples to single atom catalysts in which the active sites consist of isolated Cu-ions coordinated by N-functionalities hosted in the C-support (e.g., in the form of Cu-N4 sites alike those found in porphyrins and phthalocyanines). These catalysts' library was first submitted to electrochemical measurements (e.g., Pb underpotential deposition) to quantify their surface site density, along with complementary ex situ tests to determine their surface composition and the structure of the C-matrix. Subsequent electrochemical tests in a parallel plate cell with CO2-containing electrolyte served to establish relations among these catalytic features and those of their Cu- and C-sites.

References

[1] Lin.W, et al. Adv. Mater. (2024) 2313028.

Graphene oxide: Synthesis, assembly and film applications

Wencai Ren

Institute of Metal Research, Chinese Academy of Sciences,
72 Wenhua Road, Shenyang 110016, China

Graphene oxide (GO), an important derivative of graphene, has shown a great potential for many applications such as electronics, energy storage, membranes, and composites. In this talk, I will first introduce a green water electrolytic oxidation method for highly efficient production of GO. Then, the influence of reduction degree on the performance of GO separation membranes will be discussed and a highly stable GO-based separation membrane with superior permeability will be demonstrated. Finally, I will introduce a continuous centrifugal casting method for scalable and efficient production of highly aligned and compact GO films. With this method, a variety of GO-based functional films have also been fabricated, including reduced GO/single-walled carbon nanotube hybrid films for flexible supercapacitors with superhigh volumetric energy density and highly anisotropic GO-based nanocomposites. Importantly, we found that the defects can significantly boost the graphitization of GO-based films, which enables a graphene film with thermal and electrical conductivities comparable to those of HOPG.

References

[1] Wencai Ren, et al. Nature Nanotechnology 9 (2014) 724.

[2] Songfeng Pei, et al. Nature Communications 9 (2018) 415.

[3] Qing Zhang, et al. Science Bulletin 62 (2018) 788.

[4] K.H. Thebo, et al. Nature Communications 9 (2018) 1486.

[5] Jing Zhong, et al. Nature Communications 9 (2018) 3484.

[6] Qinwei Wei, et al. Advanced Materials 32 (2020) 1907411.

[7] Qing Zhang, et al. National Science Review 10 (2023) nwad147.

Mapping and Alleviating Fe–N–C Catalyst Degradation

Fangzhou Liu[1], Mohan Li[1], Li Wei[1], Yuan Chen[1]

[1]The University of Sydney, School of Chemical and Biomolecular Engineering, Darlington, NSW 2006, Australia

Iron–nitrogen–carbon (Fe–N–C) single-atom catalysts are promising precious metal-free catalysts for various essential reactions. However, poor stability is a roadblock to their practical applications.[1-2] First, we used a series of heterogeneous molecular catalysts with well-defined structures as model Fe–N–C catalysts to map their degradation for oxygen reduction reaction (ORR) in acidic electrolytes. Detailed characterizations identified five degradation paths and revealed the beneficial role of electron-withdrawing substituents. The direct Fe leaching from Fe-N4 sites plays a crucial role in early-stage degradation, which can be significantly suppressed by -F and -NO2 substituents. The oxidative degradation became dominant with time, forming FeOx nanoclusters on carbon nanotube substrate, which the electron-withdrawing substituents can partially alleviate.[3] Second, we showed that adding organic molecules with suitable properties as scavengers of reactive oxygen species can alleviate Fe–N–C degradation. High-viscosity scavengers (e.g., isopropanol, tert-butanol) reduce the catalytic activity. Matching organic molecule size with the pore structure of Fe-N-C catalysts other than their radical scavenging speed is more critical. The large dimethyl sulfoxide, having detrimental effects on Fe–N–C with abundant micropores, can significantly reduce the degradation of Fe–N–C with abundant mesopores and Fe–N–C with catalytic sites on carbon black surfaces.[4]

References

[1] Li Wei, et al. Current Opinion in Electrochemistry 36 (2022) 101159.
[2] Yuan Chen, et al. Carbon 1995 (2022) 272.
[3] Fangzhou Liu, et al. ACS Catalysis (2024) submitted.
[4] Mohan Li, et al. Journal of Catalysis (2024) submitted.

Technical Progresses and Prospects of Graphene @ Lithium Batteries

Xufeng Zhou[1,2] and Zhaoping Liu[1,2]

[1]Ningbo Institute of Materials Technology and Engineering, Chinese Academy of Sciences, Ningbo, China
[2]National Graphene Innovation Center, Ningbo, China

Graphene has shown great application potentials in lithium batteries, owing to its excellent electrical conductivity and unique 2D nanostructure. This talk briefly reviews the history of graphene in lithium battery applications, especially graphene conductive additives and graphene coated Al current collectors for commercial batteries. Then the technology progress of graphene in next-generation high-energy Li batteries will be introduced, including high-capacity graphene modified Si anode materials for high-energy Li-ion batteries, graphene modified metallic Li anode for Li metal batteries, graphene conductive additives that reduce Li-ion diffusion resistance in high mass loading cathodes, and graphene coated Cu current collectors that improve stability of Si anodes. Finally, the commercial application prospects and technological trend of graphene in Li batteries will be discussed.

References

[1] Rusheng Fu, Xufeng Zhou, Zhaoping Liu, et al. Energy Storage Mater. 35 (2021) 317.

[2] Wei Deng, Xufeng Zhou, Zhaoping Liu, et al. Adv. Energy Mater. 8 (2018) 1703152.

[3] Bangyi He, Xufeng Zhou, Zhaoping Liu, et al. J. Mater. Chem. A 9 (2021) 25558.

[4] Tengsheng, Chi, Xufeng Zhou, Zhaoping Liu, et al. J. Electrochem. Soc. 170 (2023) 040515.

[5] Yanan Yu, Xufeng Zhou, Zhaoping Liu, et al. Carbon 215 (2023) 118498.

New structures from high pressure densed carbon phases

Mingguang Yao[1], Jiajun Dong[1] Yucheng Shang[1], Desi Chen, Zhaodong Liu, Bingbing Liu[1]

[1]State Key Lab of Superhard Materials, College of physics, Jilin University, Changchun, Jilin province

Creating new carbon structures with excellent properties has been attracting great research interest due to their scientific importance and various applications. In this talk we will present our recent progress on the synthesis of new carbon structures from densed carbon by compressing different well-designed carbon precursors. Both crystalline and amorphous high pressure densed carbon phases were obtained, depending on the carbon precursors used and pressure conditions. These new carbon structures, including sp3-hybridized amorphous carbon and new diamond, were quenched to ambient pressure and the physical properties were studied. These new sp3-carbons exhibit superhardness and unusual electronic and optical properties, which can be further tuned by using different high pressure and temperature conditions. Our results provide efficient strategies for the synthesis of new sp3-carbon materials and solved several long-standing problems on carbon material science.

References

[1] [1] Shang YC, Yao MG*, Liu BB* et al. Nature 599, 599–604 (2021).
[2] Shang YC, Yao MG*, Liu BB* et al. Nature Communications 14, 7860 (2023).
[3] Yao MG*, et al. Advanced Functional Materials 33, 2214696 (2023).
[4] Chen DS, et al. Submitted

Investigation of pyrolysis behavior of pitch and preparation of Pitch-based functional carbon materials

Yan Song*

CAS Key Laboratory for Carbon Materials, Institute of Coal Chemistry, Chinese Academy of Sciences, Taiyuan, Shanxi, P. R. China

*Correspond Author, yansong1026@126.com

Pitch, as a raw material rich in aromatic structure, easy to graphitize, high carbon content, and low ash content, is a high-quality precursor for preparing high value-added functional carbon materials. The composition, chemical structure, and group composition characteristics of pitch are different, and the pyrolysis process during heat treatment is different, resulting in completely different pyrolysis reaction kinetics, which determines the morphology, microcrystalline structure, and application range of derived-functional carbon materials. Based on the analysis of pitch composition and structure, as well as the study of thermal reaction behavior, different types (including mesophase pitch, MCMB, carbon foam, self sintered isotropic graphite, porous carbon and carbon anode) and application of functional carbon materials were prepared in this work. The above work provides guidance for the high value-added application of pitch.

Investigating graphite oxidation via real-time offgas analysis

A. Theodosiou [a], P.R. Davies [b], A.N. Jones [a]

[a]Nuclear Graphite Research Group (NGRG), University of Manchester, Manchester, UK, M1 3BB
[b]Cardiff Catalysis Institute, School of Chemistry, Cardiff University, Cardiff, UK, CF10 3AT

Graphite is now considered as a critical mineral in the UK due to its use in batteries for electric cars, for example, as well as its use in advanced nuclear technologies. Understanding the behaviour of graphite in an oxidative environment is crucial to understanding its materials performance in these applications. Despite a relatively simple oxidation chemistry, the fundamental reaction mechanisms, particularly with relation to the kinetics, appear to be surprisingly complex; this is illustrated by the wealth of literature in this field. This work employs quantitative gas analysis (QGA) equipped directly to a tube furnace as an effective tool to monitor the off-gases produced from a range of experimental conditions in real-time. The CO and CO2 concentrations are quantified and used to derive the relevant kinetic information providing a unique insight into the underlying reaction mechanisms.

Curriculum Vitae

Dr Theodosiou is a lecturer in Nuclear Graphite Engineering and has been involved in the field of nuclear graphite since starting his PhD in 2006. He has expertise in a variety of experimental techniques which have been employed across many different projects, from fundamental understanding of graphite behavior and properties to more advanced materials research for the next generation of nuclear reactors. Dr Theodosiou has been heavily involved in the field of Advanced Nuclear, working on projects investigating the physio-chemical behaviour of Generation IV reactor candidate graphites and improving the understanding of their behaviour in reactor conditions. Dr Theodosiou has published articles on graphite oxidation, graphite microstructure and surface characterization of graphitic materials, amongst others.

Concentrated Nanocarbon Dispersions in Water Stabilized by Polymerized Ionic Liquids

John Texter

Coating Research Institute, School of Engineering, Eastern Michigan University
Ypsilanti, MI 48197, USA
Strider Research Corporation, Rochester, NY 14610-2246, USA
Email: jtexter@emich.edu

Polymerized ionic liquids (PIL) have established a new branch of materials science, and some of these new PIL-materials are finding innovative applications as thermodynamically stable nanoparticulate dispersions and as dispersing aids with controllable solubilities. PIL are produced by radical chain-polymerization and by condensation step-polymerization. Such condensation polymeric binders, polyurethanes, polyureas, and polyesters, provide biodegradability pathways for composite materials. We discuss preparations and applications of PILs derived from polymerizable surfactants that are ionic liquids.

These PIL are illustrated to be excellent stabilizers for SWCNT (single-wall carbon nanotubes), MWCNT (multiwall carbon nanotubes), hydrothermal carbon, and graphene in water. SEM (scanning electron microscopy) illustrates that imidazolium-based PIL bind randomly and reversibly from suspension onto nanocarbon surfaces. Excellent aqueous dispersion stabilization is provided by strong binding to sp2-nanocarbon surfaces by π-overlap and by very strong solvation of imidazolium bromide by water. These strong binding and osmotic-brush effects have provided means to prepare the most concentrated aqueous dispersions of nanocarbons (MWCNT,17% w/w; graphene, 6.4% w/w) reported to date via liquid-phase exfoliation in water.

Sedimentation and shear-coating produce MWCNT and graphene electrodes suitable for conventional and advanced applications and also facilitate making phase-transferable stimuli-responsive dispersions, thermally conductive hydrothermal carbon coatings (25 W/m/K), thermally conductive MWCNT coatings in the diamond range (0.5 - 3 kW/m/K); rheo-optical (amorphous-nematic transitions) graphene dispersions induced by Couette-shear; alternative coating methods derived from stimuli-responsiveness of imidazolium-anion exchange; and transformational electrode structures produced by electrospinning.

Curriculum Vitae

John Texter is Professor Emeritus of Polymer and Coating Technology at Eastern Michigan University and Managing Consultant for Strider Research Corporation. He has been Editor-in-Chief of the Journal of Dispersion Science and Technology, Associate Editor of the Journal of Nanoparticle Research, and Section Editor for Applications of Current Opinion in Colloid and Interface Science. He has worked for Eastman Kodak Company in various areas of dispersion and emulsion technology and consults through Strider Research Corporation. He received his undergraduate engineering education and his PhD in Chemistry from Lehigh University, where he studied at the Zettlemoyer Center for Surface and Coatings Research. He is an experienced lecturer, inventor, editor, organizer, and technical project manager with over 250 publications including five books, 47 issued U.S. patents, and many research and review articles. He has received numerous awards and honors and is a Fellow of the American Physical Society, the American Chemical Society, the Royal Society of Chemistry, and of the Society for Imaging Science and Technology. He has served as Chairman of the ACS Division of Colloid and Surface Chemistry, Chairman of the Chemistry at Interfaces Gordon Research Conference, Chair of the Chemistry of Supramolecules and Assemblies Gordon Research Conference, and organizer of many international symposia. He is a visiting guest Chair Professor at the School of Chemistry, Chemical Engineering, and Materials Science of Soochow University and has been a visiting professor at the Colloid and Interfaces Max Planck Institute in Potsdam (Golm), at the Graduate School of Science and Technology, Kumamoto University, and at the Fachhochschule Esslingen. His research focuses on stimuli-responsive polymers, particles, and materials and the general area of dispersion science and practice.

A Protocol for Accurately Describing Carbon Materials

M. Monthioux

Centre d' Elaboration des Matériaux et d' Etudes Structurales (CEMES), CNRS,
Université de Toulouse, France

Carbon are unique materials because of their diversity, which comes from (i) 3 possible hybridization states (sp, sp2, sp3); (ii) their ability to exist as both molecular forms and allotropic forms; (iii) their ability to transform from a phase to another under the effect of external constraints (thermal, mechanical, pressure); (iv) the existence of a metastable structure (turbostratic structure) which allows them to commonly exist under 1D, 2D, and 3D morphologies; (v) the tolerance of the graphene lattice to the presence of various types of defects; (vi) the tolerance of graphene layers to severe deformation. This situation makes the description of carbons difficult, because of the lack of a normalised procedure based on a standardised vocabulary. For instance, terms such as structure, microstructure, nanostructure, microtexture, nanotexture can be used to designate similar features, or various aspects including mere morphological ones. In addition, the existing standardized nomenclatures are often ignored, and terms such as allotrope, graphitic, graphitation are often wrongly used in the literature. All this results in the literature about carbon materials to be confusing, and not easily comparable.

This work [1] is an attempt to clarify the whole picture, by reminding the definitions of terms when they exist, and proposing a description protocol based on 4 descriptors: morphology, texture, crystallinity, structure. As defined, those descriptors are exclusive (i.e., they cover exclusively different features), and are assumed to be enough to fully describe any kind of carbon material. If the time allotted to the talk allows for it, examples of use of this description protocol for a variety of carbon material families (nanotubes, cokes…) will be provided and commented.

References

[1] M. Monthioux, Carbon Trends 14 (2024) 100325

Doping of Diamond: From Nanocrystalline to Monocrystalline CVD Diamond Films

K. Haenen [1,2]

[1] Institute for Materials Research (IMO), Hasselt University, Wetenschapspark 1, B-3590 Diepenbeek, Belgium

[2] IMOMEC, IMEC vzw, Wetenschapspark 1, B-3590 Diepenbeek, Belgium

To fully harness the superior material properties of diamond, effective doping is essential. While p-type conductivity can be relatively easily achieved through the incorporation of substitutional boron, which acts as an acceptor, known n-type dopants like phosphorus and nitrogen still result in relatively deep donor levels. Furthermore, the plasma conditions required to successfully grow doped nanocrystalline diamond films can significantly differ from those used to grow homoepitaxial crystals, and vice versa.

This presentation focuses on recent advancements in the precise control of dopant incorporation into the crystal lattice structure for both types of diamond. It examines the impact of methane concentration during plasma-enhanced microwave chemical vapor deposition on achieving desirable electrical properties, highlighting challenges related to crystalline quality and dopant concentration control [1,2]. Additionally, the presentation discusses the influence of film microstructure and composition on the Young's modulus and residual stress in thin nanocrystalline diamond [3], as well as the electrical characteristics of Schottky barrier diodes based on a combination of monocrystalline boron and phosphorus-doped layers.

References

[1] R. Rouzbahani, KH, et al., Acc. Mater. Res. (2024), accepted

[2] R. Rouzbahani, KH, et al., Carbon 221 (2024) 118923

[3] R. Mary Joy, KH, et al., Acta Mater. 264 (2024) 119548

Assessing the Compatibility and Irradiation Behaviors of Nuclear Graphite for Molten Salt Reactors

Zhoutong He

(Shanghai Institute of Applied Physics, Chinese Academy of Sciences, Shanghai, 201800)

Email: hezhoutong@sinap.ac.cn

Molten salt reactors, as an important type of Generation IV reactors, have attracted attention both domestically and internationally. With the support of the Chinese Academy of Sciences' "Advanced Nuclear Fission Energy - Thorium-based Molten Salt Reactor Nuclear Energy System" pilot project, we are committed to the development of thorium-based molten salt reactors. Nuclear graphite is the preferred moderator and reflector material for molten salt reactors, but these reactors have special requirements for nuclear graphite, and currently, there are no mature nuclear graphite materials specifically for molten salt reactors. On one hand, nuclear graphite in molten salt reactors is in direct contact with molten salt, and there is little research on its compatibility with molten salt. On the other hand, to solve the issue of molten salt infiltration into nuclear graphite, ultrafine particle graphite is generally used, and assessing the irradiation dimensional change behavior of this type of graphite has become a key issue in selecting nuclear graphite materials for molten salt reactors.

This report will first introduce the research on the service behavior of nuclear graphite in molten salt reactors carried out by Oak Ridge National Lab in the United States during the 1960s and 1970s for the development of molten salt reactors. Then, it will introduce our research over the past decade on the compatibility of nuclear graphite with molten salt, discussing the infiltration behavior of molten salt in nuclear graphite and the impact of molten salt infiltration on the microstructure, mechanical, and thermal properties of nuclear graphite. Additionally, we will introduce the ion irradiation method developed to evaluate the irradiation dimensional change behavior of new grades of nuclear graphite, addressing the bottleneck issues in the current development of new nuclear graphite grades for molten salt reactors.

Development of Circularly Designed CFRPs and Sustainable Carbon Fibers

Toshihira Irisawa[1], Izuru Shimabukuro[1], Y. Nonoguchi[2] and Koji Nakabayashi[3]

[1] Gifu University, 1-1 Yanagito, Gifu, Japan
[2] Kyoto Institute of Technology, Matsugasaki, Sakyo-ku, Kyoto, Japan
[3] KusyuUniversity, 6-1 Kasugakouen, Kasuga, Fukuoka, Japan

Carbon Fiber Reinforced Plastics (CFRP) is expected to expand its demand dramatically toward the realization of a sustainable society because of its impact on the fields of energy saving and new energy. Actually, it has already indicated that the use of CFRP in transportation equipment can be expected to improve fuel efficiency and reduce CO2 emissions due to its energy-saving effect. Moreover, the use of CFRP in wind turbines, which represent a renewable energy source, and in fuel tanks for fuel cell vehicles that use hydrogen, a clean energy source, are also being developed. However, the manufacture of virgin carbon fiber (vCF) and CFRP requires a large amount of the energy, and CO2 emissions during production are higher compared with other structural materials. In order to widely use vCF and CFRP as sustainable materials, they need to be designed as more circular materials with a low-cost process.

This study will first focus on the development of low-cost vCF with high-performance properties. And then, the development of recyclable and reusable CFRP with a circular design will be reported.

Research on the production of carbon materials using stranded resources

Koji Nakabayashi[1,2*], Hiroki Shimanoe[2], Taisei Tomaru[1], Jin Miyawaki[1,2], Seong-Ho Yoon[1,2]

[1] Interdisciplinary Graduate School of Engineering Sciences (IGSES), Kyushu University, 6-1 Kasuga-koen, Kasuga, Fukuoka 816-8580, Japan
[2]Institute for Materials Chemistry and Engineering (IMCE), Kyushu University, 6-1 Kasuga-koen, Kasuga, Fukuoka 816-8580, Japan
*Correspond Author, nakabayashi@cm.kyushu-u.ac.jp

Stranded resources are equivalent to the general concept of 'Waste material'. Although stranded resources have the potential to add value, they are stranded in the market and their potential is not being utilized. For example, stranded resources include a wide variety of resources such as fossil fuel, waste oil, waste plastics, and industrial waste, but their advanced utilization has not been determined at this point. In addition, stranded resources are generally used as fuel and emit large amounts of carbon dioxide. Carbon dioxide is one of the stranded resources. In our laboratory, we are using stranded resources to produce various carbon materials. In addition, we aim to contribute to industrial development and a carbon-circulating society by converting (upcycling) "stranded resources" into "carbon materials. In this presentation, we will outline our latest research activities. Specifically, I will introduce some research on the production of carbon nanofibers using carbon dioxide and the production of carbon fibers using stranded resources.

Scalable manufacturing technology for carbon nanotube fibers

Qingwen Li[1,2,*], Yongyi Zhang[1,2,3,*], Zhenzhong Yong[1,2,3], Kunjie Wu[1]

[1] Suzhou Institute of Nano-Tech and Nano-Bionics, Chinese Academy of Sciences, Suzhou, China
[2] Department of Materials Science and Engineering, University of Science and Technology of China, Hefei, China
[3] Division of Nanomaterials and Jiangxi Key Lab of Carbonene Materials, Jiangxi Institute of Nanotechnology, Nanchang, China

Carbon nanotube fibers are a novel type of carbon-based fiber material. Currently, the strength of carbon nanotube fibers reported in the literature is close to 10 GPa, while the strength of carbon nanotube bundles can reach 60-80 GPa. Additionally, carbon nanotube fibers exhibit high electrical and thermal conductivity, making them highly strategically valuable as super high-strength fiber materials. However, the development of carbon nanotube fiber materials has faced significant challenges, such as "how to achieve scalable and continuous manufacturing of high-performance carbon nanotube fibers". Focused on the aforementioned issue, this report mainly highlights our team's recent research advancements in areas such as "how strong are carbon nanotube fibers exactly" and "the mass production of carbon nanotube fibers".

Fig. 1 Carbon nanotube fiber

References

[1] Yongyi Zhang, et al. Science (Accepted)
[2] Yongyi Zhang, Qingwen Li, et al. Nano Research 16 (2023) 12762-12771.
[3]张永毅, 李清文等, 物理化学学报, 38 (2022) 2106034.

Graphene fibers and their multi-functional composites

Chao Gao*

Department of Polymer Science and Engineering, Zhejiang University, Hangzhou 310012, P. R. China

Graphene fibers are directly assembled from graphene sheets, promising great potential to integrate high mechanical properties and multi-functionalities. In 2011, our team proposed and invented graphene fiber by wet-spinning of graphene oxide (GO) liquid crystal followed by chemical reduction. Then, several methods have been developed to improve the properties of graphene fibers including defect engineering and plasticization stretching. We found that GO fibers showed the interesting phenomenon of reversible fusion and fission. Recently, we propose a concept of bidirectionally promoting the assembly order in graphene fibers by multiple shear-flow fields, making graphene oxide achieve synergistically concentric arrangement in the cross-section and alignment along fiber axis. These densified and orderly crystalline graphene fibers show ultrahigh modulus of 901 GPa and thermal conductivity (1660 W/(m K)). Based on their impressive properties, we further fabricated graphene fiber composites with high thermal conductivity and electromagnetic shielding effect.

References

[1] Li,P.; Gao,C.; et al. Nat. Commun. 2024, 15,409

[2] Liu ,W.; Guo,H.-W.; et al. Nat. Electron. 2022, 5, 281

[3] Ming, X.; Liu, Y.; Li, P.; et al. Adv. Mater. 2022, 34, 2201867

[4] Chang, D.; Liu, J.; Fang, B.; Xu, Z.; et al. Science. 2021, 372, 614

Advancements in high thermal conductive carbon fibers for enhanced phase change energy storage

Ting Ouyang[1*], Chong Ye [1,2,3*]

[1]College of Materials Science and Engineering, Hunan Province Key Laboratory for Advanced Carbon Materials and Applied Technology, Hunan University, Changsha, 410082, PR China

[2]Hunan Province Engineering Research Center for High Performance Pitch-based Carbon Materials, Hunan Toyi Carbon Material Technology Co., Ltd., Changsha, 410000, PR China

[3]State Key Laboratory of Advanced Design and Manufacturing for vehicle body, Hunan University, Changsha, 410082, PR China

*oyt@hnu.edu.cn, yec@hnu.edu.cn

Mesophase pitch-based carbon fibers, renowned for their high thermal conductivity (TC), exceptional modulus, and low coefficient of thermal expansion, have found extensive applications in the aerospace industry and high-tech sectors. At Hunan University, we have innovated a full-scale production process utilizing catalytically cracked oil residue, establishing a 20-ton/year production line that yields a diverse range of fiber products with tailored structures and properties (tensile strength of 1.5~3.0 GPa, tensile modulus of 200~900 GPa, and TC of 500~1000 W·m-1·K-1). Our research has concentrated on the integration of these carbon fibers into various composite materials, including carbon-based, resin-based, metal-based, and ceramic-based systems. A focal point of our study is the application of these high TC fibers in phase change energy storage systems. Phase Change Materials (PCMs) are essential for thermal energy storage, with TC dictating their efficiency in heat transfer and storage. We employed novel slurry molding and air felting techniques to construct a series of carbon fiber networks that to optimize thermal interface resistance. The obtained composites are with superior TC (45~144.8 W·m-1·K-1) and enhanced PCM leakage prevention. Our findings demonstrate that the integration of mesophase pitch carbon fibers with PCMs can substantially boost the thermal management capabilities of composite materials, particularly for energy storage applications.

Carbon Marries Silicon for High-Performance Anodes

Shichao Wu [1], Quan-Hong Yang [1]

[1] Nanoyang Group, School of Chemical Engineering and Technology,
Tianjin University, Tianjin 300072, China

Microparticulate Si, normally shelled with carbons, features higher tap density and less interfacial side reactions compared to its nanosized counterpart, showing great potential to be applied as high energy lithium-ion battery anodes. However, localized high stress generated during fabrication and particularly, under operating, could induce cracking of carbon shells and release pulverized nanoparticles, significantly deteriorating its electrochemical performance. Here we design a strong yet ductile carbon cage from an easily processing capillary shrinkage of graphene hydrogel followed by precise tailoring of inner voids. Such a structure, analog to the stable structure of plant cells, presents "imperfection-tolerance" to volume variation of irregular Si microparticles, maintaining the electrode integrity over 1000 cycles with Coulombic efficiency over 99.5%. This design enables the use of a dense and thick (3 mAh cm^{-2}) microparticulate Si anode with an ultrahigh volumetric energy density of 1048 Wh L^{-1} achieved at pouch full-cell level coupled with a $LiNi_{0.8}CO_{0.1}Mn_{0.1}O_2$ cathode.

References

[1] Fanqi Chen, et al. Natl Sci Rev, 8 (2021), nwab012.
[2] Yue Zhai, et al. J. Am. Chem. Soc., 146 (2024), 15209.
[3] Zhenshen Li, et al. Adv. Energy Mater., 13 (2023), 2300874.

Carbon-based mesoporous materials for electrochemical applications

Liang Peng[1]

[1]Department of Mechanical Engineering, City University of Hong Kong, Hong Kong 999077, China

E-mail: lianpeng@cityu.edu.hk

Carbon-based mesoporous materials are popular multifunctional materials, which have great application potentials in various fields. This is mainly because mesoporous carbon materials not only have the properties of general porous nanomaterials, such as adjustable pore size, large pore volume, high specific surface area, etc, but also possess many advantages of unique morphologies or structures. It has a broad prospect in catalysis, hydrogen storage, separation and electrochemistry. So far, many methods have been developed to regulate the morphology and structure of mesoporous materials, including spraying, drip, Stober method, colloid assisted assembly, confined assembly and solid state reaction. However, the morphology and mesoscopic structure synthesized by the above methods still cannot meet the needs of the development of the times. At the same time, the research on the synthesis mechanism is not profound enough, and the application research is also very limited. Therefore, it is of great significance to further develop a simple and powerful method to construct carbon-based mesoporous nanomaterials, to conduct in-depth research on its synthesis mechanism, and to give full play to its electrochemical application potential.

References

[1]. Peng, L.et al, Nat. Protoc., 2023, 18, 1155–1178.
[2]. Peng, L.et al, Nat. Commun., 2023, 14, 8148.
[3]. Peng, L. et al, Science Adv. 2021, eabi7403.
[4]. Peng, L. et al, Chem. 2021, 7, 1020.
[5]. Peng, L. et al, J. Am. Chem. Soc. 2019, 141, 7073.
[6]. Peng, L. et al, J. Am. Chem. Soc, 2022, 144, 15754.

Diamond Materials for Electrochemical Energy Applications

Nianjun Yang

Department of Chemistry & IMO-IMOMEC, Hasselt University
Dipenbeek 3590, Belgium

Conductive diamond owns unique features, such as a wide potential window, high stability in different solutions and under harsh conditions, tuneable surface terminations, rich surface chemistry, and the possibility to form composites with other materials. It thus has been recognized as one of the most promising electrode materials as well as the substrates to load stable catalysts. In the context, conductive diamond (e.g., heavily doped diamond) and their composites have been extensively employed for electrochemical energy applications in recent years. In this presentation, our activities about the design, synthesis, and characterization of several diamond materials will be presented, followed by their applications for the fabrication of high-performance supercapacitors and zinc-air batteries. The confinement behaviour of inert ions (here K^+) inside electrochemically expanded graphene channels grown on diamond will be highlighted. The future research directions of diamond materials for electrochemical energy storage and conversion applications will be also discussed and outlined.

Biomass wastes derived hierarchical porous carbons used for high-performance electric double-layer capacitors

Chih-Hao Chiang[1], Hao-Lun Wang,[1] Tzu-Hsien Hsieh,[2] Guan-Tin Yu,[1] Li-Ming Lu [1], Jarrn-Horng Lin[1,*]

[1]Dept. Materials Science, National University of Tainan
No. 33, Sec. 2, Shu-lin St., Tainan, Taiwan, ROC.
[2]Green Technology Research Institute, CPC Corporation,
No. 2, Tso-Nan Rd., Kaohsiung 81126, Taiwan, ROC
*E-mail: janusjhlin@mail.nutn.edu.tw

We develop a pore-size tunable and highly-pure nanosheet-like activated carbon (AC) form spent coffee grounds (SCGs) with a high specific-surface-area (2497 m2 g-1) and large pore volume (1.28 cm3 g-1). The mesopore/total-pore volume proportions (r) of the as-prepared samples, AC1200 (r = 10.25%), AC1800 (r = 20.10 %), and AC2500 (67.34%), are simply adjusted at desired temperatures (750–900 °C) using physical activation with CO2. The SCG-based electrodes for symmetric electric double-layer capacitors (EDLC) were prepared using free-standing carbon films on cellulose separator, not on Al or Cu foils, without adding any conductive additives. Interestingly, the AC2500-based electrode, a narrow mesopore-rich (mainly 2–4 nm) AC, presents a superior capacitance of 222.4 F g-1 at 0.5 A g-1 in 1.5 M LiClO4/propylene carbonate. The capacitance is superior to those of micropore-rich samples (AC1200 and AC1800) including two commercial products (YP-50 and ACS20). Furthermore, the AC2500-based EDLC exhibits a remarkable energy density of 48.3 Wh kg-1 at 627 W kg-1, and maintains a high energy density of 33.8 Wh kg-1 at 3733 W kg-1. In the present work, we demonstrate that narrow mesopore (2–4 nm) dominant and nanosheet-like carbons would conduct high-performance electrochemical behaviors.

References

[1] J. H. Lin* and S. B. Wang Environ. Technol. Innov. 2017, 8, 17-27.

[2] C.-H. Chiang, J. Chen, J. H. Lin* J. Environ. Chem. Eng. 2020, 8(4), 103929.

[3] T. H. Hsieh, H. L. Wang, G. T. Yu, G. M. Huang, J. H. Lin* J. Environ. Chem. Eng. 2021, 9, 106418.

[4] S. P. Hou, M. D. Liao, Peng, C. Peng, D. Min, X. Y. Li, J. Chen*, G. T. Yu, J. H. Lin* ACS Energy & Fuels 2021, 35(20), 16860-16869.

[5] G.-T. Yu, T.-H. Hsieh, Y.-H. Lin, J. H. Lin* ACS Sustainable Resour. Manage. 2024, 1, 76-87.

Curriculum Vitae

Prof. Jarrn-Horng Lin graduated from National Sun Yat-sen University, Taiwan, in 1989 and received his Ph. D in 1999 from the Department of Chemistry. He is a Professor at the Department of Materials Science of the National University of Tainan since 2013 and became the Chairperson of Dept. Mater. Sci. of NUTN since 2019. Additionally, he served as a Researcher and Project manager at the Research & Development Center of the Chinese Synthesis Rubber Corporation (CSRC) during 1993-2003. He founded the Carbon Society of Taiwan (CST) in 2017; served as the General Secretary of the CST from 2017-2020 and has been a standing supervisor of the CST since 2020.

His research activities focus on carbon nanotubes, general and e-conductive carbon black, polymer reinforcement, energy storage materials (mainly EDLC and sodium-ion batteries), and environmental wastes converting into carbon materials and hydrogen fuels. He has published over 50 SCI papers with a WOS h-index of 21, i10 index of 30. He has given over 25 plenary/keynote/invited lectures at various conferences and won 9 times Reward College Talents of MOST (2013-2023), Taiwan, and the Best Paper Award at the international Nano-Micro Conference in 2018, Dec. 17-20, 2018, Jeju Island, Korea. He also holds and as the chair 8 times of the cross-stair symposium on new carbon materials since 2012. Recently, he held the first international carbon conference-2023 1st international symposium on carbon materials-2023 ISCM-1, Jan.31-Feb.3, National University of Tainan, Tainan, Taiwan.

The Promotion of Emerging Carbon Based Energy Materials for Next-Generation Batteries through Lithium Bond Chemistry

Qiang Zhang, Xiang Chen

Department of Chemical Engineering, Tsinghua University, Beijing 100084, China

As one of the most important applications of the element, Li batteries afford emerging opportunities for the exploration of Li bond chemistry. In this talk, the historical development and concept of the Li bond are reviewed, in addition to the application of Li bonds for rational design of carbon based energy materials are introduced. The Li bond was described in the context of Li batteries, including discussions of sulfur cathodes, liquid electrolytes, and Li metal anodes. Such discourse on the chemistry of the Li bond can provide fruitful insight into the fundamental interactions within lithium-sulfur batteries and thus deliver a deeper understanding of their working mechanism on carbon based energy materials.

References

1) Chen X, Bai YK, Zhao CZ, Shen X, Zhang Q. Angew Chem Int Ed 2020, 59, 11192.

2) Hou TZ, Xu WT, Chen X, Peng HJ, Huang JQ, Zhang Q. Angew Chem Int Ed 2017, 56, 8178

3) Zhang R, Chen XR, Chen X, Cheng XB, Zhang XQ, Yan C, Zhang Q. Angew Chem Int Ed 2017, 56, 7764.

4) Chen X, Zhang Q. Acc Chem Res 2020, 53, 1992.

5) Chen X, Bai YK, Shen X, Peng HJ, Zhang Q. J Energy Chem 2020, 51, 1

6) Shi P, Hou LP, Jin CB, Xiao Y, Yao YX, Xie J, Li BQ, Zhang XQ, Zhang Q. J Am Chem Soc 2022, 144, 212.

7) Hou LP, Zhang X, Yan N, Chen X, Li B, Shi P, Jin C, Huang J, Zhang Q. Chem 2022, 8, 1083.

8) Hou LP, Li Y, Li Z, Zhang QK, Li B, Bi C, Chen Z, Su L, Huang J, Wen R, Zhang X, Zhang Q. Angew Chem Int Ed 2023, 62, e202305466.

9) Shi P, Fu ZH, Zhou MY, Chen X, Yao N, Hou LP, Zhao CZ, Li BQ, Huang JQ, Zhang XQ, Zhang Q. Sci Adv 2022, 8, eabq3445

10) Gao YC, Yao N, Chen X, Yu LG, Zhang R, Zhang Q. J Am Chem Soc 2023, 145, 23764-23770.

11) Cheng XB, Yang SJ, Liu Z, Guo J, Jiang FN, Jiang F, Xiong X, Tang W, Yuan H, Huang JQ, Wu YP, Zhang Q. Adv Mater 2024, 36, 2307370

Curriculum Vitae

Prof. Qiang Zhang is a full professor at Tsinghua University. He obtained his Ph.D. in chemical engineering (2009) from Tsinghua University, China, and subsequently held Research Associate/Postdoc Research Fellow positions in the Case Western Reserve University, USA, and Fritz Haber Institute of the Max Planck Society, Germany. He held the Newton Advanced Fellowship from Royal Society, UK and the National Science Fund for Distinguished Young Scholars. He is selected as highly cited researchers at 2017-2023 by Clarivate Analytics. His current research interests are advanced energy materials, including dendrite-free lithium metal anode, lithium sulfur batteries, and electrocatalysis, especially the structure design and full demonstration of advanced energy materials in working devices. His h-index is 169 now. He is the Advisor Editor of Angew. Chem., Associate Editor of J Energy Chem & Energy Storage Mater. He is sitting on the advisory board of Joule, Chem Soc Rev, Matter, Adv Funct Mater, ChemSusChem, J Mater Chem A, Chem Commun, Energy & Fuels, and so on. He is the deputy head of the expert group on energy storage and smart grid of the national key research and development plan. He has won the first prize of Natural Science of the Ministry of Education, the first prize of Fundamental Science of the Chemical Engineering Society.

INCREASING THE ELECTROCHEMICAL PROPERTIES OF SUPERCAPACITORS DUE TO GRAPHENE DEPOSITED IN A FLAME ON A NICKEL CURRENT COLLECTOR AND ACTIVE GRAPHENE-LIKE CARBON MATERIAL FROM PLANT BIOMASS

N.G. Prikhodko [1,2], M.A. Yeleuov [1,3], A.A. Abdisattar [1,3], K. Askaruly [1,3], A.T. Taurbekov [1,4], A.B. Tolynbekov [1,4], N.B. Rakhymzhan [1]

[1] Institute of Combustion Problems, Almaty, Kazakhstan
[2] Almaty University of Power Engineering and Telecommunications, Almaty, Kazakhstan
[3] Satbayev University, Almaty, Kazakhstan
[4] Al-Farabi Kazakh National University, Almaty, Kazakhstan
Email: nik1951999@gmail.com

This study presents a new approach using flame-synthesized graphene onto a metal current collector for energy storage devices. Graphene was synthesized in a premixed propane/oxygen/argon flame under atmospheric conditions at flow rates of 170/370/160 cm3/min, respectively; synthesis temperature 900-950°C, synthesis time 30-60 s; ratio C/O=0.65-0.7. This method successfully produced 5 to 10 layers of graphene on a nickel substrate with high growth rates (30–60 s). Ultrasonic treatment was used to clean the current collector from the amorphous soot structure. Next, the nickel substrate with graphene was coated with an active graphene-like carbon material obtained from barley straw and used as a current collector in supercapacitors. Studies have shown that the specific capacitance of a supercapacitor based on a current collector coated with graphene is 21% higher compared to an uncoated current collector: 218 F/g and 180 F/g at 1 A/g, respectively. At 5000 charge-discharge cycles, the specific capacity of supercapacitors remains at 98%.

This study presents a scalable and simple concept of incorporating thin layers of graphene onto a metal current collector to improve the electrochemical performance of the electrode. The possibility of continuous processing and the high growth rate of graphene contribute to the effectiveness of the method. The study demonstrates the potential of flame-synthesized multilayer graphene to improve the electrochemical properties of supercapacitor electrodes and its suitability for large-scale production, making it a promising direction for future advances in energy storage.

Curriculum Vitae

Academic degree: Doctor of Chemical Sciences

Academic title: Professor

Position: Head of the Laboratory of Synthesis of Carbon Nanomaterials in Flames,

Chief Researcher at the Institute of Combustion Problems

Educational activity: Professor at G. Daukeev Almaty University of Power Engineering and Telecommunications

Differential Design Strategies for High-Capacity Hard Carbon Anodes in Lithium, Sodium, Potassium-ion Batteries

Jilei Liu [1*]

[1] College of Materials Science and Engineering, Hunan Joint International Laboratory of Advanced Materials and Technology for Clean Energy, Hunan Province Key Laboratory for Advanced Carbon Materials and Applied Technology, Hunan University, Changsha 410082, China. Email: liujilei@hnu.edu.cn

The intricate microstructures of hard carbons together with the diverse physicochemical properties of alkali metal ions contribute to the enigmatic nature of Li/Na/K ions storage mechanisms. The current state of research still lacks comprehensive theoretical models and structural design strategies, making it challenging to optimize the performance of hard carbons for Li/Na/K-ion batteries in a targeted manner. Herein, we quantitatively investigated the extrinsic defects, crystallite parameters, and nanopores of resin-derived hard carbons and associated these microstructural parameters with the electrochemical performance. Our findings demonstrate that enabling a sufficient large graphene interlayer spacing (> 3.9 Å) could facilitate Li migration into the closed pores for clustering, resulting in satisfactory reversible capacity (497 mAh/g). Additionally, fabricating abundant closed pores with suitable graphene interlayer spacing (≈ 3.7 Å) is a compelling pathway to achieve fascinating reversible capacity in Na/K-ion batteries (530 mAh/g for Na, 520 mAh/g for K). Our work highlights the importance of the combined effects of adsorption, intercalation, and nanoclustering for achieving a reversible capacity of approximately 500 mAh/g in Li/Na/K-ion batteries, and provides clear design guidelines for the purposeful tuning of the structural characteristics to optimize their electrochemical performance.

References

[1] Jilei Liu*, et al. Adv. Mater. 2023, 35, 2209833

[2] Jilei Liu*, et al. Energy Environ. Sci. 2024, 17, 274

[2] Jilei Liu*, et al. Adv. Funct. Mater. 2024, 34, 2312994

[2] Jilei Liu*, et al. Adv. Funct. Mater. 2024, 2405174

[2] Jilei Liu*, et al. Carbon. 2023, 205, 353

Seamless 3D graphene framework with enriched topological defects for battery-related applications

Hirotomo Nishihara

Advanced Institute for Materials Research (WPI-AIMR), Tohoku University, 2-1-1 Katahira, Aoba-ku, Sendai, Miyagi, 980-8577, Japan

Graphene MesoSponge® (GMS) is a new type of graphene-based mesoporous material synthesized via template-directed chemical vapor deposition, followed by template removal and high-temperature annealing at 1800 °C [1,2]. GMS primarily consists of single-layer graphene walls with a minimal number of edge sites, exhibiting extreme elasticity (Fig. 1) [3]. Using in-house temperature-programmed desorption (TPD) up to 1800 °C, which can determine the number of edge sites in any sp2-based carbon materials, it has been revealed that GMS has fewer edge sites than single-walled carbon nanotubes (SWCNTs). Consequently, GMS outperforms SWCNTs in terms of a wide voltage window, reaching up to 4.4 V when assembled in a supercapacitor [4].

Another unique feature of GMS is its topological-defect-rich framework. From the geometric requirements, GMS necessarily includes many carbon non-hexagon rings including 5- and 7-membered rings, which are called topological defects [5]. The topological defects can play the role of anchoring sites for metal nanoparticles such as Pt [6] and Ru [7], which are useful to fuel cell and Li-O2 battery applications, respectively. Furthermore, GMS can be molded into self-standing sheets with controllable macroporosities. The optimized structure outperforms any carbon-based cathodes for Li-O2 batteries [8].

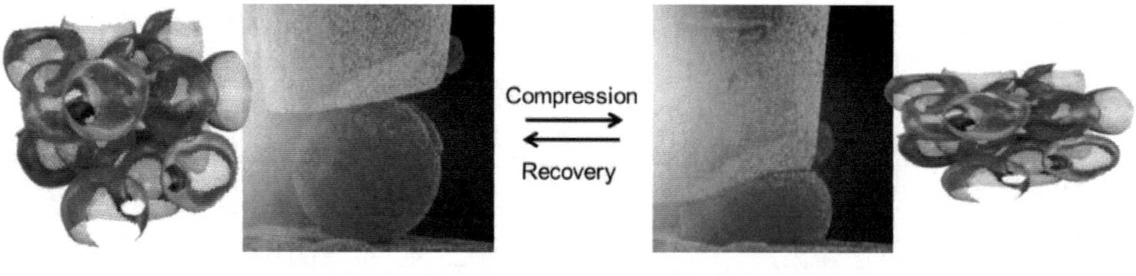

Fig. 1 Unique mechanical elasticity of GMS.

References

[1] H. Nishihara et al., Adv. Funct. Mater. 26, 6418 (2016).
[2] S. Sunahiro et al., J. Mater. Chem. A 9, 14296 (2021).
[3] K. Nomura et al., Nat. Commun. 10, 2559 (2019).
[4] K. Nomura et al., Energy Environ. Sci. 12, 1542 (2019).
[5] W. Yu et al., Adv. Sci. 10, 2300268 (2023).
[6] A. Ohma et al., Electrochim. Acta 370, 137705 (2021).
[7] Z. Shen et al., J. Phys. Chem. C 127, 6239-6247 (2023).
[8] W. Yu et al., Adv. Energy Mater. 14, 2303055 (2024).

The Development and Trend of Carbon for Lithium-ion Batteries

Xi Chen, Jianguo Ren

BTR New Material Group Co.,Ltd.

Carbon plays an important role in the development of lithium-ion batteries. This report focus on the application of carbon in lithium-ion batteries, such as artificial graphite, natural graphite, and carbon for Si-based anode. The market demand, development history, and research progress of anode will be discussed, as well as the technology trend and application capability in market of carbon in lithium-ion batteries.

Xi Chen got her Ph. D degree from Purdue University in 2020 and her major is Materials Engineering. Upon graduation, she has worked as a R&D Engineer in BTR New Material Group CO. LTD. Her main job responsibility is developing low expansion & high-performance Si-based anode materials.

Design of polymeric derived porous carbon spheres with in-situ nanoconfinement for energy storage

Jiaying Yang, Xinren Zhang, Fei Xu

[1] Center for Nano Energy Materials, School of Materials Science and Engineering, Northwestern Polytechnical University and Shaanxi Joint Laboratory of Graphene (NPU), 710072, P. R. China

Porous hollow carbons are unique in that they show high surface area, tunable morphology and tailored functionality, and thus they demonstrate great potential as key materials in rechargeable metal batteries. Nevertheless, precise manipulating the carbon pore structure with predicted high surface area and well-anchored high active guest materials remains challenging. In this report, we will first briefly show the strategies to improve the energy storage performances via high-porosity manipulation, and then with an emphasis on the in-situ pore nanoconfinement. In contrast to the conventional tedious and multistep procedures with large particle size and poor orientated distribution, we propose the interfacial micelle copolymerization with in-situ particle confinement strategy. In this context, atomic metals, nanoclusters, nanoparticles are targeted engineered into the nanospace of hollow carbon spheres, including confining in the shell micropores, embedding within the shell, encapsulating at the inner walls and hollow cavities. The key is to manipulate the driving force for orientated particle encapsulation while without disturbing the micelle interfacial copolymerization for hollow cavity formation. Then the synergistic effect of single atoms/ nanoparticles and hollow architecture as anode for metal ion batteries and as seeding/hosting interphases for dendrite-free Na and Zn metal anode was investigated. The results could be of great significance towards the durable, high-rate and energy-dense energy storage devices.

References

[1] F Xu, et. al., Matter, 3 (2020) 246-260.
[2] F Xu, et. al., Sci. Adv. 8 (2022) eabm7489.
[3] J Yang, et. al., Carbon 205 (2023) 171-179.
[4] X. Zhang, et. al, Adv Energy Mater (2024) 2401139.

New Insights into Carbide-Dichalcogenide Heterostructures: Synthesis, Superconductivity, Electrocatalysis and more

Mauricio Terrones [1*]

[1]1Department of Physics, Department of Chemistry, Department of Materials Science and Engineering and Center for 2-Dimensional & Layered Materials. The Pennsylvania State University, University Park, Pennsylvania 16802, USA

(*mut11@psu.edu)

This talk will summarize the progress in the synthesis, characterization, and computational studies of transition metal carbide (TMC) / transition metal dichalcogenide (TMD) heterostructures and will provide a perspective on future directions of the evolution of this emerging field. To begin, we will introduce TMDs and ultrathin TMDc (UThTMCs) and deposition synthesis routes towards isolating these ultrathin materials. Then, we will discuss the synthesis of TMC/TMD heterostructures, covering the partial conversion of TMCs to TMDs, and vice versa, along with experimental observations and computational predictions of the diverse set of fascinating heterostructures. In particular, we will discuss an enhancement of superconductivity in α-Mo_2C/γ'-MoC_{1-x}/MoS_2 heterostructures obtained by controlled sulfurization of thin flakes of α-Mo_2C. The heterostructures displayed a more than 50% greater Tc (6.8 K) when compared to pristine α-Mo_2C crystal (4.0 K). The reasons that lead to Tc enhancement are attributed to the interfacial strain and the presence of the metastable γ'-MoC_{1-x} phase that was stabilized by sulfurization of the α-Mo_2C, which may favor a higher Tc by increasing the frequency of relevant phonon modes and/or increasing the electron-phonon coupling. We will also summarize results related to electrocatalysis in these heterostructures and discuss challenges in this new field.

References

[1] Sredenschek, A.J., Terrones, M., et al. Heterostructures coupling ultrathin metal carbides and chalcogenides. Nature Materials 2024/4/1.

[2] Zhang, F., Terrones, M, et al. Superconductivity enhancement in phase-engineered molybdenum carbide/disulfide vertical heterostructures. Proceedings of the National Academy of Sciences 117, 19685–19693 (2020).

[3] Zhang, F., Terrones, M., et al. Carbon doping of WS2 monolayers: Bandgap reduction and p-type doping transport. Sci Adv 5, (2019).

Functional carbon materials for advanced electro-technology

Jieshan Qiu

College of Chemical Engineering, State Key Laboratory of Chemical Resource Engineering, Beijing University of Chemical Technology, Beijing, 100029, China

Email: qiujs@mail.buct.edu.cn

Functional carbon materials with tuned structure and properties are one of the key components in most of the electrochemical systems such as energy storage and conversion. In this talk, several examples will be presented to show the potential of carbon materials in the electroreduction of carbon dioxide, nitrogen, the electrocatalytic oxidation of glycerol to formic acid, and the cascade reaction for highly selective electrosynthesis of aldehyde. The intrinsic relationship between the structure and properties of the carbon materials and the catalytic performance will be analyzed, which may shed a light on the design and fabrication of carbon-supported heterogeneous catalysts with tuned interface chemistry.

References

[1] Ren YW, Li SF, Yu C, Zheng YH, Wang C, Qian BZ, Wang LS, Fang WH, Sun Y, Qiu, JS. J. Am. Chem. Soc., 146(10), 6409-6421(2024).

[2] Xie YY, Yu C, Ni L, Yu JH, Zhang YF, Qiu JS. Adv. Mater., 35(14), 2209652(2023).

[3] Guo W, Dun CC, Marcus MA, Venturi V, Gainsforth Z, Yang FP, Feng XF, Viswanathan V, Urban JJ, Yu C, Zhang QY, Guo JH, Qiu JS. Adv. Mater., 35(19), 2211603(2023).

[4] Huang HL, Son XD; Yu C; Wei QB; Ni L; Han XT; Huang HW; Han YN; Qiu JS. Angew. Chem. Int. Ed., 62(4), e20221632(2022).

[5] Wei QB; Yu C; Song XD; Zhong YP; Ni L; Ren YW; Guo W; Yu JH; Qiu JS. J. Am. Chem. Soc., 143, 6071-6078(2021)

[6] Chang JW; Yu C; Song XD; Tan XY; Ding YW; Zhao ZB; Qiu JS. Angew. Chem. Int. Ed., 2021, 60, 3587-3595

]7] Huang HL; Yu C; Han XT; Huang HW; Wei QB; Guo W; Wang Z; Qiu JS. Energy Environ. Sci., 13, 4990-4999(2020)

[8] Han XT; Sheng HY; Yu C; Walker TW; Huber GW; Qiu JS; Jin S. ACS Catalysis, 10, 6741-6752(2020).

Curriculum Vitae

Dr. Jieshan Qiu, Cheung-Kong Distinguished Professor of Carbon Science and Chemical Engineering at Beijing University of Chemical Technology, China. His research encompasses both fundamental and applied aspects of carbon materials and science, with a focus on the methodologies of producing carbon materials for energy storage and conversion, catalysis, and environment protection. He has published 900+ papers in peer-reviewed high-profile journals, including Nat. Mater., Adv. Mater., Energy Environ. Sci., PNAS, Angew. Chem. Int. Ed., J. Am. Chem. Soc., etc., with citations 66500+ times and h-index of 128 (Google scholar), and his work has been featured on 70 covers of international journals. He has been honored by 30+ prestigious awards and prizes including the 1st class award from the Education Ministry of China. He is a highly cited researcher named both by Clarivate Analytics from 2018 to 2023 and by Elsevier from 2019 to 2023. He is now the editor of three journals including Chemical Engineering Science (Elsevier) and Battery Energy (Wiley).

Porous Carbons for Gas Separation

An-Hui Lu

State Key Laboratory of Fine Chemicals, School of Chemical Engineering,
Dalian University of Technology, Dalian, 116024

Porous carbon materials have been widely used as adsorbents in gas separation because of their intrinsically good chemical-, thermal- and moisture-resistant properties.[1] For practical application, i.e., in a dynamic flow situation, the adsorption capacity of the porous carbons is determined largely by the adsorption kinetics, which is closely related to size, depth, shape and connectivity of the pores.

We have synthesized porous carbons, derived from poly(benzoxazine) system, with multiple-length-scale porosity and nitrogen-containing framework. These materials showed outstanding CO2 capture and separation capacities.[2] Moving forward in design of porous carbon materials with nanometer-scale skeleton, short diffusion paths and high microporosity, these porous carbons greatly accelerate CO2 adsorption kinetics, increasing the utilization degree of overall porosity and surface area.[3]

Moreover, we have developed a thermoregulated phase transition assisted synthesis of carbon nanoplates with over ~80% sp2 carbon, unimodal ultramicropore and controllable thickness.[4] Thin structure allows orientated growth of carbon crystallites, and stacking of crystallites in nearly parallel orientation manner are responsible for single-sized micropores. When used for gas separation from CH4, carbon nanoplates exhibit high uptakes (5.2, 5.3 and 5.1 mmol g-1) and selectivities (7, 71 and 386) for CO2, C2H6 and C3H8 under ambient conditions. The dynamic adsorption capacities are close to equilibrium uptakes of single components, further demonstrating superiority of carbon nanoplates in terms of selectivity and sorption kinetics.

References

[1] An-Hui Lu, et al., ChemSusChem, 2021, 14(6): 1428-1471.
[2] An-Hui Lu, et al., Adv. Mater., 2010, 22(7): 853-857.
[3] An-Hui Lu, et al., J. Am. Chem. Soc., 2011, 133(29): 11378-11388.
[4] An-Hui Lu, et al., Nat. Chem. Eng., 2024, 1: 411-420.

Photocatalytic regeneration of saturated activated carbons used in water treatment: a pilot scale study

Nadia El-Ouahedy[1], Kana Maiga[1], Esther Gomez[2], Yolanda Ballesteros[2], Roberto Torio[2], Conchi O. Ania[1,3*]

[1] CEMHTI, CNRS (UPR 3079), Université d' Orléans, 45071 Orléans, France.

[2] SOCAMEX SAU, Valladolid, Spain

*E-mail: conchi.ania@cnrs-orleans.fr

Despite adsorption using activated carbons is a mature technology for the removal of emerging pollutants, the cost associated to the carbon regeneration still represents a major limitation for further implementation [1]. Hence, the reactivation of saturated activated carbons is a crucial aspect in the upgrading and restructuring of existing water and wastewater treatment plants. Our studies have pointed out the ability of nanoporous carbons to generate radical oxygen species upon illumination, and induce the photooxidative degradation of pollutants adsorbed in the porosity [2,3]. This work describes the photocatalytic regeneration of spent activated carbons upon UV exposure in aqueous environments. Irradiation of the spent carbon resulted in a gradual regeneration, which efficiency depended on the saturation state of the carbon and the illumination time. A fraction of the pollutants retained in the porosity of the carbon may be released into the solution upon illumination of the carbon particles, being subsequently photodegraded. Furthermore, data of a case study on spent activated carbons from a wastewater treatment plant equipped with activated sludge technology and granular activated carbon in Almendralejo (Spain) will be presented and discussed in terms of the quality of the reactivated carbon for further reutilization at the WWTP, and the quality of the effluent generated upon the treatment [4].

Acknowledgements: This work has received funding from the European Union' s Horizon 2020 research and innovation program under the grant agreement N° 776816 (Project Ô, grant 64816).

References

[1] Roskill Report. Activated Carbon, Global Industry, Markets and Outlook, 2017, p. 350.

[2] T.J. Bandosz, C.O. Ania*, Adv. Sci. 2018, 5, 1800293-315.

[3] L.F. Velasco, R.J. Carmona RJ, J. Matos, C.O. Ania*, Carbon 2014, 73, 206-215.

[4] Patent Application (BET 23P4307 EMO/SF n° 23306872.5). Inventors: C. Macias, C.O. Ania*. Water treatment method with regeneration of activated carbon particles by UV light.

Which are the photocatalytic active sites in carbocatalysts? Examples in various photocatalytic reactions

Juan Matos [1,*], Po S. Poon [2] and Conchi O. Ania [3]

[1]Unidad de Cambio Climático y Medio Ambiente (UCCMA), Instituto Iberoamericano de Desarrollo Sostenible (IIDS), Facultad de Arquitectura, Construcción y Medio Ambiente, Universidad Autónoma de Chile, Temuco, Chile.

[2]Unidad de Desarrollo Tecnológico (UDT), Universidad de Concepción, Barrio Universitario s/n, Concepción, Chile.

[3]CEMHTI, CNRS (UPR 3079), Université d' Orléans, 45071 Orléans, France.

(12pt Times New Roman Italic)

The catalytic activity of metal-free and semiconductor-free carbon photocatalysts (herein called as carbocatalysts) will be presented and discussed. Emphasis will be placed on selected examples of such photocatalytic activity for the degradation of aqueous pollutants driven by solar irradiation. The influence of the nanoconfinement of the target pollutant inside the pores of those carbons, and that of heteroatoms doping on the photochemical reactivity will be reviewed and discussed. This allows us to introduce a new definition of a catalytic turnover frequency parameter when applied to photocatalytic reactions using carbocatalysts (PTOF) for a correct reporting of their photocatalytic activity. The PTOF parameter is described considering the textural properties of the carbon materials. In addition, we will analyze the catalytic reaction rate constants as a function of the surface density of the adsorbed molecules. The latter kinetic parameter allows the photochemical reactivity of the carbocatalysts to be normalized in terms of the frequency of the photocatalytic turnover. Correlations between the PTOF, the porosity (surface area, pore size distribution, pore volumes) and the surface chemistry of the carbocatalysts will be presented and discussed, in order to understand and better describe the photocatalytic performance of these carbon materials.

below.

Acknowledgements. This research is ascribed within the cooperation framework between CNRS (UPR 3079) and Autonomous University of Chile. J. Matos and P.S. Poon thank the financial support of the Chilean ANID projects: Anillo-ATE220014, Fondecyt-1220228, Fondef-ID23I10085, and Fondecyt-1240441.

Carbon Materials for Interfacial Solar Evaporation: Fundamental & Applications

Yan Song [1] and Jia Zhu [1,2]

[1] National Laboratory of Solid State Microstructures, College of Engineering and Applied Sciences, Jiangsu Key Laboratory of Artificial Functional Materials, Collaborative Innovation Center of Advanced Microstructures, Frontiers Science Center for Critical Earth Material Cycling, Nanjing University; Nanjing, 210093, China.

[2] School of Sustainable Energy and Resources, Nanjing University; Suzhou, 215163, China.

At the nexus of energy and water, interfacial solar evaporation is attracting worldwide attention as an emerging technology for both solar energy and water treatment. Carbon-based materials are particularly noteworthy among the potential materials for solar evaporators due to their high solar absorption, adjustable structure, easy preparation, low cost, and abundance in the Earth. In this talk, we will discuss the advancement of carbon-based materials for interfacial solar evaporation from the perspective of both fundamental and applications. We report that the solar evaporation rate can be well above the assumed limit by exploiting energy input from the environment through a three-dimensional carbon evaporator. Inspired by the transpiration process in plants, we report an artificial transpiration device with a unique design of two-dimensional water path. The energy transfer efficiency of this artificial transpiration device is independent of water quantity and can be achieved without extra optical or thermal supporting systems, therefore significantly improving the scalability and feasibility of this technology. At the end, we would like to demonstrate that this type of interfacial solar vapor generation can have direct implications in various fields such as solar desalination, zero liquid discharge, sterilization, and atmospheric water harvesting.

Controlled Synthesis and Applications of Ultralong Carbon Nanotubes

Rufan Zhang [1]

[1] Tsinghua University, Beijing 100084, China

Ultralong carbon nanotubes (CNTs) refer to the CNTs with lengths over centimeters, which have perfect structures and excellent mechanical, electrical, and thermal properties. Ultralong CNTs are ideal candidates for fabricating CNT fibers, transparent conductive films, carbon-based chips, etc. However, the biggest challenges faced with ultralong CNTs is their extremely low yields, which severely restricts their practical applications. In this report, I will talk about the controlled synthesis and high-yields production of ultralong CNTs using a substrate interception and direction strategy developed by our group. Besides, I will also talk about the property study and application exploration of ultralong CNTs in various fields.

References

[1] Qinyuan Jiang, Rufan Zhang*, Adv. Mater., 2024, doi: 10.1002/adma.202402257
[2] Fengxiang Chen, Rufan Zhang*, Sci. Adv., 2022, 8(26), eabn5882.
[3] Qinyuan Jiang, Rufan Zhang*, Nano Lett., 2023, 23(2), 523-532.
[4] Qinyuan Jiang, Rufan Zhang*, Adv. Funct. Mater., 2022, 33, 2212665.
[5] Yanlong Zhao, Rufan Zhang*, ACS Nano, 2023, 17(3), 2893-2900.
[6] Yanlong Zhao, Rufan Zhang*, ACS Appl. Mater. Interf., 2024, 16, 20949–20958.
[7] Run Li, Rufan Zhang*, SusMat, 2023, 3(1), 102-110.
[8] Qinyuan Jiang, Rufan Zhang*, Adv. Mater., 2022, 34(18), 2107062.
[9] Yunxiang Bai, Rufan Zhang*, Fei Wei*, Science, 2020, 369(6507), 1104-1106.

An Organic Electrochemical Transistor for Multi-modal Sensing, Memory and Processing

Shijie Wang [1], Wei Ma [1*]

[1] State Key Laboratory for Mechanical Behavior of Materials, Xi' an Jiaotong University, China

By integrating sensing, memory and processing functionalities, biological nervous systems are energy and area efficient. Emulating such capabilities in artificial systems is, however, challenging and is limited by the device heterogeneity of sensing and processing cores. Here, we present a universal solution to simultaneously perform multi-modal sensing, memory and processing using organic electrochemical transistors [1]. The device has a vertical traverse architecture and a crystalline–amorphous channel that can be selectively doped by ions to enable two reconfigurable modes: volatile receptor and non-volatile synapse. As a volatile receptor, the device is capable of multi-modal sensing (chemical species, temperature, light etc.). As a non-volatile synapse, it is capable of 10-bit analogue states, low switching stochasticity and good state retention. Homogeneous integration of such devices enables functions such as conditioned reflex and real-time cardiac disease diagnose via reservoir computing, illustrating the promise for future edge AI hardware.

References

[1] Wang S., Ma W.* et al., Nature Electronics, 6, 281-291 (2023).

Carbonized Polymer Dots: Syntheses, Properties and Applications

Bai Yang

State Key Laboratory of Supramolecular Structure and Materials, College of Chemistry, Jilin University, Changchun, China

In the past decade, a new exciting class of luminescence nano-building blocks, carbon dots (Cds), has been developed rapidly [1]. As a kind of CDs, CPDs are typically synthesized by the bottom-up crosslinking-polymerization and carbonization routes from many kinds of small molecules or polymers with multi-functional groups [2]. This presentation will mainly focus on the syntheses, properties and PL mechanism of CPDs. CPDs often have high luminous efficiency, which comes from the molecular state centre in CPDs [2,3]. The crosslink-enhanced emission (CEE) effect have been also confirmed in CPDs [4,5]. To give a whole summary, the category [6] and synthesis routes [7], as well as long wave fluorescence [8], room temperature phosphorescence [9] and potential applications of CPDs [10], are also briefly introduced in this presentation.

(This work was financially supported by NSFC 22035001)

References

[1] S.J. Zhu, Y.B. Song, & B. Yang, et al. Nano Research 8 (2015) 335-382

[2] S.J. Zhu, & B. Yang, et al. Angew. Chem. Int. Ed. 52 (2013) 3953-3957

[3] Y.B. Song, S.J. Zhu, & Yang B., et al. J. Mater. Chem. C 3 (2015) 5976

[4] S.J. Zhu, & B. Yang, et al. Angew. Chem. Int. Ed. 54 (2015) 14626-14637

[5] S.Y. Tao, & B. Yang, et al. Angew. Chem. Int. Ed. 59 (2020) 9826

[6] Q.S. Zeng, & B. Yang, et al. Light: Science & Applications 10 (2021) e142

[7] C.L. Xia, & B. Yang, et al. Adv. Sci. 6 (2019) 1901316

[8] S.Y. Lu, & B. Yang, et al. Adv. Mater. 29 (2017) 1603443

[9] S.Y. Tao, S.Y. Lu, & B. Yang, et al. Angew. Chem. Int. Ed. 57 (2018) 2393-2398

[10] J.J. Liu & B. Yang et al. ACS Central Science 6 (2020) 2179

Single Atom Catalysis

Dingsheng Wang

Tsinghua University, China

Single atom catalysts represent an important class of catalytic materials that provide platforms to study heterogeneous catalysis and homogeneous catalysis. The correlation between their structure and performance has attracted extensive research attention. Revealing catalytic nature and regularity of catalysts is a significant issue in chemistry and chemical engineering science and technology. Understanding catalytic active sites and centers, identifying key factors that affect catalytic performance, applying advanced achievements obtained by nanotechology in catalytic basis and applied research to enhance activity, selectivity and stability and developing and fabricating efficient and exclusively selective natural biological enzyme catalysts that can achieve catalysis at room temperature and ordinary pressure are the most challenging research issues in the era. We have developed a series of new synthetic methodology to achieve precisely controllable synthesis of nanocrystal, cluster, and single-atom catalysts, and achieved the structure regulation of catalysts at atomic scale and deeper understanding of structure-performance relationship.

Carbon nanotube based fiber bioelectronic materials and devices

Xuemei Sun

State Key Laboratory of Molecular Engineering of Polymers, Department of Macromolecular Science, and Institute of Fiber Materials and Devices, Fudan University, Shanghai, 200438, China.

Bioelectronic devices can realize real-time monitoring and precise regulation of human physiological information, which are of great significance in the fields of health monitoring and disease diagnosis and treatment. However, conventional bioelectronic devices are mainly based on rigid materials such as silicon and metal, whose mechanical properties mismatch with soft biological tissues, making it difficult to form a stable device/tissue interface. And this will lead to device failure and tissue damage under physiological environment, especially during long-term work in vivo. To this end, we develop a series of flexible fiber bioelectronic devices based on carbon nanotubes, which have unique advantages of miniaturization, lightweight, flexibility and minimally invasive implantation. Firstly, composite carbon nanotube fibers are prepared by multiscale assembly process, which has high flexibility, high conductivity and biosafety. Furthermore, fiber bioelectronic devices are constructed by introducing different active materials into fiber electrodes, including fiber biosensors that can detect neurotransmitters and glucose in vivo, fiber stimulators that can change physiological microenvironment, and fiber energy devices that can generate electricity from biological chemicals. Finally, these flexible fiber devices are implanted into the blood vessel and brain of animals through minimally invasive injection, and realize long-term monitoring of various physiological information in vivo, such as ions, dopamine and glucose. These flexible fiber devices provide a new approach for the fields of biomedical research and health management.

References

[1] Wang, L.; Xie, S.; Wang, Z.; Liu, F.; Yang, Y.; Tang C.; Wu, X.; Liu, P.; Li, Y.; Saiyin, H.; Zheng, S.; Sun, X.; Xu, F.; Yu, H.; Peng, H. Nat. Biomed. Eng. 2020, 4, 159-171.

[2] Yang, H.; Qian, Z.; Wang, J.; Feng, J.; Tang, C.; Wang, L.; Guo, Y.; Liu, Z.; Yang, Y.; Zhang, K.; Chen, P.; Sun, X.; Peng H. Adv. Funct. Mater. 2022, 32, 2204794.

[3] Qian, Z.; Yang, Y.; Wang, L.; Wang, J.; Guo, Y.; Liu, Z.; Li, J.; Zhang, H.; Sun, X.; Peng, H. Angew. Chem. Int. Ed. 2023, 62, e202303268.

[4] Wang, L.; Wan, F.; Xu, Y.; Xie, S.; Zhao, T.; Zhang, F.; Yang, H.; Zhu, J.; Gao, J.; Shi, X.; Wang, C.; Lu, L.; Yang, Y.; Yu, X.; Chen, S.; Sun, X.; Ding, J.; Chen, P.; Ding, C.; Xu, F.; Yu, H.; Peng, H. Nat. Nanotech. 2023, 18, 1085.

[5] Wang, J.; Wang, L.; Yang, Y.; Li, H.; Jiang, H.; Huang, X.; Liu, Z.; Yu, S.; Tang, C.; Chen, J.; Shi, X.; Li, W.; Chen, P.; Tong, Q.; Yu, H.; Sun, X.; Peng H. Adv. Mater. 2024, 2309862.

Non-invasive Salivary Glucose Biosensor Based on Conductive Polymer Hydrogels/carbon nanotubes

C. H. Liu[1], C. J. Liao[2], S. Gupta[1], D. W Huang[2], C. Y. Lee[1], Y. T. Lai[2] and N. H. Tai[1]

[1]Department of Materials Science and Engineering, National Tsing Hua University, Hsinchu, 30013, Taiwan, ROC

[2]Department of Materials Engineering, Ming Chi University of Technology, New Taipei City, 24301, Taiwan, ROC.

In this work, we develop a non-invasive salivary glucose biosensor based on a core-shell nanostructure of conductive polymer hydrogels/carbon nanotubes (CNTs) on carbon paper. Due tothe electrostatic interaction between pyrrole monomer and the functional groups on CNTs, the crosslinked polypyrrole (PPy) hydrogel forms in situ on the CNT surface. The 3D interconnected networks of core-shell PPy/CNTs feature high porosity and high surface area, facilitating efficient electron and ion transport, thereby leading to a highly sensitive glucose sensor in the region of 50-700 μM. Furthermore, excellent performance on selectivity tests of the as-fabricated sensor in the presence of various interferences is achieved, demonstrating a highly potential application of the sensor on non-invasive glucose measurement.

Chemical construction of magnetic molecular probe and its application in tumor diagnosis and treatment

Yonglong Hou[1,2*]

[1]Sun Yat-sen University, Shenzhen, China, 518107

[2]Peking University, Beijing, China, 100871

*E-mail: hou@pku.edu.cn

Magnetic nanostructures have important applications in the fields of information storage, biomedicine and nano-catalysis [1]. The synthesis of monodisperse magnetic nanomaterials is challenging due to their high specific surface area and magnetic dipole interaction. Based on the principle of surface-interface chemistry, we have realized the controllable preparation of magnetic nanostructures by using a series of complex precursors, they include single-component nanostructures (such as metals, metal alloys, metal oxides/carbides) and multi-component nanomaterials (heterostructures and exchange-coupled nanomagnets) . In particular, heterostructures benefit from the synergistic properties of interface coupling, improving magnetic, optical, or catalytic properties [1]。 Among the magnetic nanostructures, the photothermal conversion properties of FeCx were found, which laid a foundation for the construction of magnetic nanoprobes for the integrated diagnosis and treatment of cancer [2,3]. First, a novel multi-functional magnetic nanoprobe Fe5C2-ZHER2:342 was designed for targeted photothermotherapy of breast cancer using magnetic resonance imagin and photoacoustic imaging, PH-sensitive nano-probe PEG/Fe5C2@Fe3O4 was designed, As a reactive oxygen species (ROS) nano-reactor, T2T1 can be used as a magnetic resonance imagin for breast cancer treatment and degradation. Furthermore, the smart nano-probe DOX-ICG@Fe/FEO-PPP-FA was constructed by the modification of coordination targeting, which was used in the dual-modality (MRI and fluorescence imaging)-mediated combination therapy of human oral epidermis carcinoma, and showed a remarkable effect in diagnosis and treatment [5]. In order to increase the penetration depth of laser, a multifunctional infrared bidomain-responsive AG2S@FE2C-DSPE-PEG-IRGD magnetic molecular probe was constructed to stimulate the tumor microenvironment response through the activation of Nanase, to achieve the synergistic treatment of photothermal and chemotherapy for breast cancer [6,7]. Recent studies have shown that FeS-PEG-CAI-NPs can induce acidosis and hydroxyl free radicals, which synergize with chemical kinetics photothermic gas therapy in dynamic tumor diagnosis and treatment [8,9] . In summary, a series of multifunctional magnetic nanostructures have been synthesized, and a variety of diagnostic and therapeutic nanoprobes have been constructed.

KeyWords: Magnetic nanostructure, chemical synthesis, surface-interface interaction, tumor diagnosis and treatment, molecular probe

References

[1] Wang, S.; &Hou, Y. et al. Chem. Rev. 2022, 122, 5411–5475.
[2] Yang, C.; & Hou, Y. et al. J. Am. Chem. Soc. 2012, 134, 15814–15821.
[3] Yu, J.; & Hou, Y. et al. Adv. Mater. 2014, 26, 4114–4120.
[4] Wang, Z.; & Hou, Y. et al. Nat. Commun. 2019, 10, 4418.
[5] Wang, Z.; & Hou, Y. et al. Sci. Adv. 2020, 6, eabc8733.
[6] Wang, S.; & Hou, Y. et al. Sci. Adv. 2022, 8, eabn3883.
[7] Wang, J.; & Hou, Y. et al. J. Am. Chem. Soc. 2022, 144, 19884–19895.
[8] Sun, Z.; & Hou, Y. et al. J. Am. Chem. Soc. 2023,145, 11019–11032.
[9] Wang, J.; & Hou, Y. et al. Adv. Mater. 2023, 2307006.

Curriculum Vitae

Yonglong Hou, Sun Yat-sen University

Chair Professor, Dean of the School of Materials, Royal Society of Chemistry member (FRSC) , Chinese Chemical Society member (FCCS). It is mainly engaged in the chemical controllable synthesis of multifunctional magnetic nanostructures, and has developed a general preparation method of monodisperse magnetic nanomaterials, and explored the application of magnetic nanoparticles in the diagnosis and treatment of important diseases such as tumors. So far, more than 260 academic papers have been published and cited more than 29,000 times. There are 21 patents filed and 16 patents granted. In 2019, he won the second-class National Natural Science Award. He was awarded the National Innovation Award, the Beijing Mao Yisheng Youth Science and Technology Award, and the Chinese Chemical Society-Royal Society of Chemistry Youth Chemistry Award. He was supported by the National Outstanding Youth Science Foundation, and was selected as a distinguished professor of the Minister of Education. More than 100 times at international and various bilateral meetings. He is currently an international consultant editor of journals including Advanced Science, Advanced Healthcare Materials, Advanced Therapeutics and National Science Reviews, as well as a Chinese Chemical Society director and deputy secretary-general.

Low-frequency noise in carbon nanotube thin-film transistors and its impact on analog integrated circuits

Yutaka Ohno

[1] Institute of Materials and Systems for Sustainability,
Nagoya University, Furo-cho, Chikusa-ku, Nagoya, Janan

Flexible electronics is a promising field for the application of carbon nanotube (CNT) thin-film transistors (TFTs), which possess high mobility and mechanical flexibility. While analog front-end circuits using CNT TFTs have been demonstrated [1], there is a need to reduce low-frequency noise (<1 kHz) for more accurate measurements of biological signals such as ECG and EEG. Here, we investigated the low-frequency noise characteristics of CNT TFTs on flexible substrate. It will be shown that the noise power coefficient, which is related to traps at/near the interface between the CNT and gate oxide, depends on field concentration to the one-dimensional material. The physics of the noise behavior and its impact on analog integrated circuits such as sensing amplifiers will be discussed.

References

[1] T. Kashima, et al. Research Square (2020) DOI:10.21203/rs.3.rs-68702/v1

Carbon nanotube-based high-performance bioelectronics

Youfan Hu

Center for Carbon-Based Electronics, School of Electronics, Peking University, Beijing 100871, China

The continuous advancements in wearable and flexible electronics are expected to enable advanced bioelectronics with the capability of health assessment, physical activity tracking, and even personalized therapy. An intimate interface between the electronics and organisms for high fidelity information capture and high-performance electronics to support advanced functions are the key enablers of these applications. Here, we show several examples of our recent progress to construct such bioelectronic systems for physiological information monitoring from human body surface, which integrate flexible sensors, carbon nanotube based interface circuits and memory. The carbon nanotube-based electronics demonstrate great performance to enable advanced functions, such as in-situ signal processing and information storage, revealing a great possibility in advanced health monitoring, medical diagnostics, etc.

References

[1] G. Long, W. Jin, F. Xia, Y. Wang, T. Bai, X. Chen, X. Liang, L.-M. Peng*, Y. Hu*, Nature Communications, 13, 6734 (2022).

[2] L. Xiang, Y. Wang, F. Xia, F. Liu, D. He, G. Long, X. Zeng, X. Liang, C. Jin, Y. Wang, A. Pan, L.-M. Peng, Y. Hu*, Science Advances, 8, eabp8075, 2022.

[3] L. Xiang, X. Zeng, F. Xia, W. Jin, Y. Liu, Y. Hu*, ACS Nano, 14, 6449, 2020.

[4] H. Zhang, Y. Liu, C. Yang, L. Xiang, Y. Hu*, L.-M. Peng*, Advanced Materials, 30, 1805408, 2018.

[5] L. Xiang, H. Zhang, G. Dong, D. Zhong, J. Han, X. Liang, Z. Zhang, L.-M. Peng, Y. Hu*, Nature Electronics, 1, 237, 2018.

[6] H. Zhang, L. Xiang, Y. Yang, M. Xiao, J. Han, L. Ding, Z. Zhang, Y. Hu*, L.-M. Peng*, ACS Nano, 12, 2773, 2018.

Flexible carbon nanotube thin film transistors for optoelectronic neuromorphic devices and CMOS circuits

Jiaqi Li[1,2], Shuangshuang Shao[1,2], Min Li[1,2] and Jianwen Zhao[1,2*]

[1] Key Laboratory of Semiconductor Display Materials and Chips, Division of Nanodevices and Related Nanomaterials, Suzhou Institute of Nano-Tech and Nano-Bionics, Chinese Academy of Sciences, No. 398 Ruoshui Road, Suzhou Industrial Park, Suzhou, Jiangsu Province, 215123, PR China

[2] School of Nano-Tech and Nano-Bionics, University of Science and Technology of China, No. 398 Ruoshui Road, Suzhou Industrial Park, Suzhou, Jiangsu Province, 215123, PR China

Email: jwzhao2011@sinano.ac.cn

Semiconducting SWCNTs (sc-SWCNTs) are promising semiconductor materials for flexible printed thin film transistors (TFTs) because of their excellent electrical properties, solubility, flexibility, high chemical and physical stability, and low process temperatures[1-9]. In this talk, I will show some new results in our group about flexible printed carbon nanotube TFTs, (stretchable) optoelectronic neuromorphic devices, logic gates (NOR and NAND gates, CMOS or CMOS-like inverters) and ring oscillators. Furthermore, flexible printed SWCNT TFT devices and circuits with the recorded irradiation resistance and strong self-repair ability can be achieved by optimizing the designed device structures and using high-capacitance dielectric materials[1-9].

References

[1] Zebing Wang, et al. ACS Nano 18 (2024) 14298-14311

[2] Nianjie Zhang, et al. Nano Letters (2024) DOI: 10.1021/acs.nanolett.4c01691

[3] Nianzi Sui, et al. Advanced Science (2024) 2401794

[4] Tanghao Xie, et al. Advanced Functional Materials (2023) 2303970

[5] Suyun Wang, et al. Nano Energy 115 (2023) 108698

[6] Min Li, et al. Small 19 (2023) 2207311

[7] Penghui Guo, et al. Carbon 115(2023) 118453

[8] Xin Wang, et al. Advanced Materials 34 (2022) 2204066

[9] Maguang Zhu, et al. Nature Electronics 3 (2020) 622-629

Oral Presentations

A new discovery on the surface activity of carbon black.

Jian Chen *, Mao-Yuan Hu , Rui Li, Wei Jiao, Fu-Lin Tang, Da-Shun Li

Key Laboratory of material corrosion and protection, Sichuan University of Science & Engineering, Zigong, China, 643000

*Email: jchenzg@aliyun.com

It is well known that the particle size, structure, and surface activity of carbon black are the main factors influencing the reinforcement of rubber. The smaller the particle size of carbon black and the higher its structure, the thicker the bonding layer formed with rubber, leading to better reinforcement performance. However, the exact location of the active sites on carbon black’ s surface remains uncertain. We conducted gradient high-temperature heat treatment on carbon black and found that even after completely removing the functional groups on the surface of carbon black, it still exhibited good reinforcement properties. Therefore, we believe that the activity of carbon black is determined not by the functional groups on its surface, but by the edges of the oligomeric layer structures. By measuring the surface energy of carbon black treated at different temperatures using inverse gas chromatography, we observed that the surface free energy of carbon black gradually increased with higher heat treatment temperatures. Through STM, TEM, and other tests, we characterized the size and structure of the local oligomeric layer on the surface of carbon black after heat treatment at different temperatures. Additionally, using density functional theory (DFT) and first-principles calculations (VASP) to calculate the electron cloud density of the oligomeric layer and step parts, we found that the flat layer in the structure is an inert surface, while the step part is an active surface.

Keywodes: carbon black; activity; oligomeric layer;

References

[1] Raos G., Application of the Christensen-Lo model to the reinforcement of elastomers by fractal fillers, Macromol. Theor. Simul, 12, 17-23 (2003)

[2] Dannenberg E. M., Bound rubber and carbon-black reinforcement, Rubber Chem. Technol., 59, 512-524 (2016)

[3] Zhang S.Y., Zhong R.P., Hong R.Y., et al. On factors affecting surface free energy of carbon black for reinforcing rubber, Nanotechnol Rev, 9, 170-181 (2020)

Grafting CNTs on carbon fabrics with enhanced mechanical and thermal properties for tribological applications of carbon fabrics/ phenolic composites

Beibei Wang[1,2], Qiangang Fu[1], Hejun Li[1], Lehua Qi[1], Yewei Fu[1]

[1] Shaanxi Province Key Laboratory of Fiber Reinforced Light Composite Materials, Northwestern Polytechnical University, Xi' an 710072, China

[2] School of Materials Engineering, Xi' an Aeronautical Institute, Xi' an 710077, China

Email:15249271900@163.com

To improve the tribological properties of carbon fabrics-reinforced phenolic composites, carbon nanotubes (CNTs) were in situ grown on carbon fabrics by chemical vapor deposition. Results showed that the thermal conductivity and tensile strength of the CNT-modified composites were enhanced by approximately 21% and 45%, respectively, owing to the high thermal conductivity of CNTs and the improved interfacial adhesion between carbon fabrics and resin matrix. The friction coefficient of the composites modified by CNTs increased from 0.12 to 0.14, while the wear rate decreased from 3.7 10-14 m3 (N m)-1 to 2.8 10-14 m3 (N m)-1. CNT-grafted composites exhibited a higher friction coefficient and greater wear resistance than CNT-free composites; this can be attributed to the improved interfacial bonding of the composites by the incorporation of CNTs. Therefore, grafting CNTs on carbon fabric is an effective method to improve the thermal, mechanical, and tribological properties of carbon fabrics-reinforced phenolic composites. Most importantly, the carbon fabrics/phenolic composites reinforced by CNTs hold a great potential in producing high-performance polymer composites for tribological application.

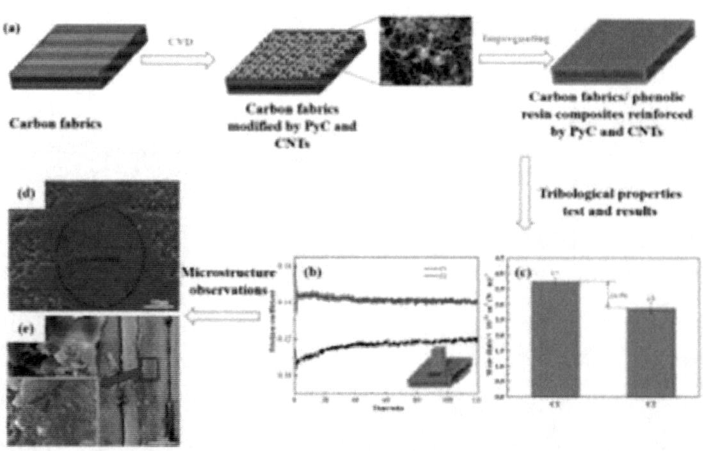

Fig. 1. Schematic of the preparation procedure and tribological properties of carbon fabric/phenolic composites reinforced by CNTs.

References

[1] Wang B., Fu Q., Li H., Qi L., Fu Y. Grafting CNTs on carbon fabrics with enhanced mechanical and thermal properties for tribological applications of carbon fabrics/phenolic composites. Carbon. 139, 45-51 (2018).

Efficient Controlled Growth of Single-Wall Carbon Nanotubes and Their Derivatives

Feng Zhang, Chang Liu

Shenyang National Laboratory for Materials Science, Institute of Metal Research, Chinese Academy of Sciences, 72 Wenhua Road, Shenyang, China

Single-wall carbon nanotubes (SWCNTs) with uniform structures are highly desired for fabricating high-performance electronics. Great efforts have been devoted to controlled synthesizing structure-specific SWCNTs1. However, there is a conflict between the growth efficiency and structural uniformity of SWCNTs prepared by chemical vapor deposition, which seriously limits their application exploration. We tried to figure out the reason by systematically investigating the thermodynamic and kinetic factors from a chemical reaction view.2 It was found that kinetic parameters should be matched with the catalyst with suitable size and composition3, small-size metal nanoparticles with high thermal stability are critical for achieving efficient growth of structural uniform SWCNTs. To obtain SWCNT-based derivatives with fascinating properties, SWCNT was respectively used as a nanoreactor and template. SWCNTs were cut by defect-induced etching and used as reactors to prepare WO3 nanowire@SWCNTs with excellent radiation resistance4. Free-standing SWCNT@h-BN films were efficiently prepared by a defect-induced epitaxial growth mode5. This provides new insights and a platform for exploring the application of SWCNTs.

References

[1] Peng-Xiang Hou, et al. Advanced Funcational Materials 32 (2022) 2108541
[2] Feng Zhang, et al. Carbon 209 (2023) 118011
[3] Xin Li, et al. ACS Nano 16 (2022) 232-240
[4] Xue Zheng, et al. Nanoscale 15 (2023) 3931-3939
[5] Changping Yu, et al. Materials 16 (2023) 1864

Tuning structure and electron transport properties of graphene by chemical modification using ion-beam irradiation

K. Takai[1,2], Y. Zhao[1] K. Nakamura[1], H. Yoshimoto[1] and T. Nishimura[2]

[1]Dept. Applied Chemistry, Hosei University, 3-7-2, Kajino, Koganei, Tokyo, Japan
[2]Research Center of Ion Beam Technology, Hosei University, 3-7-2, Kajino, Koganei, Tokyo, Japan

2D materials like graphene are promising for future electronic devices [1-3]. To modify graphene to tune its carrier density and electron mobility, we have modified it by irradiating it with a Au- or I-ion beam at 200 keV with doses of 1013–1014 cm−2 using a NaCl sacrificial layer as a quantitative chemical modification method [4]. After removing the sacrificial layer, ion-irradiated graphene was evaluated by Raman spectroscopy, electrical conductivity measurements, and Rutherford backscattering spectrometry. Both carbon vacancies and heavy atoms (Au, I) are introduced into graphene by ion-beam irradiation; the introduced heavy atoms act as charged impurities, which results in the Raman D band and a downshift of the Fermi energy caused by hole carrier doping. More precise knowledge of the influence of heavy-ion beam irradiation of graphene on electron transport properties will lead to the development of a new class of materials for electronics and spintronics.

References

[1] Y. Ishiguro, et al. J. Phys. Chem. C 124 (2020) 27176-27184
[2] Y. Ishiguro, et al. Nanotechnology 34 (2023) 275701
[3] Y. Zhao, et al. Nuclear Inst. and Methods in Phys. Res. B 543 (2023) 165095
[4] Y. Zhao, et al. Carbon Rep. 3, (2024) 37-46

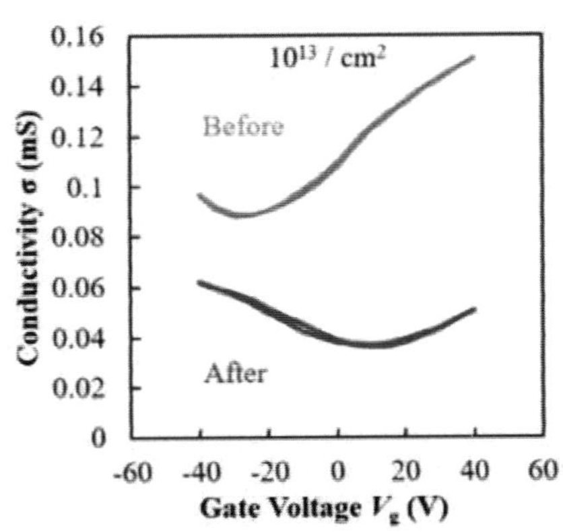

Fig.1 Conductivity of graphene before / after Au+ irradiation.

β-resin cross-linking Graphitization for Highly Thermally Conductive Graphene Film

Zheng Wang[1,2], Hui Jia[1], Dong Jiang[1] and Cheng-Meng Chen[1]

[1] CAS Key Laboratory of Carbon Materials, Institute of Coal Chemistry, Chinese Academy of Sciences, Taiyuan 030001, China

[2] University of Chinese Academy of Sciences, Beijing 100049, China

Highly thermally conductive graphene films (GFs) have become essential to meet efficient heat dissipation demand of electronic devices. To date, high thermal conductivity has been achieved in only in-planes. However, traditional GFs usually have low through-plane thermal conductivity and weak inter-layer interaction, due to the disordered sheet alignment and frail interfacial adhesion. Herein, a novel strategy is proposed to manipulate the structure of graphene oxide (GO) films via vacuum self-assembly coupled with β resin from coal-based mesophase asphalt (CNS). The rGO surface is thoroughly hybridized with the β resin, manifesting periodic structures with highly crystalline. Additionally, the lower layer porosity and higher interfacial strength through transforming conventional physical adhesion into covalent bonding ensure the structural stability and in-plane thermal conductivity (≈899 W m−1 K−1) of GFs. For the GCNS2800 with the thickness of 55 μm, the through-plane thermal conductivity is 3.83 W m−1 K−1 and the peel force reaches 202 gf, approximately 4 times and 3.5 times those of the GFs prepared by the traditional assembly method, respectively. This work not only paves the way for the technological application of highly thermal conductive graphene films but also provides a general strategy to efficiently improve the synthesis and properties of high interlayer bonding strength graphene-based films.

Applications of Vertical Graphene Thin-Flim Materials for Bio-electronics, Sensors and Electrochemical Energy Devices

Qinhua Zhong [1], Zhuang-Qiang Gao [2] and Xin Zhao[1]

[1]Shenzhen Yick Xin Tech Ltd. Co., Shenzhen City, China
[2]School of Medicals, Shenzhen University, Shenzhen City, China

Vertical Graphene (VG) thin-films materials, comprising of carbon nanosheets, have been mass produced by specialized plasma enhanced chemical vapor deposition (PECVD) via catalysts-free nucleation and crystal growth processes. This complex material, having unique hierarchy 3D morphology, exhibits rapid manipulable and long-range charge transport feature at graphene-electrolyte interface, complying with surface-confined quantum-state electrochemistry theory, making it crucial for molecular electronics sensors. Graphene nanosheets can be modified with a range of biomolecules, opening up applications in medical clinic chemistry analysis, including enzyme, nucleic-acids and immunoassays. Biochips made of VG materials demonstrates high sensitivity and limited of detection of IL-10 antigen at 1pg/ml. The VG thin-film up-grown on carbon-nanotube cloth/papers also showed high performance as electrochemical energy device electrodes.

'Pre-dissolving carbon' in metal catalysts for growing high-crystallinity vertically aligned carbon nanotube arrays

Zhong-Hai Ji[1], Lili Zhang[1*], Chang Liu[1*], Hui-Ming Cheng[1,2]

[1]Shenyang National Laboratory for Materials Science, Institute of Metal Research (IMR), Chinese Academy of Sciences, 72 Wenhua Road, Shenyang, 110016, PR China

[2]Faculty of Materials Science and Energy Engineering/Institute of Technology for Carbon Neutrality, Shenzhen Institute of Advanced Technology, Chinese Academy of Sciences, Shenzhen 518055, PR China

*E-mail: zhangll@imr.ac.cn; cliu@imr.ac.cn.

Direct growth of vertically aligned carbon nanotube (VACNT) arrays with high crystallinity and controllable structures are highly important for their thermo- and electrical-related applications. There have been tradeoffs between the crystallinity and length/diameter of VACNT arrays, due to the lack of efficient synthesis methods and in-depth understanding of the growth mechanism. Here, we propose a 'pre-dissolving carbon' and high-throughput catalyst design strategy to boost catalyst activations and to address the synthetic barrier for growing high-crystallinity VACNT arrays with controllable wall number, diameter and length. In this approach, C2H2/H2 plasma was used to pre-dissolve carbon in metal catalyst nanoparticles, which improved the carbon diffusion and incorporation efficiency for array growth while reduced the formation of defects. Highly crystalline (IG/ID=12) double wall-VACNT (DW-VACNT) arrays with large diameters (average diameter, d=6.6 nm) and single wall VACNT (SW-VACNT) arrays with a height of~1.309 mm were then obtained tuning C2H2/H2 plasma pretreatment time and catalyst compositions. Therefore, high thermal diffusivity (~151 mm2/s) and high electrical sheet conductance (~9.2×10-3 S/□) were simultaneously achieved by the DW-VACNT arrays.

Molecular and ionic transport through graphene membranes

Pengzhan Sun [1]

[1]Institute of Applied Physics and Materials Engineering, University of Macau, Macau SAR, China

It is widely believed that, despite being one-atom thick, graphene and other 2D crystals is completely impermeable to all gases and liquids. In this talk, I will present my recent research1-4 on the topic "Molecular transport through angstroporous 2D crystals."

Using monocrystalline container made from atomically flat graphite, which is tightly sealed with graphene, we have achieved measurements that put the permeation limit through 2D materials at 8–9 orders of magnitude lower than previously, such that we would discern (but did not observe) just a few helium atoms per hour crossing micrometer-size membranes. This detection limit is also valid for all other gases tested, except for hydrogen. Hydrogen shows noticeable permeation, even though its molecule is larger than helium. The mechanism of this anomalous observation is proposed and later corroborated by experiments. To make the generally "impermeable" graphene not only "permeable" but also highly "selective", we have developed a perforation technique which involves a short-time exposure of the graphene membrane to low-energy electrons. Using the same monocrystalline containers, we are able to study gas transport through the created individual graphene pores with an effective size of about one missing carbon ring. Helium and hydrogen permeate easily through these pores whereas larger molecules such as xenon and methane are blocked. Permeating gases experience activation barriers that increase quadratically with the kinetic diameter, and the transport process crucially involves surface adsorption.

References

[1] P. Z. Sun, et al. Nature 2020, 579, 229.
[2] P. Z. Sun, et al. Nat. Commun. 2021, 12, 7170.
[3] P. Z. Sun, et al. PNAS 2023, 120, e2300481120.
[4] Z. F. Wu, P. Z. Sun, et al. Nat. Commun. 2023, 14, 7756.

Construction and differential growth mechanism of uniform and controllable CNTs and CNWs by ZIF-67 precursor

Zhaofan Zhou[1], Jinhua Lu[1] and Hejun Li[1]

[1]State Key Laboratory of Solidification Processing, Shaanxi Key Laboratory of Fiber Reinforced Light-Weight Composites, Northwestern Polytechnical University, Xi' an, Shaanxi, P.R. China

Carbon nanotubes (CNTs) and carbon nanowires (CNWs) have been successfully derived from ZIF-67 nanoparticles which belongs to metal-organic frameworks on carbon cloth surface. Affected by the atmosphere, ZIF-67 nanoparticles derived ZIF-67@CNTs in a melamine environment while derived ZIF-67@CNWs in a C2H5OH environment. SEM, TEM, Raman, and so on were used to explore the construction of the ZIF-67@CNTs/CNWs modification layer on the carbon fiber surface. The macroscopic morphology of ZIF-67@CNTs/CNWs is influenced by ZIF-67 concentration. When the growth temperature rase, CNTs and CNWs' graphite degrees significantly improved. CNTs and CNWs both adhere to the traditional tip-growth mode, whereas their growth mechanisms differ. "Bamboo-like" CNTs are formatted due to the different precipitation rates of graphite sheets (GSs), which are caused by catalyst particles in the solid and liquid phases. Solid CNWs are constructed with stacked GSs produced by single-phase catalyst particles. It is the phase uniformity of catalyst particles that ultimately determines the catalytic product.

Keywords: Carbon nanotubes (CNTs); Carbon nanowires (CNWs); Carbon fibers (CFs); ZIF-67

Fabrication of high-quality graphene films and mechanism of graphitization promoted by lattice defects

Qing Zhang[1]

[1] Institute of Materials Research, Tsinghua Shenzhen International Graduate School, Tsinghua University, Shenzhen, People' s Republic of China.

Fabricating highly crystalline macroscopic films with extraordinary electrical and thermal conductivities from graphene sheets is essential for applications in electronics, telecommunications, and thermal management. High-temperature graphitization is the only way known to date for the crystallization of all types of graphitic materials, where defects are gradually removed with increasing the temperature. However, when using graphene-based nanosheets as precursors including graphene oxide, reduced graphene oxide and pristine graphene, even a long-term graphitization at 3000 °C can only produce graphene films with small crystallites and abundant structural disorders, which limit their conductivities. Here, we show that high-temperature defects in graphene lattice can substantially accelerate the graphitization of graphene films, enabling ideal AB stacking as well as over 100, 64 and 28 times improvements in crystallite size, electrical and thermal conductivities, respectively, from 2000 to 3000 °C. This process is realized by heteroatom doping, which retards the in-plane lattice restoration of graphene and retains high-temperature defects which could promote carbon atom migration and rearrangement during graphitization. With this approach, a highly ordered crystalline graphene film similar as highly oriented pyrolytic graphite is fabricated, whose electrical and thermal conductivities (~2.0×104 S cm-1; ~1.7×103 W m-1 K-1) are improved ~6 and 2 times, respectively, compared to those of the graphene films fabricated by graphene oxide. Such graphene film exhibits a superhigh electromagnetic interference shielding effectiveness of ~90 dB at a thickness of 10 m, outperforming all the synthetic materials of comparable thickness. This work not only paves the way for the technological applications of highly conductive graphene films but also provides a general strategy to efficiently improve the synthesis and properties of other graphitic materials such as graphene fibers, carbon nanotube fibers, carbon fibers, polymer-derived graphite, and highly oriented pyrolytic graphite.

References

[1] Zhang Q., Wei Q.W., Huang K., Liu Z.B., Ren W.C., et al. Defects boost graphitization for highly conductive graphene films. Natl. Sci. Rev., 2023, 7: nwad147.

Graphene-based hierarchical structure for significant enhancement in thermal conductivity of carbon fiber/epoxy composites with high mechanical reinforcement

You Zeng [1,2,*], Xinyang Sun [1,2], Han Wang [1,2]

[1]Shenyang National Laboratory for Materials Science, Institute of Metal Research, Chinese Academy of Sciences, Shenyang 110016, P. R. China

[2]School of Materials Science and Engineering, University of Science and Technology of China, Shenyang 110016, P.R. China

*E-mail address: yzeng@imr.ac.cn (Y. Zeng)

In order to obtain high-performance carbon fiber reinforced polymer composites (CFRP) with enhanced thermal conductivity and high mechanical properties, we have fabricated carbon fiber (CF) laminated composites by utilizing graphene-based hierarchical structures as functional layers and infiltrating with epoxy. High-quality CVD graphene foam (GF) and graphene nanosheets (GNPs) were used and laminated with CF fabrics to construct high-efficiency thermal conduction pathways. Thermal conductivity and mechanical properties of the CF/graphene laminated composites were investigated in detail. We found that, with addition of the graphene hierarchical networks (as low as only 3.0 vol%), the CF/graphene laminated composites exhibited remarkable enhancement in through-plane and in-plane thermal conductivity, increasing by 158.8% and 28.8% in comparison with that of CFRP. Meanwhile, the tensile, flexural, compressive, and interlaminar shear strength for the laminated composites were also increased by 19.3, 13.1 and 8.8% comparing with that of CFRP. Such remarkable reinforcement in thermal and mechanical properties is mainly attributed to the highly-conductive graphene hierarchical structures, strong interfacial interactions, and high-efficiency stress transferring. The CF/graphene laminated composites have great potential to be used as multi-functional materials in the fields of aerospace, transportation, and electrical industry.

Atomic Origin of the Axis-Symmetrical Breaking of a Growing Single-Wall Carbon Nanotube

Lili Zhang[1], Thomas W. Hansen[2], Jakob B. Wagner[2,*], Chang Liu[1,*] Hui-Ming Cheng[1,3]

[1] Shenyang National Laboratory for Materials Science, Institute of Metal Research, Chinese Academy of Sciences, 72 Wenhua Road, Shenyang 110016, China,

[2] National Centre for Nano Fabrication and Characterization, Technical University of Denmark, 2800, Kgs. Lyngby, Denmark,

[3] Institute of Technology for Carbon Neutrality, Shenzhen Institute of Advanced Technology, Chinese Academy of Sciences, 1068 Xueyuan Road, Shenzhen 518055, China

Corresponding author(s): zhangll@imr.ac.cn, jabw@dtu.dk, cliu@imr.ac.cn

The chirality of a single-wall carbon nanotube (SWCNT) is usually considered to be closely related to the atomic configuration of catalyst seed nanoparticles. Hence, a change in the local atomic structure of the catalyst may generate a local defect on the tube and then break the symmetry in the tube chirality. However, the axis symmetry and its variation during the early stage of SWCNT growth remains unexplored and its origins at the atomic scale are elusive. Here, we report the direct atomic-scale visualization of both the axis-symmetrical growth and its alteration processes of a SWCNT by using in situ environmental transmission electron microscopy [1-3]. Our results reveal that in contrast to the continuous growth of a straight tube, the .formation of topological defects on the sidewall of a SWCNT acts as a buffer for stress release, inherently breaking its axis-symmetrical growth (Figure 1). Atomic-level details reveal the importance of the tube/catalyst interface and how the atom rearrangement of the solid-state platinum catalyst around the interface influences the final tubular structure. Theoretical simulations confirm that the active sites responsible for trapping carbon dimers and providing enough driving force for carbon incorporation and asymmetric growth are the low-coordination step edges.

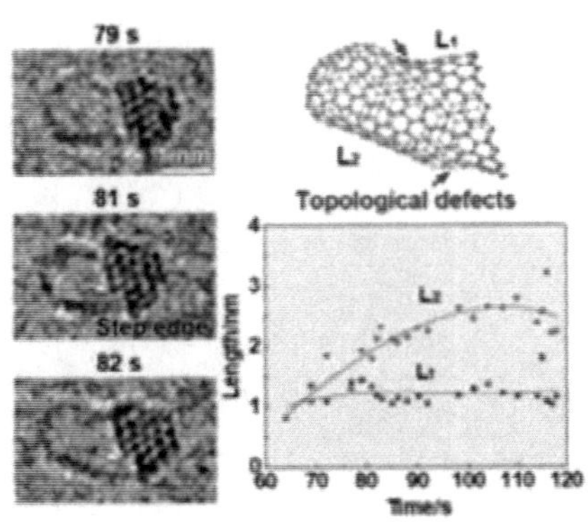

References

[1] L. Zhang, et al. Adv. Sci. (2023) 2304905.
[2] R. Ma, et al. ACS Nano 16 (2022) 16574-16583.
[3] Y. Wang, et al. Sci Adv 8(2022) eabo5686.

Interfacial enhancing of carbon nanotube fiber with Cu shell for future lightweight cables

Zhao-Qing Gao [1,2], Peng-Xiang Hou [1,2,*], Chang Liu [1,2,*], and Hui-Ming Cheng [1,2,*]

[1]Shenyang National Laboratory for Materials Science, Institute of Metal Research, Chinese Academy of Sciences, Shenyang, 110016, China

[2]University of Science and Technology of China, No.96, JinZhai Road Baohe District, Hefei, Anhui, 230026, P.R.China.

Wet-spun carbon nanotube fibers (CNTFs) have been considered as one of the most promising candidates as next generation cables due to their excellent high strength, electrical conductivity, ampacity, and low density [1]. However, wet-spun CNTFs as the lightweight cables are still facing two critical issues to be solved. One is to reliably connect CNTFs with Cu electrodes as conducting components in electronic and electrical systems. The other is to further improve electrical conductivity of CNTFs, which generally metallizes CNTFs. However, one conjoint question in the above two issues is to improve interfacial bonding strength between CNTFs and metals. Here, we selected and optimized intermediate metal materials to tackle this issue based on the understanding in binding energy, affinity, work function and Bader charge differences between carbon nanotubes (CNTs) and metal atoms. In consequence, CNT/Cu core/shell fibers with good interfacial bonding and good electrical properties are prepared, and a strong connection between CNTFs and a Cu matrix was welded by a commercial eutectic Sn-37Pb alloy showing mechanically strong and electrically conductive joints.

References

[1] Z.Q. Gao, L.L. Xu, X.Y. Jiao, P.-X. Hou, C. Liu, H.-M. Cheng. ACS Nano 2023, 17, 18290–18298.

The role of water on the electrochemical synthesis of graphene oxide

Jiaqi Guo [1,2], Songfeng Pei [1,2], Kun Huang [1,2], Qing Zhang [1,2], Jinmeng Tong [1,2], Hui-Ming Cheng [1,2,3], Wencai Ren [1,2]

[1] Shenyang National Laboratory for Materials Science, Institute of Metal Research, Chinese Academy of Sciences, 72 Wenhua Road, Shenyang 110016, P. R. China.

[2] School of Materials Science and Engineering, University of Science and Technology of China, 72 Wenhua Road, Shenyang 110016, P. R. China.

[3] Shenzhen Institute of Advanced Technology, Chinese Academy of Sciences, Shenzhen 518055, P. R. China

With the development of graphene applications, graphene oxide (GO) has become an important basic chemical material needing mass production for industrial applications. Electrochemical (EC) synthesis of GO, despite having shown its advantages in environmental-friendliness, safety, and low cost, is also suffering from its drawbacks on insufficient and non-uniform oxidation as well as low yield. Whether these drawbacks can be effectively solved is the key to determining the practicality of EC methods for industrial production. Here, we found that water plays a key role for the EC synthesis of GO, which is indispensable for EC oxidation of sulfuric acid-graphite intercalation compound (SA-GIC) but can lead to deintercalation of SA-GIC at the same time. Moreover, both the EC oxidation by water electrolysis (OWE) and deintercalation of SA-GIC caused by water absorption (DIWA) are significantly influenced by the diffusion of water from the electrolyte to the interior of electrode. The competition between these two processes determines whether SA-GIC can be uniformly oxidized. Based on these understandings, we developed a liquid membrane electrolysis (LME) method to precisely control the water diffusion pathway to achieve a dynamic equilibrium between DIWA and OWE, which leads to a high GO yield of 180% with monolayer ratio >99% and average later size ~13 m, solving the inherent inhomogeneous oxidation issue of the existing EC methods. Moreover, LME enables precise control on lateral size and oxidation degree of GO. Remarkably, pure water can be used to produce GO. The realization of homogeneous EC oxidation by LME paves the way to industrial production of GO with overwhelming advantages over the Hummers' methods in terms of production efficiency, cost, eco-friendliness and safety.

Large-area transfer of two-dimensional materials free of cracks, contamination and wrinkles

Li Lin[1,2]

[1]School of Materials Science and Engineering, Peking University, Beijing 100871, P. R. China

[2]Beijing Graphene Institute, Beijing 100095, P. R. China

Owing to its fascinating properties, graphene has exhibited great potentials in the applications of high-performance electronics and flexible devices. The availability of graphene films on a wide range of substrates forms the basis for graphene applications. For instance, the integration of graphene with silicon-based technologies, requires graphene on silicon-based substrate with outperforming carrier mobilities. Although large-area graphene films have been successfully produced based on chemical vapor deposition methods, graphene films were only produced on limited archetypal substrates, such as metal foils. Reliable after-growth transfer techniques are critical for layering 2D materials onto arbitrary substrates for further applications. During the transfer of graphene, polymer-based transfer medium is usually introduced for avoiding crack formation, which, however, introduces unavoidable contamination on graphene surface. Traditional transfer routes, including etching and bubble-based delamination, involve the aqueous solution-based reactions or processing, which would cause doping in graphene and improve the complexity in the design of industrial equipment for batch transfer.

Here, we have achieved the crack, contamination and wrinkle-free transfer of graphene films over large area and designed transfer machines for automatic transfer with highly improved production capability: (1) we achieved all-dry transfer of graphene wafers, including the mechanical delamination of graphene from Cu surface and exfoliation of polymer from graphene surface, and all the transfer process were conducted without using water or organic solvent; (2) Transferred graphene wafers are free of cracks, contamination, and wrinkles, and the ultraclean surfaces provide the carrier mobilities of transferred graphene up to 14,000 cm2 V-1 s-1; (3) we designed and built the automatic transfer systems including automatic spin-coater, delaminator and laminator for the batch transfer of graphene films, enabling an improved production capability of 300 pieces (4-inch graphene wafer) per days.

References

[1] Yixuan Zhao, et al. Nat. Commun., 13 (2022), 4409.

[2] Sensheng Chen, et al. Adv. Mater., 36 (2024), 2308950.

[3] Zhaoning Hu, et al. Adv. Mater., 35 (2024), 2300621.

One-Side Hydrogenation of Single-Layer Graphene: Etching or only Hydrogen Chemisorption?

T. Fournier[1,2], S. El-Masaoudi[2], K. Cruz,[3] B. Lassagne[2], I. Gerber[2], C. Crespos[4], J.-M. Leyssale[4], F. Andriambelaza[4], M. Monthioux[1], G. Paredes[3], K. Gough[5], J.-M. Pouméról[1], P. Puech[1], F. Piazza[3]

[1]CEMES, CNRS, Université de Toulouse, France
[2]LPCNO, CNRS, INSA, Université de Toulouse, France
[3]Institut des Sciences Moléculaires (ISM), CNRS, Université de Bordeaux, France
[4]Laboratorio de Nanociencia, PUCMM, Santiago, Dominican Republic
[5]Department of Chemistry, Universidad de Manitoba, Winnipeg, Canada

Many claims for the successful one-side hydrogenation of a single-layer graphene (1LG) can be found in the literature. However, while calculation works are many, experimental evidences of hydrogenated 1LG/SiO2 are scarce and supported by barely convincing data.

After a thorough investigation of various hydrogenation conditions using the hot-filament (W) process, we were able to discriminate between the conditions preferably leading to either graphene etching or graphene hydrogenation. Based on our experiments supported by calculations in which H · radicals are added one by one, it appears that the stability of one-side hydrogenated 1LG is very low, meaning that obtaining such a material without creating nano-holes corresponds to a narrow condition range.

Raman spectroscopy (with visible light) is a key characterization method for this, however, it is not always interpreted accurately. For instance, for a genuine one-side hydrogenated 1LG, the ID'/ID ratio should be low (< 1/13), otherwise etching should be suspected. Carrying out annealing to remove the chemisorbed hydrogen atoms is a good way for figuring this out: after genuine hydrogenation, annealing is able to recover a low D-band, whereas, after etching, it is not. In some cases, completely removing the D band is even possible, as a signature that the majority of the "defects" having been removed are hydrogen atoms formerly grafted on one side (top side) of the 1LG, switching the sp3 hybridization of carbon atoms back to sp2. For a one-side fully hydrogenated 1LG, the opening of the electronic gap leads to optical transparency, and the standard Raman signal vanishes. Therefore, other techniques (IR, EELS, TEM, STM, TERS, devices) need to be considered. The questions remain to know where the H atoms are located (to edge or to in-lattice carbon atoms), and how the small data variations are interpreted.

Controlled Synthesis of Two-Dimensional Diamane from Graphene under Extreme Conditions

Yabin Chen [1,2]

[1]Advanced Research Institute of Multidisciplinary Sciences,
Beijing Institute of Technology, Beijing 100081, China
[2]Beijing Institute of Technology, Zhuhai Beijing Institute of Technology (BIT), Zhuhai 519088, China
E-mail: chyb0422@bit.edu.cn

Bulk diamond crystals have attracted much attentions these years as the desired candidate of next-generation semiconductors, owing to their superior thermal and electrical properties. However, the fragile characteristic and non-adjustable band gap limits its potential applications in various fields, and the two-dimensional diamond, named diamane emerges and can offer the novel insights, especially for the flexibility and sizeable band gaps. Herein, we will report our recent progress on the controlled synthesis and properties of two-dimensional diamane from graphene under high temperature and high pressure. Few-layer graphene can transform to diamond-like sp3 structure under high pressure, and its band gap can be opened and modulated with pressure, evident from the in-situ optical and electrical transport measurements. Moreover, combined with laser-heating approach, the ultrathin diamane can be successfully quenched to ambient condition, and the minimum thickness can approach to several nanometers. The lattices exhibited cubic or hexagonal structures based on HRTEM and SAED results. The optical band gaps can be regulated according to the ratio of sp2 and sp3 carbon, that leads by the distinct extreme conditions. We believe these results can shed a light on the structural design and controlled preparation of novel carbon nanomaterials.

Confined Synthesis of Graphene Nanoribbons inside Single-Walled Carbon Nanotubes

Kunpeng Tang[1], Yingzhi Chen[1], Huiju Cao[1], Wendi Zhang[2], Kecheng Cao[2], Lei Shi[1]

[1] Sun Yat-sen University, Guangzhou, China
[2] Shanghai Tech University, Shanghai, China

Unraveling the chemical reaction process still remains a challenge. The 1D nanospace of carbon nanotube is an ideal nanoreactor for synthesis of nonexistent or unusual materials, e.g., previously we have synthesized ultra-long linear carbon chains [1,2]. Thanks to the confinement of the carbon nanotube and Raman Spectroscopy, the evolution (or reaction) of precursor molecules under high temperature or high pressure can be studied in detail now. Here, our recent results discovered that some precursor molecules with/without hexatomic rings can be converted into graphene nanoribbons [3-7], whereas some aromatic molecules can be transformed into linear carbon chains. These unusual phenomena are beyond the conventional chemical reaction. In addition, many kinds of 1D elemental or compound chains can be obtained by using the same confined synthesis, which paves the way to create a library of novel 1D materials that do not exist or are yet to be explored.

References

[1] Lei Shi, et al. Nature Materials 15 (2016) 634–639
[2] Weili Cui, et al. Carbon 219 (2024) 118784
[3] Hans Kuzmany, et al. Carbon (2021) 171, 221–229
[4] Yifan Zhang, et al. Nano Research (2022) 15, 1709–1714
[5] Haoyuan Zhang, et al. Nano Research (2023) 16, 10644–10651
[6] Yanghao Feng, et al. Nano Research (2024) doi:10.1007/s12274-024-6571-z
[7] Kunpeng Tang, et al. in preparation (2024)

Wafer-Scale Single-Crystal 2D Materials Growth on Insulating Substrates

Bo Tian

Nanyang Technological University (NTU), Singapore
E-mail: botianlab@gmail.com

Two-dimensional (2D) materials have attracted considerable interest in the science community in recent years. Until today, high-quality 2D material films have been grown on metal foils by the chemical vapor deposition (CVD) method. However, nanodevice applications usually require 2D material on insulating substrates. To date, growing high-quality 2D layered materials, such as graphene and hexagonal boron nitride (hBN), on insulators is still challenging owing to the lack of suitable metal catalysts, imperfect lattice matching with substrates, and other factors. Therefore, developing a generally applicable approach for realizing high-quality 2D layers on insulators remains crucial.

In this presentation, I will discuss our past study on the epitaxial growth of wafer-scale single-crystal graphene at the interface between Cu(111) and Al2O3(0001) using a multi-cycle plasma etching-assisted CVD and carbon diffusion method. I will also introduce our recent work on the new strategy for the universal non-epitaxial synthesis of wafer-scale single-crystal 2D materials on diverse insulating substrates. This strategy has been utilized to synthesize large-area single-crystal monolayer, bilayer, and trilayer graphene and hBN on various insulators. These explorations are expected to be extended to the future CVD synthesis of wafer-scale 2D materials and multilayer 2D van der Waals heterostructures on insulating substrates.

Biography:

Dr. Tian is a Postdoctoral Research Fellow at Nanyang Technological University (NTU). He received his BSc in physics at Xiamen University in 2015. He obtained his PhD in Materials Science and Engineering from King Abdullah University of Science and Technology (KAUST) in 2022. He was selected as the winner of the Outstanding PhD Thesis Award of the Carbon Journal Prize by Elsevier in 2023. Following his doctoral studies, Dr. Tian served as a research associate and visiting scientist at KAUST from 2023 to 2024. His research interests span physics, CVD growth, 2D materials, van der Waals heterostructures, and artificial intelligence areas.

Metallic-enriched SWCNTs grown from CoxRe1-x by quasi-static CVD method

Xin Li[1,2], Feng Zhang[1,2], Lili Zhang[1,2], Zhong-Hai Ji[1,2], Peng-Xiang Hou[1,2,*], Zi-Wei Xu[3,*], Lin-Quan Ping[1,2], Hui-Ming Cheng[1,4], Chang Liu[1,2,*]

[1]Shenyang National Laboratory for Materials Science, Institute of Metal Research, Chinese Academy of Sciences, Shenyang, 110016, China

[2]University of Science and Technology of China, No.96, JinZhai Road Baohe District, Hefei, Anhui, 230026, P.R.China.

[3]Schol of Materials Science and Engineering, Jiangsu University, Zhenjiang, 212013, P.R. China

[4]School of Materials Science and Engineering, Shenzhen Institute of Advanced Technology, Chinese Academy of Sciences, Shenzhen, 518055, P.R. China.

Carbon nanotubes (CNTs)

The growth of high-quality single-wall carbon nanotubes (SWCNTs) with single conductivity in narrow diameter distribution is urgently demanded for the fabrication of high-performance electronic devices. However, it is very difficult to directly synthesize m-SWCNTs from traditional metal catalysts due to their high activity. Herein, we design and prepare high-melting-point CoxRe1-x unlimited solid solution catalysts with a hexagonal-closed-packed structure and uniform size. It is found that the CoxRe1-x functions as active catalytic phase for m-SWCNT growth, under a proper low carbon and hydrogen feeding and low temperature quasi-static CVD growth conditions. Re functions to inhibit catalysts' agglomeration and enlarge the growth rate difference for SWCNTs with different chirality. As a result, the grown m-SWCNTs have a very narrow diameter distribution centered at 1.1 ± 0.2 nm and a high metallic content of > 80%. Our findings open up the possibility of the chirality-controlled growth of SWCNTs through controlling its growth rate for practical nanoelectronic applications.

References

[1] Li, X., Zhang, F., Zhang, L., Ji, Z. H., Zhao, Y. M., Xu, Z. W., Wang, Y., Hou, P. X., Tian, M., Zhao, H. B., Jiang, S., Ping, L. Q., Cheng, H. M., & Liu, C. Kinetics-Controlled Growth of Metallic Single-Wall Carbon Nanotubes from CoRex Nanoparticles. ACS Nano 2022, 16 (1), 232-240.

[2] Zhang, F., Zhang, L., Jiang, H., Li, X., Liu, F., Ji, Z.-H., Hou, P.-X., Guo, S., Cheng, H.-M., Kauppinen, E. I., Liu, C., & Ding, F. (2023). Growth of high-density single-wall carbon nanotubes with a uniform structure using a CoRu catalyst. Carbon, 209, 118011.

[3] Ji, Z.-H.; Zhang, L.; Tang, D.-M.; Chen, C.-M.; Nordling, T. E. M.; Zhang, Z.-D.; Ren, C.-L.; Da, B.; Li, X.; Guo, S.-Y.; Liu, C.; Cheng, H.-M., High-throughput screening and machine learning for the efficient growth of high-quality single-wall carbon nanotubes. Nano Research 2021.

[4] Zhang, F.; Sun, J.; Zheng, Y.; Hou, P.-X.; Liu, C.; Cheng, M.; Li, X.; Cheng, H.-M.; Chen, Z., The importance of H2 in the controlled growth of semiconducting single-wall carbon nanotubes. Journal of Materials Science & Technology 2020.

Preparation of Coal Tar Pitch-Based Multidimensional Carbon Materials and Their Light Metal Anode Performance

Jian Xiao [1,*], Yunpeng Zhao [1], Jingpei Cao [1], and Jieshan Qiu [2,*]

[1]Key Laboratory of Coal Processing and Efficient Utilization, Ministry of Education,
China University of Mining & Technology, Xuzhou 221116, Jiangsu, China
[2]State Key Laboratory of Chemical Resource Engineering, College of Chemical Engineering,
Beijing University of Chemical Technology, Beijing 100029, China.
Email: jxiao@cumt.edu.cn (J. Xiao), qiujs@mail.buct.edu.cn (J. S. Qiu)

Light metals represented by Li, Na, and Zn have high theoretical specific capacity, low mass density/redox potential, and are considered as the most promising anode materials for next generation of high-energy density batteries. However, due to the high reactivity, dendritic growth, and volume expansion of these metal anodes, the electrochemical performance is poor with significant safety hazards, which hinders the practical application of light metal batteries. Therefore, improving the safety of the anodes while taking into account the stability of the interface, so as to achieve high electrochemical performance, is the key to promoting the commercialization of light metal batteries. In this thesis, starting from coal tar pitch, 1D CrO0.78N0.48 nanoseed-coupled carbon fibers [1], 2D high fluorine content stepped carbon nanosheets [2], 2D semi-ionic C-F dominated edge-rich carbon sheets, and 3D periodical FO-CNTs/CNFs composite paper [3] are prepared based on electrospinning, template confinement, liquid-phase fluorination, and plasma treatment technologies, respectively, so as to construct stable lightweight metal anodes. The results indicate that the reasonable design of physical structure and the optimal control of chemical properties of coal tar pitch-based carbon materials are of great significance for the construction of high energy density and high safety lightweight metal anodes.

References

[1] Jian Xiao, Nan Xiao, Jieshan Qiu, et al. Small, 16 (2020), 2003827.
[2] Jian Xiao, Nan Xiao, Jieshan Qiu, et al. Adv. Energy Mater., 12 (2022), 2103123.
[3] Jian Xiao, Nan Xiao, Jieshan Qiu, et al. Adv. Funct. Mater., 32 (2022), 2111133.

Elucidating the intrinsic core-shell structure of carbon nanospheres from glucose hydrothermal carbonization

Kui Wang[1,2], Jefferson W. Tester [2]

[1]School of Chemistry and Chemical Engineering,
Hefei University of Technology, Hefei 230009, China
[2]Smith School of Chemical and Biomolecular Engineering,
Cornell University, Ithaca 14853, NY, United States

This study explores the core-shell structure formation mechanisms of primary carbon nanospheres (PCNs) through hydrothermal carbonization (HTC) of glucose at 200 °C, focusing on key phase-change polymerization reactions. Colloidal carbon nanoparticles in the aqueous phase filtrate self-assembled into secondary carbon nanospheres (SCNs) with intrinsic hollow structures during room temperature storage. FTIR results revealed similar functional groups on PCNs and SCNs surfaces due to esterification reactions during HTC cooling. XPS and 13C NMR analyses identified HMF aldol condensation and etherification as dominant reactions for PCNs, while esterification and aldol condensation with levulinic acid were dominant for SCNs. The hypothesis suggests that PCNs initially formed hollow microframeworks but collapsed due to consumption of encapsulated organics, resulting in hydrophobic cores. These cores grew through aggregation (linear) and surface reactions (exponential), internalizing hydrophilic surfaces into hydrophobic cores, forming the final core-shell structure of PCNs.

Preparation of graphitized mesocarbon microbeads as anode material for lithium-ion battery by conductive graphene induction

Lantao Liu, Weiwei Pang, *

Petrochemical Research Institute, PetroChina Company Limited,102206 Beijing, P. R. China

Graphitized mesocarbon microbeads (GMCMBs) is considered to be a promising anode material of Li-ion batteries due to its unique structure and superior electrochemical performance. Limited by the complicated preparation process and the difficult control of structural design, GMCMBs has not been well developed. In this study, we have developed a new efficient and rapid method for the preparation of MCMBs, which is induced by conductive graphene through π-π interaction to accelerate the process of nucleation, growth and coalescence of MCMBs. Simultaneously, conductive graphene could act as a defect maker and provide an ion transport channel to significantly improve the Li-ion storage performance of GMCMBs. In the Li-ion half-cell test, conductive graphene-added samples exhibit good performance in cycling and rate tests. This conductive graphene-induced method provides a new avenue for developing anode materials for industrialized Li-ion batteries.

129Xe NMR Study of Adsorption Interaction between Porous Carbons and Xe Molecules

Minghao Li[1], Keiko Ideta[2], Koji Nakabayashi[1,2], Seong-Ho Yoon[1,2], Jin Miyawaki[1,2]

[1]Interdisciplinary Graduate School of Engineering Sciences (IGSES), Kyushu University, 6-1 Kasuga-koen, Kasuga, Fukuoka 816-8580, Japan
[2]Institute for Materials Chemistry and Engineering (IMCE), Kyushu University, 6-1 Kasuga-koen, Kasuga, Fukuoka 816-8580, Japan

Since the 1980s, 129Xe-NMR method has been widely used to evaluate pore structure of materials. However, how to apply the 129Xe-NMR method to investigate porous carbon materials is still confusing. This is partly due to the structural disorder and heterogeneity of surface properties of porous carbon materials, and partly due to the lack of reliable control over the density of adsorbed Xe [1].

In this study, we evaluated the interaction between porous carbon materials and the adsorbed Xe molecules by using a homemade temperature- and pressure-controlled 129Xe-NMR system. By an extrapolation to an ordinate in plots between the surface-area-normalized Xe adsorption amount and 129Xe-NMR chemical shift of the adsorbed Xe molecules to remove the influence of the Xe-Xe interaction, ultimate interaction information between the porous carbon and one Xe molecule was estimated. It was also confirmed that the interaction between the pore surface and the adsorbed Xe molecules depended on a parameter related to the pore size of the porous carbons regardless of the amount of oxygen-containing surface functionalities.

This work was supported by JSPS KAKENHI Grant Number JP19K05213.

References

[1] P.A. Simonov, et al. Carbon 37 (1999) 591–600.

Investigating the relationship between nitrogen content and the structural and electronic properties in nitrogen-carbon materials.

Mamta Devi[1], Prakash Pandey[1] and Swati Sharma[1]

[1]School of Mechanical and Materials Engineering, Indian Institute of Technology, Mandi, HP, 175075, India.

Nitrogen incorporation into carbon lattices alters crystal structure and electronic properties in N-C materials, ranging from graphitic carbon nitride (g-CN) to nitrogen-doped carbon. With nitrogen fractions spanning from over 50% to parts per billion, these materials exhibit diverse N/C ratios. This study investigates how this chemical transition affects the crystal structure and electronic properties of the resulting materials. Waste-derived g-CN serves as the initial material, with its N/C atomic ratio sequentially reduced from 0.52 to 0.10 through heat treatment. A combination of experimental techniques and computational simulations elucidates the range of N/C ratios where: (i) g-CN maintains its compound nature with a layered structure, (ii) the material exhibits a random distribution of nitrogen within a turbostratic carbon framework, and (iii) the material transforms into N-substituted graphite-like material, predominantly featuring graphitic nitrogen. At each stage, the electronic characteristics of the material change contingent upon the specific crystal structure associated with a given N/C ratio. The findings presented in this study provide valuable insights, allowing scientists to predict the structure of any N-C material based solely on its N/C ratio. Furthermore, this research sheds light on the thermal degradation behavior of g-CNs, which holds significance for various practical applications.

Microstructural Representations and Reaction Mechanisms during Multiple High-Temperature Processes of Complex Amorphous Carbon Materials

Qifan Zhong[1,2], Zihan You [1], Liuzhou Zhou [1] and Yu Zhou

[1] School of Metallurgy and Environment, Central South University, Changsha 410083, China;
[2] National Engineering Research Center of Low-carbon Nonferrous Metallurgy, Changsha 410083, China;

The complex amorphous carbon materials (CAC) containing petroleum coke, calcined petroleum coke, and coal tar pitch are crucial feed stocks for artificial carbon and graphite products. Lack of exploration of microstructural transformation and reaction mechanisms of CAC during multiple high temperature processes have hindered the development of preparations and applications of graphite products. Here, the microstructural features and functional groups were investigated based on high-resolution transmission electron micrographs (HRTEM) analysis. The atomistic representations of CAC incorporating experimental parameters were firstly constructed for comprehensive observation and simulation via the self-developed automated modeling strategy [1]. The reaction mechanisms of CAC during pyrolysis, combustion, gasification, coking, calcination, graphitization, and molten electrolysis were explored via combining reactive force field molecular dynamics (ReaxFF MD), density functional theory (DFT) calculation, and advance detection methods. The molecular structural evolution, as well as the removal and catalytic mechanism pathways of the impurities (S, N, Na, V, Fe, and etc.) in CAC were ascertained [2].

References

[1] Qifan Zhong, et al. Carbon 129 (2018) 790-802.
[2] Lei Tang, et al. Carbon 188 (2022) 336-348.

Capacitance of porous carbon materials unraveled by graphene mesosponge

Rui Tang [1], Takashi Kyotani [2] and Hirotomo Nishihara [2]

[1]College of Materials Science and Engineering, Hunan University, Changsha 410082, China
[2]Institute of Multidisciplinary Research for Advanced Materials, Tohoku University, Sendai 980-8577, Japan

The understanding of the capacitance of porous carbon materials is hindered by the structural complexity. On one hand, there are controversial opinions on the low areal capacitance of porous carbon materials. On the other hand, the cyclic voltammetry curves are elusive, being a "butterfly-shaped" one for some porous carbons but a rectangular-shaped one for others. To resolve the above puzzles, we synthesized graphene mesosponges with different graphene stacking layers (1–3 layers) and nitrogen doping amounts (0.1–4.5 wt.%) by a templating method [1,2]. These graphene mesosponges have uniform mesoporous structures with the average pore sizes (7 nm) replica the size of nanosized Al2O3 templates, therefore suitable to be used as model materials for the investigation of capacitive behavior. Our results suggest that the capacitance of porous carbons is less affected by the graphene stacking layers but strongly affected by the doped nitrogen. Nitrogen doping enhances the density of states of porous carbons and the quantum capacitance is improved as a result. Consequently, the total capacitance is improved and the shape of the cyclic voltammetry curve is transformed from a "butterfly-shaped" one to a rectangular-shaped one.

References

[1] Rui Tang, et al. Electrochim. Acta 429 (2022) 141009
[2] Rui Tang, et al. Small (2023) 2308066

Structure control of carbon films by addition of graphene oxide in the precursor film

Wataru Takarada[1,2], Takehito Yokoyama[2] and Masatoshi Shioya[2]

[1] Institute of Textile Science and Technology, School of Science and Technology, Shinshu University, Ueda, Japan

[2] Department of Materials Science and Engineering, School of Materials and Chemical Technology, Tokyo Institute of Technology, Meguro, Japan

Highly oriented graphite materials have excellent thermal conductivity and are used as thermal diffusion films. The authors aim to establish a novel method for the preparation of highly oriented graphite materials from polymeric films. Some polymer films can obtain carbon films by carbonizing at high temperatures, but in many cases the carbonization yield is low, and the orientation of the resulting carbon films is also low. In this study, we tried to improve the orientation of carbon films by addition of graphene oxide (GO) in the precursor film. A solution of phenol resin and alkali-treated lignin with 10 wt% GO was used as the precursor solution. The precursor film was formed by the solution-casing method, and carbon films were prepared by carbonizing at 1000 degree C. No orientation was observed in the carbon film prepared wuithout GO, but in the carbon film prepared with GO, the graphite layer tended to be oriented parallel to the film surface. It was also confirmed that the addition of GO increased the electrical conductivity of the carbonized film. These results indicate that carbon crystals had grown from GOs that were oriented during the preparation process of the precursor film.

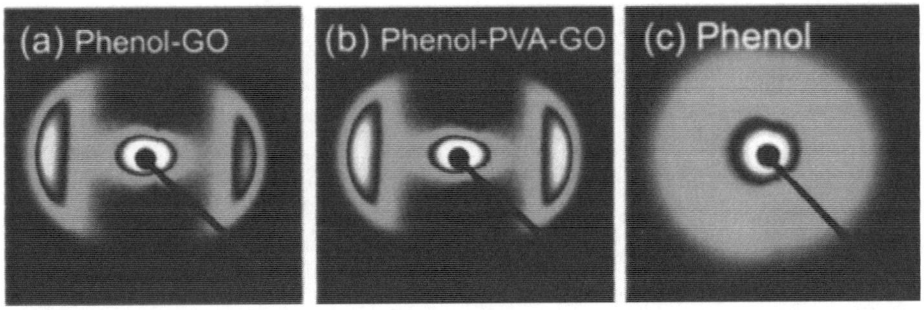

Figure 1. Edge view WAXD patterns of carbon films from (a) phenolic resin with GO, (b)phenolic resin and PVA mixtures with GO and (c) neat phenolic resin.

Controllable preparation and application of Coal-based functional carbon materials in energy storage

Xiaoxiao Qu[1], Baolin Xing[1] and Chuanxiang Zhang[1]

[1]College of Chemistry and Chemical Engineering, Henan Polytechnic University, Jiaozuo 454000, China

Henan Polytechnic University is the first mining university in Chinese history, which was established in 1909, and has a history of more than one hundred and ten years. Based on the demands of environmental protection and the strengths of our university in the research of traditional energy, that is mining and coal, we provide new ideas from traditional energy to new energy and energy storage technology. Thus, in our work, we employed coal and its derivatives (humic acid and coal tar pitch) as precursors to prepare various carbon materials with different dimensions, including 0D coal-based carbon quantum dots, 1D coal-based carbon fiber and carbon nanotube, 2D coal-based graphene and carbon nanosheets, 3D coal-based activated carbon, graphite, and hard carbon for energy storage application in supercapacitor, lithium-ion battery, and sodium-ion battery.

Influence of Oxygen-containing Surface Functional Groups on Fluorination of Porous Carbons

Xiazhe Zhai[1], Keiko Ideta[2], Yoshiyuki Hattori[3], Koji Nakabayashi[1,2], Seong-Ho Yoon[1,2], Jin Miyawaki[1,2]

[1] Interdisciplinary Graduate School of Engineering Sciences (IGSES), Kyushu University, 6-1 Kasuga-koen, Kasuga, Fukuoka 816-8580, Japan

[2] Institute for Materials Chemistry and Engineering (IMCE), Kyushu University, 6-1 Kasuga-koen, Kasuga, Fukuoka 816-8580, Japan

[3] Faculty of Textile Science and Technology (FTST), Shinshu University, 3-15-1 Tokida, Ueda, Nagano 386-8567, Japan

Ozonation and fluorination are common methods for modifying the surface properties of carbon materials, transitioning between hydrophilicity and hydrophobicity. However, the influence of hydrophilic surface functional groups on carbon surface modification through fluorination remains unexplored. This study investigates the impact of oxygen-containing functional groups introduced by the ozonation treatment of porous carbons on the fluorination treatment. It was found that the oxygen-containing surface functional groups were effectively removed during the initial stages of the fluorination treatment. In addition, through fluorination, the surface was converted from hydrophilic to hydrophobic regardless of the presence of the oxygen-containing surface functional groups, leading to a significant decrease in pore filling factor of water vapor. Furthermore, the results revealed that there were two types of oxygen-containing surface functional groups, the easily-removal and hardly-removal ones by the fluorination treatment. The former was found to be associated with high hydrophilicity, while the latter was low hydrophilic or hydrophobic. This research demonstrates a potential of fluorination for flexible control over the surface characteristics of porous carbon materials, offering insights into tailoring their properties for various applications.

This work was supported by JSPS KAKENHI Grant Number JP23H01691.

Solvated-metal-ion cointercalation chemistry in graphite

Zheng-Long Xu

[1]Department of Industrial and Systems Engineering, The Hong Kong Polytechnic University, Kowloon, Hung Hom, Hong Kong SAR, China

Reversible intercalation of guest ions in graphite has been considered as the key feature in rechargeable battery systems. Evidently, graphite has been widely used as anode materials in Li ion battery (LIB) owing to the reversible Li de/intercalation in graphite. On the road to explore graphite as universal anodes for post Li-ion batteries, such as Na-ion batteries, the conventional intercalation chemistry had to be revisited, and recent findings demonstrate that a new intercalation chemistry involving the insertion of solvated ions, designated as co-intercalation, could overcome some of the obstacles in the conventional intercalation of graphite. Accordingly, remarkable progresses have been recently attained in high energy and power Na ion batteries with graphite anodes undergoing co-intercalation reactions. More recently, extensive accomplishments have been continuously made in other post Li-ion batteries with graphite anodes, such as Ca ion batteries (CIBs), by exploiting the co-intercalation reactions. This talk will present our interesting findings on the solvated metal ions intercalation into graphite for post-Li-ion battery technology.

References

[1] Y Yi, et al. Deciphering anion-modulated solvation structure for calcium intercalation into graphite for Ca-ion batteries, Angew Chem. Int. Ed. 2024, e202317177.

[2] J. Park, † Z.L. Xu, † G. Yoon, † S.K. Park, J. Wang, H. Hyun, H. Park, J. Lim, Y.J. Ko and K. Kang, Stable and High power calcium ion batteries enabled by calcium intercalation in graphite, Adv. Mater. 2020, 32.4: 1904411.

[3] Z.L. Xu, et al. Tailoring sodium intercalation in graphite for high energy and power sodium ion batteries, Nature Communications, 2019,10, 2598.

Efficient combustion of fixed coal layer in the advanced chamber of the low-power boiler

Z.A. Mansurov, B.A. Urmashev, A.B. Kudasheva, Z.M. Azizov, M.Zh. Khasimov

[1]Department of Industrial and Systems Engineering, The Hong Kong Polytechnic University, Kowloon, Hung Hom, Hong Kong SAR, China

To increase the efficiency of small-capacity boilers when burning coal in a fixed bed, an improved boiler design is proposed [1,2]. The improved design of the boiler has the following features: instead of fire grate, a solid metal plate is used, avoiding the loss of fine coal particles in the ash tray; The air flow is controlled by means of vertical air distributors in the coal layer having radial openings in a horizontal direction. In addition, the exclusion of air supply through the grate will: reduce the airspeed of flow rate by 2-2.5 times, because of the time spent in the heat exchanger is prolonged; The carrying away of particularly fine coal particles is avoided and the combustion of these particles emits additional heat. Realization this hypothesis will lead to a significant reduction in the emission of harmful substances into the environment, thus significantly increasing the efficiency of the heating system.

Objective: To improve the efficiency of small-capacity industrial boilers combustion solid fuels (coal) in a fixed layer by improving the design, parameters, and operating modes. Tasks: 1. Mathematical modeling of the movement of hot gases after the region of coal combustion processes in a fixed coal bed on a laboratory sample of a low-power boiler. 2. Experimental study of the flow of air into the fixed coal layer by physical simulation. 3.The hydrodynamic model is obtained in the form of regression equations from two factors influencing the completeness of combustion area coverage. 4. Comparative evaluation of fixed coal layer combustion in an advanced small-capacity boiler laboratory sample with a classical combustion system.

Design and operation of advanced laboratory sample of low-power boiler. According to the objective of improving the efficiency of low-power industrial boilers 0.21-0.50 kcal/h burning solid fuels (coal) in a fixed layer, pre-developed laboratory sample of the heating device of limited size for setting parameters and modes of combustion process. For the study, a room with an area of 12.96 m2 was selected and a household boiler with a capacity of up to 10 kW. To make a comparative assessment of the stationary coal combustion according to the existing and proposed scheme, a space has been provided in the boiler design for the installation of a standard grate or blind

Efficient combustion of fixed coal layer in the advanced chamber of the low-power boiler

Z.A. Mansurov, B.A. Urmashev, A.B. Kudasheva, Z.M. Azizov, M.Zh. Khasimov

[1]Department of Industrial and Systems Engineering, The Hong Kong Polytechnic University, Kowloon, Hung Hom, Hong Kong SAR, China

tile. he standard grate and deaf steel plate had the same external dimensions of 190x230 mm. The blank flat plate was 10 mm thick and had a 45 mm diameter hole in the center. Through the opening of the plate passed the base of the air injection pipe, which had a connection to the fan of the injection air.

The test was carried out by combustion of coal in one boiler with a change of air supply: classical way through the grille of the grate; without grate by means of injectors into the coal layer. Research for different coal combustion methods was conducted on different days, by selecting the same ambient temperature. And in each method the same quantity of coal (10 kg) of the same brand (Shubarkol) with the same granulometric compositions was used. During the combustion, the temperature conditions of the walls of the water boiler, the water and the air were determined with the help of the Fluke measuring instrument (changes in the indoor air temperature and water in the tank were recorded by time). The coal combustion test was carried out in a single boiler with a change of air supply by: the classic method through the grille of the grate; without the grate using the superchargers to the coal layer.

References

1.Khazimov M.Zh., Urmaschev B.A., Idrishev K.Zh., Khazimov K.M. Vodogreyny solid fuel boiler/ Utility model patent 3368. It was announced on 05/29/2018. Publ. 16.11.2018. byul. No.43.

2.Khazimov M.Zh., Berdimurat A.D., Kudasheva A.B., Sagyndykova Zh.B., Khazimov K.M., Urmaschev B.A., Rahman Sh. Industrial Low-power Water Heating Boiler / Patent for the invention RK 35521. The application was made on 08.09.2020. Pub. 02/18/2022, Office No. 7.

Molecular Evolutionary Growth and Assembly of Ultrapure Semiconducting Carbon Nanotubes

Zhenxing Zhu[1], Fei Wei[1]

[1]Beijing Key Laboratory of Green Chemical Reaction Engineering and Technology, Department of Chemical Engineering, Tsinghua University, Beijing 100084, China

Email: zxing@mail.tsinghua.edu.cn

Carbon nanotubes (CNTs) are the promising candidates for the novel integrated electronics. Whereas, it' s a challenge to mediate their bandgap or chiral structures from a vapor liquid-solid growth process. Here, we demonstrate the molecular evolution of ultralong semiconducting CNT (s-CNT) arrays based on an interlocking between the atomic assembly rate and bandgap of CNTs. Rate analysis verified the different evolutionary rates of metallic (m-) and s-CNTs as length increased. Quantitatively, a nearly ten-fold faster decay rate of m CNTs, led to a spontaneous purification of 99.9999% s-CNTs at a length of 154 mm and wafer scale ultralong CNTs with the maximum length up to 650 mm were achieved within a precisely designed micro reactor. Transistors fabricated on them delivered a high current of 1.4 A/µm with an on/off ratio around 108and mobility over 4000 cm2/V s.

Furthermore, ordered chirality-consistent CNT tangles and coils were synthesized by entangling an individual monochromatic ultralong CNT under the effect of "acoustic-induced vortices" . Each as-synthesized high-density CNT tangle was made of one self-entangled CNT with the length over 100 mm, tandem with a (1~10)x104 µm2

area range controlled by a minimum consumed energy model. Devices fabricated with the large-area CNT tangles

exhibited unique photoelectronic properties with an on/off ratio up to 103~106at 4 mA on-state

current, which is also the highest output current record so far among single-CNT-based transistors. Our self-purification strategy based on molecular evolution offers more freedom to in-situ control the CNT purity and provides a robust method to synthesize perfectly entangled condensate over a wide length scale.

References

[1] Zhu et al., ACS Nano 15, 5129 (2021).

[2] Zhu et al., Nat. Commun. 10, 4467 (2019).

[3] Zhu et al., Adv. Funct. Mater. 2109401 (2021).

[4] Zhu et al., Adv. Sci. 2205025 (2022).

[5] Zhu et al., Adv. Sci. 2003078 (2021).

Favorable Component Regulation of Coke for High-Performance Graphite Blocks

Chuanjun Tu, Run Li, Yanliu Liu, Jiao Tan

College of Materials Science and Engineering, Hunan University, Hunan Province, Changsha, China

High performance carbon graphite materials have been widely applied in various fields, such as aerospace, energy, transportation, and chemical engineering. Normally, carbon graphite blocks are prepared via either conventional kneading or self-sintering approaches. However, most kneaded carbon blocks suffer drawbacks of phase separation and interface defects, and the self-sintering carbon block faces the challenge of poor thermal-shock resistance, brittleness and easy cracking. To solve those problems, a highly active interface transition layer is created on the surface of the raw materials, ensuring good contact of various components and minimizing the formation of pores[1]. Sulfonated graphene is employed as the reinforcement phase to intensify the interactions between the aggregate and the binder, thus obviously improving the structure integrity of the whole carbon materials[2]. Moreover, to clarify the specific contribution of each component in green petroleum coke, the coke is divided into several fragments via solvent-extraction. When the dichloromethane-soluble compounds are removed, the obtained green coke demonstrates a higher self-bounding and self-sintering capacity[3-4]. The dichloromethane-soluble-toluene-insoluble components, which display characteristic of high content of naphthene, oxygen-containing groups, and plasticizer, are also beneficial to the carbonization of coke[5].

References

[1] Ping Liu, et al. Sustainable Materials and Technologies 40 (2024) e00984.
[2] Chuanjun Tu, et al. Carbon 148 (2019) 378-386.
[3] Ping Liu, et al. Materials Letters 309 (2022) 131387.
[4] Ping Liu, et al. Materials Letters 336 (2023) 133826.
[5] Jiao Tan, et al. Fuel 343 (2023) 127912.

Study on oxidation rates and gas product for matrix fuel element under high temperature steam oxidation

Guma Yeli[1], Shaochang Hao and Bing Liu

[1]Institute of Nuclear and New Energy Technology, Tsinghua University, Beijing, China

The high-temperature gas-cooled reactor fuel elements are produced by uniformly mixing TRISO-coated fuel particles with matrix graphite. Under extreme accident conditions, water or water vapor may enter the reactor core, causing rapid oxidation of the fuel elements above 1000°C. In this study, steam oxidation of graphite fuel element was conducted at 800°C, 900°C, and 1000°C under different water vapor partial pressure. The coupled mass spectrometry results indicate that the main gas products at these temperatures are H2, CO and small amount of CO2. By combining graphite fuel element weight loss and the production of gas product, the reaction rates were obtained. The surface microstructure and internal three-dimensional pore structure of graphite fuel elements have been characterized. These data provide insight into the oxidation behavior of matrix fuel element under water ingress accident condition.

Laser extreme manufacturing of nanocarbon materials from low rank carbonaceous materials (Coal, tar, pitch, and beyond)

Xining Zang [1]

[1] Mechanical Engineering, Tsinghua University, Beijing, China, 100084

The policy-driven and strategic reduction in the use of fossil fuels for electric power generation is renewing the interest in the potential for coal and other heavy hydrocarbon materials (HCMs) to be used as feedstocks for carbon-based products. The present work shows that optimal selection of initial HCMs based on molecular constituents and processing parameters are essential in tuning the material for a particular and targeted electronic application. Combining the selection of feedstock chemistry (H:C and aromatic content) and controlling variable laser treatment parameters (laser power, speed, and focus) lead to full control over the H:C ratio, sp2 concentration, and degree of graphitic stacking order of the products [1,2]. Furthermore, ultrafast laser can induce shock wave pressure leading to effective and speedy tuning of phase change during the carbon conversion comparing to traditional hexagonal press and Maximum multianvil cells. The broad intertunability of these factors results from a wide distribution of carbon material crystallinity from amorphous, to highly graphitic, and to sp3 dominated nano-diamonds. We will also showcase a few examples of extreme manufacturing of HCMs based devices for sensing and environmental applications [3-5].

References

[1] X Zang, et al., Science Advances, 2020, 6, eaaz5231.
[2] X Zang, et al., ACS Nano, 2022, 16, 2, 2101–2109
[3] X Zang, et al., Matter, 2022, 5, 430-447.
[4] J Yu, et al., Advanced Science, 2023, 2302479
[5] C-C Liu, et al., Nano Energy, 2023, 108634

Multiscale Structural Engineering of Carbon Aerogels for Broadband Electromagnetic Wave Absorption

Gaofeng Shao

School of Chemistry and Materials Science, Nanjing University of Information Science & Technology, Ningliu Road No.219, Nanjing 210044, P.R. China

This presentation encapsulates our latest research on the multiscale structural engineering strategy that optimally tailors carbon aerogels to effectively manage electromagnetic wave absorption properties. Herein, we introduce a series of hierarchical carbon aerogels, created through processes such as self-assembly, freeze casting and wet-spinning[1,2]. Our findings reveal that the strategic regulation of macroscale dimensions along with microscale porous architecture and nanoscale domains significantly enhances microwave reflection and scattering, facilitates electron migration and hopping, and encourages defect-induced polarizations. The optimized carbon aerogels demonstrate an impressive absorption bandwidth while utilizing less than 1wt% of the material. Additionally, incorporating these aerogel into a polysilsesquioxane host-guest aerogel system not only augments their practical application scope but also imparts attributes like superhydrophobicity, thermal insulation, and anti-freezing capabilities, hence adeptly navigating complex electromagnetic environments[3].

References

[1] Gaofeng Shao(1st), et al. Advanced Materials 2020, 32, 1907176.

[2] Gaofeng Shao*, et al. Nano-Micro Letters 2022, 14, 107.

[3] Gaofeng Shao(1st), et al. ACS Materials Letters 2022, 4, 1787-1797.

Graphene oxide reinforced epoxy nanocomposites coatings with superior corrosion resistant and mechanical performance

Ding Nan, Xin Li, Jie Chen, Dongxia Huo

College of Chemistry and Chemical Engineering, Inner Mongolia University, Hohhot, China

Carbon materials, such as graphite, graphene oxide (GO) and porous carbon nanotube, have been became key materials for high-performance coatings for the protection of metals from corrosion. Our research focuses on the application of GO reinforced epoxy nanocomposite coatings for metals protection. We have successfully synthesized monolayer GO using modified Hummer' s method from natural graphite. Monolayer modified graphene oxide (MGO) was prepared by covalent grafting and non-covalent methods. Subsequently, MGO were integrated with epoxy resin to fabricate nanocomposite coatings exhibiting exceptional corrosion resistance. By harnessing the distinctive structure of MGO alongside the outstanding properties of epoxy resin, we attained coatings characterized by high strength, superior corrosion resistance, and exceptional weathering resistance. Furthermore, our study revealed the impact of MGO content in the epoxy coatings on their performance and evaluated their potential applications in various environmental conditions. Our findings provide fresh insights and methods for further research and application in the field of metals protection.

References

[1] Xin Li, et al. Surf. Coat. Tech. 485 (2024) 130883

[2] Ding Nan, et al. Mater. Today Sustain. 21 (2023) 100282

Hard Carbons Derived from Different Condensation Degree COPNA Resin and Their Sodium Ions Storage Performances

Qi Han, Ying Zhou, Yuhang Xie, Jiahui Ding, Kening Xing, Chunlei Wang*

Dalian University of Technology, Dalian, China
*Corresponding author e-mail: clwang-chem@dlut.edu.cn

In recent years, sodium-ion batteries (SIBs) are gaining popularity because of their abundance, worldwide resources, low cost, which were considered as the most promising alternative to lithium-ion batteries presently the dominant battery technology. Hard carbons with an expanded interlayer distance (>0.335 nm) and a tunable porous structure have been regarded as the most promising anode material for SIBs, e.g. [1]. Based on the energy structure of China, which is rich in coal, poor in oil, and low in gas, the application of coal-based heavy organic matter to prepare hard carbon as an anode material for SIBs is an effective way to realize the utilization of heavy resources. COPNA resin, a thermosetting resin with a three-dimensional network of polynuclear aromatic compounds crosslinked by electrophilic substitution reaction, is an excellent binder for carbon materials and a potential precursor for advanced carbon materials, e.g. [2]. In this paper, COPNA resin with different condensation degree was prepared by cross-linking coal liquefied residue (CLR) with trioxane (TOX) under the catalysis of 50 % wt. H2SO4, and the hard carbon anode was then obtained by curing and carbonization. The results showed that the sodium ions storage performance of the hard carbon derived from COPNA resin with high condensation degree after carbonization at 1300 °C exhibited a reversible capacity of 316.43 mAh g-1 with an initial Coulombic efficiency of 71.64 %.

Keywords: coal liquefied residue; COPNA resin; hard carbon; sodium ion battery; condensation degree

Acknowledgments: This work was supported by the National Natural Science Foundation of China (No. 22178053).

References

[1] Sun N, et al. J. Advanced Energy Materials, 12 (2022) 2200715.
[2] Fang W, et al. J. Fuel, 320 (2022) 123916.

Heavy Organic Matters Based Hard Carbon Anodes via Mechanochemical Method and Their Sodium Storage Performance

Yuhang Xie, Ying Zhou, Qi Han, Kening Xing, Jiahui Ding and Chunlei Wang*

Dalian University of Technology, Dalian, China

*Corresponding author e-mail: clwang-chem@dlut.edu.cn

Hypercrosslinked polymers (HCPs) are a promising precursor for hard carbon anode materials for sodium ions batteries, but their high synthesis cost and strict reaction conditions limit their commercialization[1]. In this work, to avoid the use of expensive organic solvents, mechanochemical methods were used to prepare low cost heavy organic matters based HCPs through Friedel-Crafts reaction, with ferric chloride or concentrated sulfuric acid as catalysts[2]. The results showed that the hypercrosslinking reaction catalyzed by ferric chloride had a faster reaction rate (completed in 10 minutes) than that of concentrated sulfuric acid. When concentrated sulfuric acid was used as a catalyst, besides significantly reducing of the reaction time, sulfonation side reaction was restrained also compared to traditional solvent methods. The hard carbon anode material can be obtained by directly carbonizing the heavy organic matters based HCPs prepared by mechanochemical method. When using ferric chloride as a catalyst, anthracene oil(AO)-based hard carbon materials (carbonized at 1300 ℃) exhibit a reversible sodium storage capacity of 306.8 mA h g-1 at a current density of 0.03 A g-1. When using concentrated sulfuric acid as a catalyst, the coumarone resin(CR) based hard carbon material exhibits a reversible sodium storage capacity of 304.6 mA h g-1 at a current density of 0.03 A g-1. This work successfully explored a green, economic and promising hard carbon preparation process, has important research value.

Keywords: Mechanochemistry, Hypercrosslinked polymers, sodium ion battery, Hard carbon materials.

Acknowledgments: This work was supported by the National Natural Science Foundation of China (No. 22178053).

References

[1] Zhang, S. L. et al. J. Mater. Chem. A. 3: 1896-1901 (2015)

[2] Lee, J. M., Kurihara, T. Chem. Mat. 32: 7694-7702 (2020).

PEG Template Hypercrosslinked Coal Liquefaction Asphalt Based Closed-Pore Hard Carbon Anode for Sodium Ions Batteries

Kening Xing, Ying Zhou, Jiahui Ding, Yuhang Xie, Qi Han, Chunlei Wang*

Dalian University of Technology, Dalian, China

Sodium ion batteries (SIBs) are considered the most competitive candidate for lithium-ion batteries (LIBs) in many cost sensitive application scenarios due to their merits in abundant resources and low-cost. Hard carbon (HC) is considered as the most promising commercial anode material for SIBs. Recent studies have shown that closed or sieving pores play a crucial role in improving the low voltage (<0.1V) plateau capacity of hard carbon anodes in SIBs. However, due to the complexity of the structure of hard carbon and the unclear mechanism of closed cell formation, there is currently a lack of simple and effective closed cell construction strategies[1], which seriously hinders the development of high platform capacity hard carbon anodes[2]. In this work, hard carbons with sieving pores was prepared through Friedel-Crafts knitting of coal liquefaction asphalt (CLR) with polyethylene glycol (PEG) as porogen, FeCl3 as catalyst and dimethoxymethyl (FDA) as crosslinking agent. The sample CLR:PEG-100:20-1300 exhibit a high sodium storage capacity of about 335 mAh/g and an initial Coulombic efficiency of 60%. The proposed method of chemical regulation through the pore-forming strategy is simple and practical to achieve high sodium storage energy density.

Acknowledgments: This work was supported by the National Natural Science Foundation of China (No. 22178053).

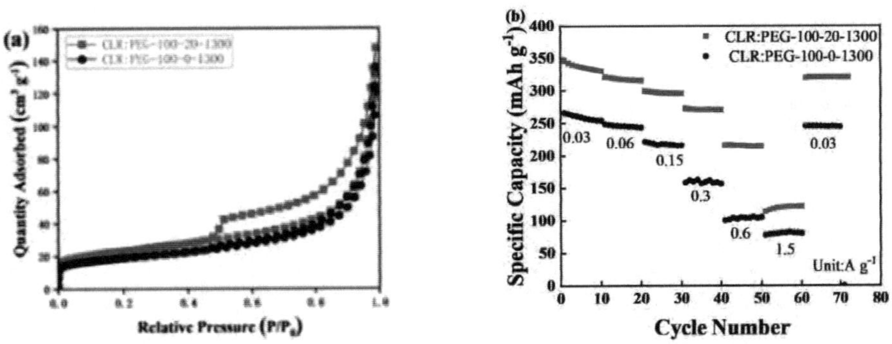

Fig (a) nitrogen adsorption isotherm (b) The magnification performance diagram at different current densities

Inducing inherent oxidation resistance in carbon fibers by incorporation of oxidation inhibitors in the precursor

Bhavika Chouhan[1], Bikash Raut [1], Swati Sharma [1]

[1] School of Mechanical and Materials Engineering, Indian Institute of Technology Mandi, Kamand, Mandi, 175075, HP, India

The utility of carbon fibers (CFs) and carbon/ carbon (C/C) composites is limited by their low thermal stability in oxygen-rich environments above 400 0C. A thermal barrier coating is commonly used on manufactured parts to prevent their oxidation. Here we report on a novel approach to achieve inherent oxidation protection up to 620 0C in electro-spun CF mats by incorporating SiO2, an oxidation inhibitor in the CFs, achieved by a controlled incorporation of polydimethylsiloxane (PDMS) directly into the CF precursor. The concentration of PDMS is optimized such that the mechanical properties of the fibers remain within the usable range. In addition, the effect of boric acid dip-coating on the SiO2-containing CFs is investigated. The modified CFs are characterized for their chemical composition and bonding, and the oxidation behavior is studied via Thermogravimetric Analysis (TGA). Based on the data, a mechanism for the oxidation resistance of such fibers is proposed. This scalable, one-step, and green method has tremendous potential in various commercial as well as research applications.

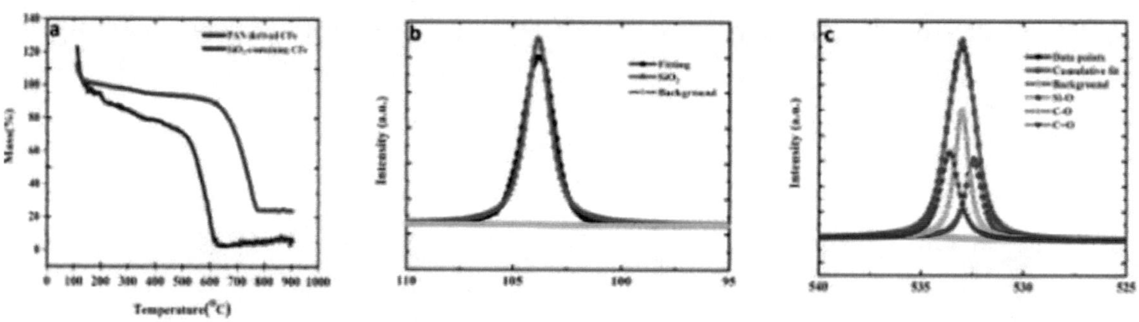

Figure 1: (a) TGA plot of untreated PAN-derived CFs vs SiO2-containing CFs, X-ray photoelectron spectroscopy graph showing deconvoluted (b) Si2p spectra and (c) O1s spectra, confirming the presence of SiO2

Upcycling waste wind turbine blades into flash graphene

Jiahui Lei[1], Qunxing Huang[1]

[1]State Key Laboratory of Clean Energy Utilization, Institute for Thermal Power Engineering, Zhejiang University, Hangzhou, 310027, China

Wind turbines, designed with a lifespan of approximately 20-25 years, are approaching the end of their operational life and will soon require replacement [1]. It is estimated that by 2034, ~225,000 tons of waste wind turbine blades (WWTB) will be retired annually [2]. WWTB are mainly made of carbon fiber reinforced polymer (CFRP) composites and glass fiber reinforced polymer (GFRP) composites [1]. Most recycling technologies concentrate on recycling carbon fiber, typically through mechanical recycling of CFRP [3]. However, disruptive separation steps often degrade the fibers' mechanical properties, significantly diminishing their value [4]. In this work, an approach to upcycling CFRP products is presented. This method employs flash Joule heating (FJH), a highly efficient technique that uses minimal electrical energy to produce high-quality turbostratic graphene called flash graphene (FG) [5]. The FG obtained from CFRP is characterized by Raman spectroscopy and has a high-intensity 2D band with a low-intensity D band. The XRD of FG shows a (002) peak that occurs at a slightly lower 2θ (~25.95°) compared to graphite and the FG is turbostratic with an interlayer spacing of ~3.43 Å. The results demonstrate a workable path for reusing waste wind turbine blades through upcycling of CFRP into high-value FG.

References

[1] Lichao Ge, et al. Journal of Analytical and Applied Pyrolysis 171 (2023) 105964

[2] Wangmi Chen, et al. Journal of Analytical and Applied Pyrolysis 169 (2023) 105859

[3] Ming-Yuan Shen, et al. Composites Part B 264 (2023) 110878

[4] Lichao Ge, et al. Journal of Analytical and Applied Pyrolysis 170 (2023) 105919

[5] Duy X. Luong, et al. Nature 577, 647–651 (2020)

Developing Strategy of Low-cost Mesophase Pitch-based Carbon Fibers with High Performance

Jianxiao Yang[1], Yue Liu[1], Yuanshuo Peng[1] and Jinshui Liu[1,2]

[1] Hunan University, Changsha 410082, China

[2] Hunan Toyi Carbon Material Technology Co., Ltd, Changsha 410221, China

Mesophase pitch-based carbon fibers (MPCFs) are considered to be promising structural and functional materials in applications such as aerospace, electronics, and advanced manufacturing industry due to their high specific Young' s modulus and high thermal conductivity. Herein, some research cases for developing the low-cost MPCFs with high performance have been investigated in our work. The spinnable mesophase pitch (MP) was synthesized through the co-carbonization with coal tar pitch (CP) and petroleum pitch (PP). The MPCFs prepared from co-carbonized MP showed both good mechanical properties, like CP-based fibers, and, ultrahigh thermal conductivity, like PP-based fibers [1]. Accelerated stabilization was successfully demonstrated under effective plasma irradiation-assisted modification (PIM) of pitch fibers (PFs). The PIM treatment could obviously introduce more oxygen-containing groups into PFs, which was remarkably efficient in shortening the stabilization time of PFs with a faster stabilization heating rate, as well as in preparing the corresponding MPCFs with higher performance [2]. The boron carbide (B4C) nanoparticles doping MP was synthesized by the in-situ doping method, the B4C was evenly dispersed in PFs to provide active sites of oxygen, resulting in a homogeneous stabilization, and ameliorating the splitting microstructures of MPCFs [3].

References

[1] Jianguang Guo, et al. Energy & Fuels 34 (2020) 6474-6482

[2] Yuanshuo Peng, et al. Materials 14 (2021) 6382

[3] Yue Liu, et al. Molecules 27 (2022) 5132

Impact of Various Carbon Fiber Types on the Properties of Pb-C Composites

Lin You Cheng, Yang Tzu Ching, Shu-Huei Hsieh*

National Formosa University, No. 64, Wenhua Rd., Huwei Township, Yunlin County 632301, Taiwan

Email: shhsieh@nfu.edu.tw

Lead-carbon cloth composites with a good Pb/C interface and 100% Pb coverage have been produced to enhance the charging reaction and mitigate the voltage leakage effect of the carbon additives in lead-acid batteries.

Various types of carbon cloth, including activated, structural, flame-retardant, and recycled carbon fibers, are used to make the composite plates. Liquid lead has high surface tension1, resulting in the poor wetting of carbon materials. When we used modified carbon cloth, we achieved 100% success rate in making lead-carbon cloth composite plates using a mold designed by ourselves and a hot press2.

The electrochemical properties in 1.28M sulfate solution show that the charge transfer resistance of the composite is much lower than that of pure lead, except for the air-oxidized flame-retardant carbon fibers. Using the lead-activated carbon cloth composite plates enhances the high-rate cycling performance of lead-acid batteries. Recent experimental results reveal that composite plates used as electrodes to form a symmetric cell with recycled carbon cloth have higher capacity than those with activated carbon cloth. Further battery tests will be conducted to investigate the effect in lead-acid batteries.

References

[1] E. Dujardin, T. W. Ebbesen, H. Hiura, K. Tanigaki, Science, 265,5180 (1994).

[2] S.H. Hsieh*, C.T. Hsieh, and C.H. Xiao, J.Taiwan Institue of Chemical Engineers 152(2023), P105175.

Changes of structure and compressive strength of carbon fiberswith fatigue

Motoki Tanaka[1], Daisuke Kimura[1,2], Haruki Kobayashi[3], Wataru Takarada[1,4], and Masatoshi Shioya[1]

[1]Department of Materials Science and Engineering, School of Materials and Chemical Technology, Tokyo Institute of Technology, Meguro, Japan
[2]Nano Carbon Device Research Center, National Institute of Advanced Industrial Science and Technology (AIST), Tsukuba, Japan
[3]Faculty of Materials Science and Engineering, Kyoto Institute of Technology, Kyoto, Japan
[4]Institute of Textile Science and Technology, School of Science and Technology, Shinshu University, Ueda, Japan

Carbon fiber is used in various fields as a reinforcing material for composite materials, but the relationship between fatigue and physical properties and its structural dependence have not yet been clarified. In this study, single fiber compressive strength tests and void structure analyses were conducted on carbon fibers with different fatigue cycles to investigate how compressive strength changes with fatigue and how it is related to structure. As a result, the decrease in compressive strength due to fatigue was confirmed and correlations with structural parameters were also obtained.

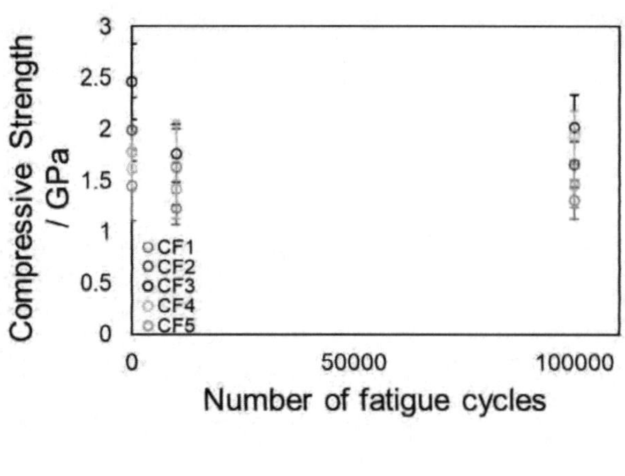

Figure 1. Compressive strength at different number of fatigue cycles

Fatigue behavior in CFRP Composites: Unravelling the Complex Relationship of Cyclic Loading and Structural Performance

Prabhjot[1], Ravinder kumar [2], Sahendra P. sharma [1,2*]

[1]School of Chemical Engineering and Physical Sciences, Lovely Professional University, Jalandhar- Delhi G.T Road, Phagwara, Punjab 144411 India,
[2]Division of Research and development, Lovely Professional University, Jalandhar- Delhi G.T Road, Phagwara, Punjab 144411 India
*sahendras@gmail.com

Cyclic loading-induced fatigue stands as the primary factor behind the performance degradation observed in carbon fiber-reinforced polymer (CFRP) composites and structures. This degradation, in turn, adversely impacts damage tolerance, consequently diminishing the overall service life. The fatigue behavior of composite materials involves a multifaceted process encompassing damage initiation and propagation. The intricacy of the fatigue damage mechanism in CFRP composites primarily results from the material's heterogeneous composition, variations in material constituents, service conditions, distribution of cyclic loads and their complicated interplay stacking. This article offers an overview of the fundamental aspects and dominant factors influencing the fatigue response of fiber-reinforced polymer (FRP) composites. Both experimental and computational studies were carried out on commercially available T700 carbon fibers in order to understand the response of the carbon fiber composite fatigue behavior. Moreover, the article delves into the damage mechanisms associated with fatigue failure in FRP composites like interface failure, fiber-matrix debonding and fiber breakage.

Keywords: Carbon fiber composite, Fabrication, Fatigue testing, Computational study

Calcium compound-assisted catalytic graphitization to improve crystallinity of PAN-based carbon fibers

Sora Lee,[1] Sungho Lee [1,2*]

[1] Carbon Composite Materials Research Center, Institute of Advanced Composite Materials, Korea Institute of Science and Technology, 92 Chudong-ro, Bongdong-eup, Wanju-gun, Jeonbuk 55324, Republic of Korea.

[2] Department of Nano Material Engineering, KIST School, University of Science and Technology.

Graphitized carbon fibers (CFs) are appealing materials because of their high thermal and electrical conductivities and tensile modulus. These attributes derive from their crystalline structures that develop during heat treatments of up to 3000 ℃ [1]. In spite of the costly thermal processes, there is a structural limit for achieving these sought-after properties for polyacrylonitrile-based CFs. Herein, the preparation of polyacrylonitrile-based CFs with highly developed microstructures via calcium-assisted heat treatments of up to 2700 ℃ is reported. The catalytic graphitization mechanisms of PAN-based CFs using Ca at various HTTs were investigated. The hydrothermally treated CFs in a solution of calcium carbonate (CaCO3) were dissolved in distilled water and heat-treated to 2700 ℃. The heat treatment of the calcium compounds at 1600 ℃ to incorporate CFs resulted in an interlayer spacing of 0.3360 nm, which was unachievable solely through heat treatment at 2700 ℃. In addition, the achieved tensile modulus and electrical conductivity of 480 GPa and 1.7 X 103 S/cm, respectively, were significantly higher than those of pure carbon fibers. The calcium penetrating the internal structure of the CFs aligned the non-uniform graphene structure and developed the graphite structure of the CFs by acting as catalysts, even at low temperatures.

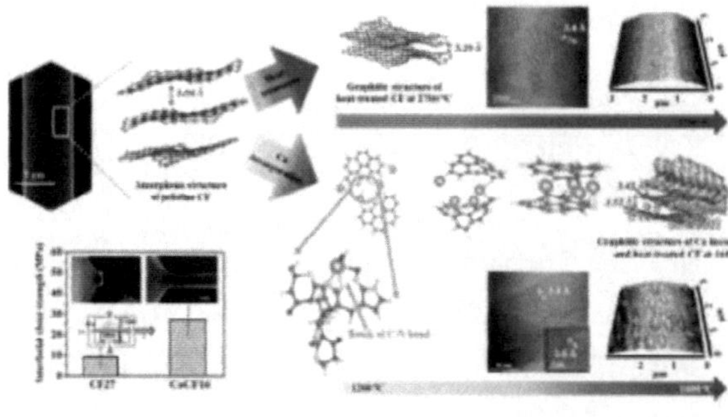

References

[1] Ōya A, Ōtani S. Effects of particle size of calcium and calcium compounds on catalytic graphitization of phenolic resin carbon. Carbon. 17 (1979) 125-129.

Preparation and oxidation ablation behavior of TaxHf1-xC (x=0.2, 0.25, 0.33, 0.5) single phase solid solution ceramic powders and its modified C/C composites

Zongyu Liu[1,2], Zhijun Dong[1,2], Hui Zhu[1,2], Guanming Yuan[1,2], Ye Cong[1,2], Baoliu Li[1,2], Jianguang Guo[1,2], Qin Zhang[1,2], Jiang Zhang[1,2], Xuanke Li[1,2]

[1]The State Key Laboratory of Refractories and Metallurgy, Wuhan University of Science and Technology, Wuhan, Hubei 430081, PR ChinaHubei Province Key

[2]Laboratory of Coal Conversion & New Carbon Materials, Wuhan University of Science and Technology, Wuhan, Hubei 430081, PR China

This study achieved controllable preparation of TaxHf1-xC ceramic powders with ideal single-phase solid solution structure characteristics via liquid phase precursor method using TaCl5, HfCl4 and phenolic resin as metal and C sources. On this basis, TaxHf1-xC modified C/C composites were prepared by precursor impregnation cracking process (PIP). The conversion process from TaxHf1-xC precursor to single-phase solid solution ceramic powders, the oxidation behavior of ceramic powders and the ablation behavior of C/C-TaxHf1-xC composite materials were studied. The results indicate that the organic precursors of TaxHf1-xC occur successively carbon thermal reduction reactions of Ta2O5 and Hf6Ta2O17, and solid solution reactions of HfC and TaC during the pyrolysis process. As the solubility of HfC increases, the temperature required to prepare TaxHf1-xC ceramic powders also increases. After pyrolysis in the temperature range of 1800 ℃ to 2000 ℃, both HfC and TaC in TaxHf1-xC (x=0.2, 0.25, 0.33, 0.5) undergo sufficient solid solution, forming a single-phase solid solution with a particle size of 150-300 nm. It can be found that the oxidation resistance of TaxHf1-xC ceramic powders and the anti-ablation ability of C/C-TaxHf1-xC composite materials are closely related to the composition of its oxidation products. In contrast, TaxHf1-xC (x=0.25) ceramic powder has better oxidation resistance.

Functional design and performance study of carbon-based composites used for lithium metal batteries

Shuzhang Niu[1], Xin Wang[1] and Peigang Han[1]

[1] School of New Materials and New Energies, Shenzhen Technology University, Shenzhen 518118, China

Lithium metal has a high theoretical specific capacity (3860 mAh g-1) and the lowest REDOX potential (−3.04 V vs. Standard hydrogen electrode), which is considered to be a high potential anode material. However, the high chemical activity of lithium metal is easy to form dendrites, which not only consumes a large amount of electrolyte, but also may puncture the battery separator leading to short circuit and safety issues. By using physical evaporation deposition (PED) technology, an Ag@C60 interface layer with a lithiophilic-lithiophobic gradient is constructed on the copper collector, which can regulate the uniform lithium-ion flow and form Li-Ag alloy, effectively reduce the nucleation overpotential and inhibit the formation of lithium dendrites. In addition, GO-BQ@LiTFSI (GBL) artificial interface layer is designed on the separator, which is prepared by combining graphene oxide with a liquid organic eutectic mixture of 1, 4-benzoquinone and lithium bis (trifluoromethane) sulfonamide salts. Due to the excellent properties of these functional materials, the GBL interface layer has abundant lithiophilic active sites, good ionic conductivity, excellent thermal stability and electrolyte compatibility, and can form a dense and LiF-rich solid electrolyte interface (SEI) layer.

References

[1] Xu, L.; Zhang, Q. C.; Niu, S. Z.; Wang, X.; Lian, Q.; Huang, Y. L.; Shi, R.; Amini, A.; Cheng, C. Adv. Energy Mater. 2023, 13, 2204214.

[2] Niu, S.; Zhang, S.-W.; Li, D.; Wang, X.; Chen, X.; Shi, R.; Shen, N.; Jin, M.; Zhang, X.; Lian, Q.et al. Chem. Eng. J. 2022, 429, 132156.

[3] Wang, X.; Huang, R.-q.; Niu, S.-z.; Xu, L.; Zhang, Q.-c.; Amini, A.; Cheng, C. New Carbon Mater 2021, 36 (4), 711.

Structural and interfacial engineering of carbon materials for advanced potassium-ion batteries

Biao Zhang

Department of Applied Physics, the Hong Kong Polytechnic University.
Email:biao.ap.zhang@polyu.edu.hk

Potassium-ion batteries (PIBs) are promising candidates for large-scale energy storage because of the potential advantages in cost and sustainability. Carbon materials play vital roles in boosting the stability and energy density of PIBs. We systematically tailor the microstructure of hard and soft carbon through controlling the carbonization temperature to establish the correlation between the surface functionality/graphitization degree and K ion storage performance. We also reveal that the electrolyte chemistry could significantly affect the cyclic stability by forming a robust solid electrolyte interphase (SEI). Moreover, the K ion insertion behavior greatly depends on the solvation structure of the electrolyte. By tuning the molecular structure of ether solvent, we successfully inhibit the detrimental solvent co-intercalation while preserving the benefits of ether in SEI construction.

Structure Regulation and Electrochemical Performances of Hard Carbon Anode Materials for Sodium-Ion Batteries

Fang-Yuan Su, Li-Jing Xie and Cheng-Meng Chen

Institute of Coal Chemistry Chinese Academy of Sciences, 27 South Taoyuan Road, Taiyuan, Shanxi, China

The nanotexture of precursors has the critical effect on the microstructures and electrochemical performances of obtained hard carbons. The inherited morphologies from precursors, the micro- and nanostructural optimization of hard carbons, including porosity, interlayer distance, defects, and more, plays a key role in enhancing the electrochemical performances of hard carbons. Therefore, the optimal utilization and controllable conversion of precursors at the molecular level have become a burgeoning area of research to develop hard carbons for advanced sodium-ion batteries.

In our research, we used precursors with different molecular structures, such as starch as a representative of ester heterocycles, and lignin and pitch as representatives of aromatic rings, as model precursors to prepare hard carbons. The thermochemical transformation process and reaction mechanism of solid-phase carbonization were thoroughly investigated. The structural evolution mechanisms of different precursor molecules during thermal processes, such as cross-linking, carbonization, and activation, were explored. The influence of precursor molecular structure, process methods, and conditions on the structural characteristics and energy storage performance of the final hard carbon material was understood [1-3]. These research provides important scientific basis for the rational selection, structural design, and process development of precursor materials for high-performance sodium-ion battery anodes.

References

[1] Li-Jing Xie, et al. Advanced Energy Materials 11 (2021) 2101650
[2] Ming-Xin Song, et al. Energy Storage Materials 51 (2022) 620-629
[3] Yilin Wang, et al. Advanced Materials, 2024, 2401249

Design and Modulation of Novel Carbon Materials

Kun Ni[1]

[1]Department of Materials Science and Engineering, School of Chemistry and Materials Science, University of Science and Technology of China, Hefei, China, 230026.

The development of research on novel carbon materials is confronted with the challenge of precisely manipulating structural characteristics and interfacial properties. The reporter has achieved a series of innovative results in addressing this issue: 1) Utilized neural network potential method to achieve million-level structure search and potential energy surface depiction for fullerene-derived carbon structures, with efficiency improved by three orders of magnitude compared to density functional theory methods, and for the first time proposed charge injection as a universal means of carbon structure regulation; 2) Identified that the interlayer interaction of graphene provides a new pathway for hot electron relaxation, elucidating the synergistic control mechanism of twist angle and pressure on the electronic structure; 3) Explored the interaction between ions and carbon using graphene as a model material, and established a multiscale structure-activity relationship between pore gradient structures and ion storage performance. The reporter's series of work provides a broader space for the structural design and property study of novel nanocarbon materials. As the (co-)first author or (co-)corresponding author, the reporter has published more than 20 papers in journals including Nature, Advanced Materials (5), Physical Review Letters, Advanced Functional Materials (3), etc., with over 1200 citations. The reporter will further focus on the control of structure-activity relationships in novel carbon materials in the future, developing a new structural design paradigm of "construct-on-demand", aiming to broaden the development space in the field of carbon materials.

References

[1] Fei Pan; Kun Ni; Tao Xu; Huaican Chen; Yusong Wang; Ke Gong; Cai Liu; Xin Li; Miao-Ling Lin; Shengyuan Li; Xia Wang; Wensheng Yan; Wen Yin; Ping-Heng Tan; Litao Sun; Dapeng Yu; Rodney S. Ruoff; Yanwu Zhu. Nature 2023, 614: 95-101

[2] Kun Ni; Fei Pan; Yanwu Zhu. Adv. Funct. Mater. 2022, 32(42): 2203894

[3] Kun Ni; Jinxiang Du; Jin Yang; Shujuan Xu; Xin Cong; Na Shu; Kai Zhang; Aolei Wang; Fei Wang; Liangbing Ge; Jin Zhao; Yan Qu; Kostya S. Novoselov; Pingheng Tan; Fuhai Su; Yanwu Zhu. Phys. Rev. Lett. 2021, 126(2): 027402

Design and preparation of multifunctional two-dimensional layered film barrier for lithium sulfur batteries

Xiang Long, Zhi-Hong Luo, and Jiao-Jing Shao

School of Materials and Metallurgy, Guizhou University, Guiyang, China

Two dimensional materials have structural characteristics such as anisotropy and large aspect ratios, which give them good film-forming properties. Some two-dimensional materials have unique properties, for example, two-dimensional clay has excellent lithium ion transport ability, and modified two-dimensional clay can further regulate the distribution of lithium ions; Two dimensional graphene has excellent electron mobility and high mechanical strength; The high conductivity and catalytic activity of two-dimensional MXenes have been widely confirmed. The above properties make two-dimensional materials have great potential for application in the field of electrochemical energy storage. The authors have been engaged in the research of preparation and controllable assembly of two-dimensional materials in liquid phase environment, as well as their application in electrochemical energy storage devices. In recent years, our team has mainly focused on the issue of the shuttle effect in high-energy density lithium sulfur batteries. Starting from the functional modification of the separator, a multifunctional two-dimensional layered film has been designed and prepared as the barrier for lithium sulfur batteries. Via surface modification and size control of two-dimensional nanochannels, on the basis of promoting efficient adsorption and catalytic conversion kinetics of polysulfides by the barrier layer, the selective transport of lithium ions by the barrier layer is further improved, and the effective suppression of polysulfide shuttle effect is achieved, with the goal of obtaining high energy density, high safety, and long life lithium sulfur batteries.

References

[1] Min Zheng, Zhihong Luo, Jie Wu, Jiao-Jing Shao*, Guangmin Zhou, et al., Composites Part B-Engineering, 2023, 110898, 1359-8368.

[2] Xiang Long, Zhi-Hong Luo, Zhi-Yuan Han, Guang-Min Zhou, Wei Lv, Jiao-Jing Shao*, et al., Energy Storage Materials, 2022, 52, 120–129.

Selective synthesis of soft and hard carbons from a single precursor through tailor-made stabilization for anode in sodium-ion batteries

Gwanwon Lee[1†], Kyoung Bin Min[2†], Min Eui Lee[3], Youn-Ki Lee[3], Hae Ri Lee[3], Sung-Soo Kim[1,3], Se Youn Cho[1,3], Han-Ik Joh[4], Young-Kwan Kim[2*], and Sungho Lee[1,3,5]**

[1]Convergence Research Center for Recyclable Air mobility, Materials and Platform (RAMP), Korea Institute of Science and Technology (KIST),92 Chudong-ro, Bongdong-eup, Wanju-gun, Jeollabuk-do, Republic of Korea

[2]Department of Chemistry, Dongguk University-Seoul, 30 Pildong-ro, Jung-gu, Seoul, Republic of Korea

[3]Carbon Composite Materials Research Center, Korea Institute of Science and Technology (KIST),92 Chudong-ro, Bongdong-eup, Wanju-gun, Jeollabuk-do, Republic of Korea

[4]Department of Energy Engineering, Konkuk University, 120 Neungdong-ro, Gwangjin-gu, Seoul, Republic of Korea

[5]Department of Nano Material Engineering, KIST School, University of Science and Technology, 217 Gajeong-ro, Yuseong-gu, Daejeon, Republic Korea

(*kimyk@dongguk.edu, **sunghol@kist.re.kr)

A major factor in classifying carbon materials is their degree of graphitization, which is defined as a structural transition from an amorphous to an ordered crystalline state. Previous studies have shown that graphitization depends on the intrinsic properties of precursors for carbon materials. Thus, these studies on graphitization have been mainly limited to modulating precursors [1]. Herein, we demonstrated the degree of graphitization that can be governed by a pre-treatment process using polyethylene (PE) as a single precursor. To demonstrate the effect of pre-treatments on graphitizability, thermal oxidation or sulfuric acid treatments were used as the pre-treatment and they diverged from a single precursor into soft and hard carbons, respectively. We also confirmed the different structures of thermally and acid-treated PEs at the early stage of graphitization based on the formation patterns of carbon cluster ions in LDI-TOF-MS. In addition, the distinct electrochemical performances of PE-derived soft and hard carbons were investigated by applying them as anode in sodium ion batteries, respectively. We believe that this study provides fundamental insights and practical tools to develop tailor-made carbon materials from a single and common precursor for energy storage application.

References

[1] A. Oberlin, Carbon, 22 (1984) 521-541

Bifunctional role of nitrogen-doped carbon quantum dots for CoS2/carbon composites in performance enhancement of Li-ion batteries

Hae Ri Lee[1,2], Yun Sik Kim [1], Seon Yeong Lee[1], U Hyeok Son[1], Sungho Lee[2], and Han-Ik Joh [1,*]

[1] Konkuk University, 120 Neungdong-ro, Gwangjin-gu, Seoul, Republic of Korea
[2] Korea Institute of Science and Technology (KIST), 92 Chudong-ro, Wanju-gun, Jeollabuk-do, Republic of Korea

CoS2 has been studied as an anode material of Li-ion batteries (LIBs) due to high electronic conductivity, thermal stability, and high theoretical capacity. However, CoS2-based anode suffered from large volume changes, shuttle effect of lithium polysulfides. These damages can be alleviated by the utilization of porous carbon supports.

Cobalt-based Metal-Organic Frameworks (Co-MOFs) have been applied to form CoS2/C composites due to high porosity and enriched heteroatom dopants [1]. However, ultramicropore-dominant porous nature of Co-MOFs hindered diffusion of Li ions during cycling.

In this study, the porosity and the content of heteroatom doping was simultaneously controlled by simply coating N-doped carbon quantum dots (NCQDs), which derived from N-methyl-2-pyrrolidone. We confirmed the mild acidity of NCQDs solution, which is capable partial-etch of ZIF-67 structures, one of Co-MOFs. Furthermore, loading of the NCQDs allowed to maintain nitrogen contents in CoS2 embedded N and S co-doped mesoporous carbon composites (CoS2/NSC) obtained after the carbonization followed by sulfurization. The CoS2/NSC pre-treated with 2.8 mL NCQD solution (CoS2/NSC-2.8) exhibited enhanced rate performances (392 mAh/g at 2.0 A/g) and cycling stability. The performance improvement might be attributed to broader pore sizes with retained heteroatom contents of CoS2/NSC-2.8 compared to those of CoS2/NSC without pre-treatment.

References

[1] Q. Wang, et al., Small, 11 (2015) 2511-2517

Electrocatalytic properties of carbon nanotube-based metal-free nanocomposites

Hironori Ogata[1], Nene Kosaku [1], Moeka Taniguchi [1]

[1]Department of Chemical Science and Technology, Faculty of Bioscience and Applied chemical and Graduate School of Engineering, Hosei University, 3-7-2, Kajino-cho, Koganei, Japan
Email: hogata@hosei.ac.jp

In renewable and green energy technologies, three simple reactions dominate the energy conversion and storage process: hydrogen evolution reaction(HER), oxygen reduction reaction(ORR), and oxygen evolution reaction (OER). Precious metal-based electrocatalysts are needed to facilitate these three reactions. However, precious metal-based catalysts are expensive, and their use is limited by their limited storage capacity. Therefore, nonprecious metal-based catalysts are currently being developed. Recently, it was reported that electrocatalysts prepared by adsorbing fullerenes on single-walled carbon nanotubes (SWNTs) exhibit even better multifunctional(ORR, OER, and HER) electrocatalytic activity over a wide range of pH values, from acid to alkaline, and better long-term stabilities than those of commercial precious metal-based catalysts[1-2]. However, the detailed mechanism of multifunctionality as an electrocatalyst has not been fully elucidated, and fabrication conditions of the composites for each function have not been optimized.

In this study, composite films of hydroxylated fullerenes[3-4] with different numbers of hydroxyl groups and single-walled carbon nanotubes(SWNTs) were prepared by reprecipitation method to clarify how the number of hydroxyl groups (x) added to fullerene and the mixing ratio of hydroxylated fullerene and SWNTs affect the electrocatalytic activity of hydrogen generation reaction (HER). A linear sweep voltamentry (LSV) curve was measured using the composite film as an electrode film, and the Tafel slope and exchange current density (j0) were calculated according to Tafel's equation. It was found that the HER electrocatalytic activity was improved by adsorbing hydroxylated fullerenes onto SWNTs. In addition, the HER electrocatalytic activity of both samples with x of 10 and 36 was improved by compounding with SWNTs, was improved depending on the mixing ratio of C and SWNTs.

Detailed experimental results on evaluation of electrocatalytic activity using these will be reported at the conference.

References

[1] Rui Gao, et al. J. Am. Chem. Soc. 2019, 141, 29, 11658–11666.
[2] Kun Guo et al., Green Energy & Environment 9, 2024, 7-27.
[3] Chiang Long Y et al., J. Org. Chem. 1994, 59, 14, 3960–3968.
[4] K. Kokubo, et al., ACS Nano 2008, 2, 2, 327–333.

Fabrication of non-graphitizing carbon electrodes for Sodium-ion Batteries

Jyoti Shikhar[1*], Swati Sharma[1]

[1]School of Mechanical and Material Science, Indian Institute of Technology Mandi,
Kamand, Himachal Pradesh, India.
*E-mail: D22090@students.iitmandi.ac.in

Sodium-ion batteries (SIBs) are emerging as a promising alternative to conventional lithium-ion batteries (LIBs) due to the abundant availability and non-toxicity of sodium compared to that of lithium. Non-graphitizing carbons (NGCs) play a crucial role as an anode material for SIBs. Their d002 spacing is >0.335 nm, which is more suitable for intercalation/ de-intercalation of sodium ions compared to pure graphite. Certain NGCs are reported to have a high thermal shock resistance, crack propagation resistance, a high capacity and low sodiation potential, leading to high energy density. In this work, preparation of a variety of NGCs from different organic precursors with the goal of obtaining inexpensive anode materials for SIBs is carried out. These NGCs are derived from natural and petrochemical waste materials such as lignin and pitch. Attempts are made to further increase the inter-layer spacing while still maintaining a reasonable electrical conductivity by tuning the carbonization parameters, as well as by incorporating heteroatoms such as Oxygen, Nitrogen and Sulphur into the carbon backbone. Additionally, the microstructure of the prepared electrode materials is thoroughly characterized, and their electrochemical performance is investigated via cyclic voltammetry. Finally, the fabrication of free-standing electrode of NGCs for SIBs will be discussed.

Keywords: Non-graphitizing carbon, Na-ion battery, Anode fabrication, Lignin;

Structural design of carbonaceous porous materials for lithium-sulfur battery and their electrochemical behaviors

Zhan Liang

East china university of science and technology, Meilong Road, shanghai, China

The ultra-high theoretical energy density of lithium-sulfur batteries (2600Wh/kg) makes them one of the most promising energy storage systems after lithium-ion batteries. When used the commercial carbon black/sulfur (CB/S) cathode, ether electrolyte and polypropylene separator, the "shuttle effect" of lithium polysulfide (Li2Sx, x=3~8) intermediate products between cathode-separator-anode is bound to occur in the charging and discharging process. The slow redox kinetic behavior of Li2Sx intermediate products between cathode-separator-anode electrode will affect the effective utilization of sulfur cathode in a long cycle. In this regard, based on the structural designability, composability and modifiability of carbonaceous porous materials, separator coating materials with different nanostructures and chemical properties have been constructed, and exhibit excellent electrochemical performance. The relevant experimental results have been published in academic journals such as Advanced Materials, Advanced Energy Materials, and ACS Nano.

References

[1] Weiqi Yao, Liang Zhan*, et al. Advanced Materials 35 (2023) 2212116
[2] Weiqi Yao, Liang Zhan*, et al. Advanced Materials 34 (2022) 2106370
[3] Xiang Li, Liang Zhan* et al. Advanced Energy Materials 14 (2024) 2303389
[4] Xin Zhang, Liang Zhan* et al. Advanced Functional Materials, 33 (2023) 2302624
[5] Xiang Li, et al. Liang Zhan* ACS Nano 17 (2023) 1653–1662

Porous Carbons for Positive Electrode of Zn-Carbon Rechargeable Battery

M. Yoshikawa, H. Fujimoto, Z. Ogumi, and T. Abe

Kyoto University, Gokasho, Uji, Kyoto 611-0011, Japan

Zinc is widely applied in multivalent energy storage systems because it is a safe, inexpensive, abundant resource, and attractive high capacities (820mAh/g) and low redox potential (−0.76 V vs standard hydrogen electrode). We are working on the development of a novel Zinc-Carbon rechargeable battery. Porous carbon materials prepared with different activation methods were evaluated as positive electrodes. Higher capacities were obtained with chemically activated carbons than with steam-activated carbons. KOH-activated carbon which has BET surface area over 3,000m2/g showed the highest capacity of over 400mAh/g-carbon. The cycle performances were measured when KOH-activated carbon was cathode and ZnCl2 was electrolyte, 33% decrease in capacity was observed after 100 cycles. It is thought that the reason is not the carbon of the positive electrode but the deterioration of the zinc of the negative electrode. In addition, the discharge capacity tended to be higher for chlorine ion-containing electrolytes such as ZnCl2 and NH4Cl. The discharge capacity of Zinc-Carbon rechargeable batteries tended to be proportional to the meso pore volume and the surface acidic functional groups of activated carbon. KOH-activated carbon has higher meso pore volume and surface acidic functional groups than other activated carbons, and provide high capacity in combination with as acidic electrolyte.

CUCUMBER PEEL DERIVED POROUS CARBONS FOR ELECTROCHEMICAL ELECTRODES

M. Nazhipkyzy [1,2,*], A. Seitkazinova[1,2], G. Kurmanbayeva[1], M. Yeleuov [1,3], A. Talgatkyzy [2], Z. Mansurov[1,2]

[1] Institute of Combustion Problems, Almaty, 050012, Kazakhstan
[2]Al-Farabi Kazakh National University, Almaty, 050038, Kazakhstan
[3] Satpayev University, Almaty, 050043, Kazakhstan
*Email: meruyert.nazhipkyzy@gmail.com

Owing to our inevitable energy necessity, alternate energy resources have to be implemented in our energy storage systems to enervate the energy demands. On the consideration of the above statement, activated high surface area three-dimensional (3D) nanoporous carbon derived from the cucumber peel bio-waste has been studied as electrode material for supercapacitor with high energy density. The nitrogen adsorption/desorption isotherms revealed that the activated nano-porous carbon (ACP600/800) exhibits a high specific surface area of 2333 m2 g−1. The fabricated symmetric cell using the activated porous carbon in aqueous electrolyte exhibits a high specific capacitance of 286 F/g at 1 A/g.

When the current density rose to 2 A/g, the ACP-800 maintained its gravimetric capacity at 279 F/g, indicating a capacity retention rate of 84.3%. Similarly, the ACP-700 exhibited a capacitive retention rate of 93.7%. This notable capacity retention can be attributed to their expansive surface area and intricate porous structure. These characteristics facilitate improved penetration of the electrolyte and the diffusion of electrolyte ions within the electrode, especially under high current density conditions.

References

[1] A. Gopalakrishnan, S. Badhulika, Ultrathin Graphene-like 2D porous carbon nanosheets and its excellent capacitance retention for supercapacitor. Journal of Industrial and Engineering Chemistry. 68 (2018) 257-266. http://doi.org/10.1016/j.jiec.2018.07.052

Acknowledgment

The authors acknowledge the Science Committee of the Ministry of Education and Science of the Republic of Kazakhstan (Grant No AP19677415 Obtaining nanocomposites from food waste and creating electrode materials based on them for supercapacitors).

CVD-coated carbon xerogels with various nodule size for high performance Na-ion battery negative electrode

Berke Karaman [1], Hélène Tonnoir [1,2], Da Huo [2], Bryan Carré [1], Alexandre F. Léonard [3], Jimena Castro Gutiérrez [4], Marie-Laure Piedboeuf [1], Alain Celzard [4,5], Vanessa Fierro [4], Carine Davoisne [2], Raphaël Janot [2], Nathalie Job [*1]

[1] Department of Chemical Engineering - NCE (Nanomaterials, Catalysis, Electrochemistry), University of Liège, 4000 Liège, Belgium

[2] Laboratoire de Réactivité et Chimie des Solides - LRCS, UMR7314 CNRS, Université de Picardie Jules Verne, Amiens, France

[3] Department of Chemical Engineering - CARPOR, University of Liège, 4000 Liège, Belgium

[4] Université de Lorraine, CNRS, Institut Jean Lamour - IJL, 88000 Épinal, France

[5] Institut Universitaire de France - IUF, 75231 Paris, France

E-mail : Nathalie.Job@uliege.be

Carbon xerogels (CXs [1]) are used as hard carbons for Na-ion battery (NIB) electrodes. These materials are made of microporous carbon nodules, the size of which can be tuned (from 20 nm to 2 μm); they are obtained by pyrolysis at 800°C of resorcinol-formaldehyde resins. However, when used in NIBs, their high specific surface areas (~600 m^2/g) leads to low Initial Coulombic Efficiency (ICE) due to electrolyte decomposition. To cope with that issue, Chemical Vapor Deposition (CVD) by ethylene cracking at 650-700°C is used to fill or mask the micropores. In the case of non-coated materials, large nodules (2 μm) lead to high Na+ storage capacity (248 mAh/g at C/20 cycling rate) and ICE (up to 80 %); quite notably, and contrarily to what is usually stated in literature, those are independent on the specific surface area measured by gas sorption. More probably, it depends on the surface that can really be accessed by the electrolyte, which could exclude part of the micropores volume when the nodules are large. In the case of the coated samples, the CVD-deposited carbon layer blocks the micropores; as a result, the specific surface area drops and both reversible capacity and ICE increase up to 298 mAh g-1 and 84 % [2], respectively, which is remarkable for carbons processed at such a low temperature (< 1000°C). Finally, electrochemical response to speed cycling depends on the nodule size. These carbon-carbon composite materials show great promise as materials for Na-ion batteries.

References

[1]Job N., Pirard R., Marien J., Pirard J.-P. Porous carbon xerogels with texture tailored by pH control during sol-gel process. Carbon, 42, 619-628 (2004).
[2]Karaman B., Tonnoir H., Huo D., Carré C., Léonard A.F., Castro Gutiérrez J., Piedboeuf M.-L., Celzard A., Fierro V., Davoisne C., Janot R., Job N. CVD-coated carbon xerogels for negative electrodes of Na-ion batteries. Carbon, revised version submitted.

Sieving carbon anodes power the high-energy sodium-ion batteries

Jun Zhang [1,2]

[1]NaCun (Tianjin) Technology Co., Ltd., Tianjin 300192, China
[2]Nanoyang Group, Tianjin Key Laboratory of Advanced Carbon and Electrochemical Energy Storage, School of Chemical Engineering and Technology, Tianjin University, Tianjin 300072, China

The design and development of high-capacity non-graphitic carbon anodes is widely accepted as the key challenge that impedes the commercialization of sodium-ion batteries (SIBs)[1]. Here, we present a vivid image of sieving carbons as the ideal configuration for superhigh sodium storage, featuring with the tightened pore entrance and enlarged pore body[2]. This talk is started by revisiting the use of commercial carbon molecular sieve (CMS) as the anode in SIBs. The unique pore entrance size of CMS[3], which is designed to precisely separate nitrogen and oxygen, play a key role in sieving the solvated molecules and achieving the high-efficiency de-solvation. Compared to CMS, sieving carbons are featured with highly developed and tunable nanopore[4]. The customized sieving carbons are studied as model anodes to unveil the controversial sodium storage mechanism related to the low-pltential plateau, and guide the design of high-capacity carbon anodes with record-high reversible capacity over 500 mAh/g and high initial coulombic efficiency over 90 %[5]. Sieving carbons are critical for accelerating the commercialization of SIBs as practical anodes, and further offers an ideal model to design and develop high-energy carbon anodes for metal-ion batteries not limited to SIBs[6].

References

[1] Siwei. Zhang, et al. Energy Storage Mater., 2016, 3: 18.
[2] Jun. Zhang ‡ , et al. Energy Environ. Sci., 2017, 10(1): 370.
[3] Kaikai. Li, et al. Nat. Commun., 2019, 10(1): 725.
[4] Qi. Li ‡ , et al. Natl. Sci. Rev., 2022, 9(8): nwac084.
[5] Qi. Li, et al. Adv. Energy Mater., 2022, 12(37): 2201734.
[6] Yue Chu, et al. Adv. Mater., 2023. 35(31): p. 2212186.

advanced multielement (o, n, s) modified strategies for pitch-based carbon anodes toward both lithium and sodium storage

Jian Wang

Taiyuan University of Technology, China

Pitch is a commonly used industrial by-product precursor in the production of anode materials for lithium-ion batteries (LIBs) and sodium-ion batteries (SIBs). However, achieving efficient utilization of waste pitch poses a challenge for recycled materials. To address this issue, we used simple multielement modification and obtain the pitch-based carbon doped with oxygen (O), nitrogen (N), and sulfur (S). Our work reveals that pitch with high softening point tends to form the crosslinked structure during pre-oxidation. Moreover, in proximity to the defects induced by oxygen functionalities, the introduction of nitrogen and sulfur can jointly impact on the diffusion kinetics of lithium/sodium, and it is demonstrated by enhancing lithium adsorption and sodium desorption from the defects. In comparison to the conventional one-step carbonization approach, the modified anodes significantly improve the rate performance of the LIBs (61.11 mAh g-1 to 222.90 mAh g-1 at 5 A g-1) and the initial Coulombic efficiency (ICE) of SIBs (41.74% to 66.40%). Our study explores the link between softening points and oxygen functionalities, as well as the effect of introducing various heteroatoms on the properties and diffusion kinetics of LIBs and SIBs, providing a reference for other materials to obtain anodes with high compatibility.

A gas diffusion layer based on synergistic effect of whisker-carbon nanotubes and carbon black for proton exchange membrane fuel cells

Ting-Ting Yao [1], Yu-Ying Li [1,2], Xiao-Fang Zhang [1], Yu-Ting Liu [1,2] and Gang-Ping Wu [1]

[1] CAS Key Laboratory of Carbon Materials, Institute of Coal Chemistry, Chinese Academy of Sciences, 27 Taoyuan Road (South), Taiyuan 030001, China

[2] Center of Materials Science and Optoelectronics Engineering, University of Chinese Academy of Sciences, No.1 Yanqihu East Rd, Huairou District, Beijing 100049, China

The superior mass transfer capacity of the gas diffusion layers (GDLs) is the key to the design and fabrication of high-performance proton exchange membrane fuel cell (PEMFC) for various applications. However, challenges remain in obtaining low-cost and high-performance PEMFC. The conventional GDLs easily generate abrupt pore gradient change and interfacial resistance between the microporous layer and the carbon paper substrate, resulting in mass transport limitation, and ultimately leading to lower output power for the PEMFC. Here we show a versatile and low-cost approach to obtaining thickness-controlled substrate-free gas diffusion layers (sfGDLs) with bimodal pore distribution and super-hydrophobic surface (i.e. contact angle of 153.96°) via non-solvent induced phase inversion method. The cell power density of sfGDL with thickness of 230 μ m and mesopores of 85.41% exhibited maximal limiting current density of up to 6.31 A cm-2 and peak power density of 1.60 W cm-2 at high humidity, which was much higher than that of sfGDLs reported in the literature under counterpart test conditions [1,2,3,4,5]. In this work, the distinguishing performances verified that the synergistic effect of the appropriate thickness and abundant mesopores can significantly promote the mass transfer efficiency of membrane electrode assembly, ensuring gas transport to catalyst layer and excellent cell performance.

References

[1] Chen-Yang Y. W, et al. J. Power Sources 173(1) (2007) 183-188.
[2] Kattamanchi, S, et al. Materials Research Express 6(10) (2019) 105502.
[3] Hung, T. F., et al. J. Power Sources 184(1) (2008) 165-171.
[4] Li, Y.-Y., et al. Int. J. Hydrogen Energy 55 (2024) 675-682.
[5] Lee, H.-F., et al. RSC Advances 8(40) 2018 22506-22514.

A gas diffusion layer based on synergistic effect of whisker-carbon nanotubes and carbon black for proton exchange membrane fuel cells

Ting-Ting Yao [1], Yu-Ying Li [1,2], Xiao-Fang Zhang [1], Yu-Ting Liu [1,2] and Gang-Ping Wu [1]

[1] CAS Key Laboratory of Carbon Materials, Institute of Coal Chemistry, Chinese Academy of Sciences, 27 Taoyuan Road (South), Taiyuan 030001, China

[2] Center of Materials Science and Optoelectronics Engineering, University of Chinese Academy of Sciences, No.1 Yanqihu East Rd, Huairou District, Beijing 100049, China

The superior mass transfer capacity of the gas diffusion layers (GDLs) is the key to the design and fabrication of high-performance proton exchange membrane fuel cell (PEMFC) for various applications. However, challenges remain in obtaining low-cost and high-performance PEMFC. The conventional GDLs easily generate abrupt pore gradient change and interfacial resistance between the microporous layer and the carbon paper substrate, resulting in mass transport limitation, and ultimately leading to lower output power for the PEMFC. Here we show a versatile and low-cost approach to obtaining thickness-controlled substrate-free gas diffusion layers (sfGDLs) with bimodal pore distribution and super-hydrophobic surface (i.e. contact angle of 153.96°) via non-solvent induced phase inversion method. The cell power density of sfGDL with thickness of 230 μ m and mesopores of 85.41% exhibited maximal limiting current density of up to 6.31 A cm-2 and peak power density of 1.60 W cm-2 at high humidity, which was much higher than that of sfGDLs reported in the literature under counterpart test conditions [1,2,3,4,5]. In this work, the distinguishing performances verified that the synergistic effect of the appropriate thickness and abundant mesopores can significantly promote the mass transfer efficiency of membrane electrode assembly, ensuring gas transport to catalyst layer and excellent cell performance.

References

[1] Chen-Yang Y. W, et al. J. Power Sources 173(1) (2007) 183-188.
[2] Kattamanchi, S, et al. Materials Research Express 6(10) (2019) 105502.
[3] Hung, T. F., et al. J. Power Sources 184(1) (2008) 165-171.
[4] Li, Y.-Y., et al. Int. J. Hydrogen Energy 55 (2024) 675-682.
[5] Lee, H.-F., et al. RSC Advances 8(40) 2018 22506-22514.

Study on effect of carbon nanomaterials as conductive additives in EDLCs

Zhanibek Ayaganov [1], Annie Ng [2] and Vladimir Pavlenko [1,2]*

[1] Institute of Combustion Problems, Almaty, Kazakhstan
[2] Department of Electrical and Computer Engineering, School of Engineering and Digital Sciences, Nazarbayev University, Astana, Kazakhstan

Effects of various carbon nanomaterials using as conductive additives (5 wt%) in activated carbon-based electrodes for electrical double layer capacitors (EDLCs) were investigated. A variety of carbon nanomaterials including carbon black (CB), single-walled carbon nanotubes (SWCNT), graphene nanoplatelets (GNP), fullerenes (C60), and activated carbon nanofibers (ACNF) are considered in this work as they provide a diverse combination of intrinsic properties such as specific surface area, conductivity, micro-mesoporous structures, and geometry of nanomaterials. The results show that each carbon additive has its own strengths and limitations for different performance parameters considered. A reduction in capacitance at high current densities and material degradation during long-term cycling are observed for CB samples, which could be attributed to the high dispersibility of CB causing limited interparticle voids and thus increasing diffusion resistance. In contrast, the use of ACNFs significantly reduces the diffusion resistance and improves the cyclic stability compared with CB based samples due to the advantages of larger specific surface area and superior micro-mesoporous structures of ACNFs. In particular, EDLCs with ACNFs demonstrate an increase in specific capacitance and energy density by 99 % and 76 %, respectively compared to CB based EDLCs at a current density of 20 A/g [1].

References

[1] Zhanibek Ayaganov, et al. Journal of Energy Storage 78 (2024) 110035

Fabrication of biomass-based functional carbon materials for energy storage

Xiaomin Yang[1], Zhimin Chen[1], Wei Li[1], Ting Lv[1], Xiaofeng Wang[1] and Jieshan Qiu[2]

[1]College of Chemistry, Jilin University, Changchun 130012, China
[2]College of Chemical Engineering, Beijing University of Chemical Technology, Beijing 100029, China

Biomass-based functional carbon materials (BFCs) with renewability, flexible structural tunability and diverse physicochemical properties have shown encouraging and bright prospects in the fields of energy storage and conversion [1]. Our group has been engaged in high efficiency utilization of biomass and achieved a series of innovative research results in the synthesis, structural regulation, and energy storage applications of BFCs [2-5]. The formation mechanism of hierarchical pore structure of rice husk-based porous carbon was revealed via the method of biomass component separation, and the relationship between pore structure and electrochemical performance was established [2]. A low-temperature dehydration carbon-fixation method was developed to regulate the structure of biomass-based porous carbon, which can significantly improve its volumetric energy density and yield [3]. The soft-hard heterostructure porous carbon was prepared by the composite technology of soft carbon and hard carbon, in which biomass and pitch were used as hard carbon and soft carbon precursors, respectively [4, 5]. The synergistic effect of soft carbon and hard carbon ensures the composite material exhibiting enhanced electrochemical performance for high mass loading supercapacitors. The self-discharge of supercapacitors has been effectively suppressed. Our research results greatly increase the practical application potential of BFCs in energy storage.

References

[1] Xiaomin Yang, et al. Materials Science & Engineering R-Reports 154 (2023) 100736.
[2] Xiaomin Yang, et al. Carbon 161 (2020) 432-444.
[3] Xiaomin Yang, et al. ChemSusChem 15 (2022) e2021019.
[4] Xiaomin Yang, et al. Carbon 196 (2022) 136-145.
[5] Xiaomin Yang, et al. Small (2024) 2310645.

Rationally designing closed-pores hard carbon structures to boost the low-voltage plateau capacity of sodium-ion batteries

Xiaoqi Yu, Xiaoyu Xu, Zipeng Jiang, Renlu Yuan, Ang Li, Xiaohong Chen, Huaihe Song*

State Key Laboratory of Chemical Resource Engineering, Beijing Key Laboratory of Electrochemical Process and Technology for Materials, Beijing University of Chemical Technology, Beijing 100029, PR China

The theoretical capacity of hard carbon anodes comprises a high-pressure slope region and a low-pressure plateau region. The high low-voltage plateau capacity makes hard carbon one of the strongest candidate materials for sodium-ion batteries. However, the mechanism of sodium storage in hard carbon remains controversial. Establishing a link between microstructure and electrochemical properties provides strong evidence for elucidating how hard carbon stores sodium. Therefore, in this study, phenolic resin-derived hard carbon (PRHC) anodes with a rich closed-pores structure were synthesized, and it was demonstrated that the presence of a closed-pores structure significantly increased the sodium storage capacity. This hard carbon anodes exhibited a reversible capacity of 352.16mAh/g at 20mA/g, with the platform capacity accounting for 78.58% and being 276.73mAh/g. As predicted, as the number of closed-pores decreases, the platform capacity gradually diminishes. The platform capacity gradually decreased to 106.85mAh/g and accounting for 58.35% of the total capacity. It is evident that the closed-pores structure of hard carbon promotes platform capacity and provides a reasonable framework for understanding the mechanism of sodium storage in hard carbon anodes.

Keywords: Phenolic resin-derived hard carbon, Closed-pores structure, Low-voltage plateau capacity, Sodium storage mechanism

Facile synthesis of S-doped Co2P nanoparticles on a P, S co-doped carbon substrate by using ion exchange resin for hydrogen evolution reaction

Xinran Yang[1], Koji Miyake[1,2], Yoshiaki Uchida[1] and Norikazu Nishiyama[1,2]

[1]Division of Chemical Engineering, Graduate School of Engineering Science, Osaka University, 1-3 Machikaneyama, Toyonaka, Osaka 560-8531, Japan

[2]Innovative Catalysis Science Division, Institute for Open and Transdisciplinary Research Initiatives (ICS-OTRI) Osaka University Suita, Osaka 565-0871, Japan

Producing green hydrogen through water electrolysis is considered to be the optimal approach to achieving future carbon neutrality goals[1-2]. Nevertheless, the actualization of this method has been hindered by its enormous energy barrier, which has plagued scientists over several decades[3-4]. Herein, we developed a facile method to synthesize an S-doped Co2P anchored P,S co-doped carbon catalyst via an ion-exchange process and one-step pyrolysis. Through prolonged ion exchange, cobalt ions were uniformly immobilized onto the organic carbon framework, ensuring a uniform distribution of Co2P nanoparticles and robust interactions between the resultant product and carbon substrate. The agglomeration of the Co2P nanoparticles during pyrolysis was prevented, thus generating a larger number of exposed active sites. In addition, doping with S can efficiently regulate the electronic states of the Co sites, resulting in an improved ΔGH^*. The best catalyst exhibited outstanding HER activity, with an extraordinarily low overpotential of 161 mV to achieve a current density of 10 mA cm−2 in 0.5 M H2SO4. Compared with the mechanically stirred sample and the sample without S atoms, the catalyst activities of the sample subjected to ion exchange are improved by 37.8 % and 54.2 % respectively.

References

[1]J. Ding, H. et al, Small 2022, 18, 2204524.
[2]Dubouis N. et al,Chem Sci 2019;10:9165–81.
[3]Yao J. et al,Small Methods 2020;4:2000494.
[4]Sun Z. et al, Int J Hydrogen Energy 2023;48:366–92.

Mechanism understanding of metal-free carbon-based electrocatalysis with model catalysts

Yangming Lin

Fujian Institute of Research on the Structure of Matter, Chinese Academy of Sciences, Fuzhou, Fujian, 350002 P. R. China

Carbon materials have attracted significant attention as promising metal-free catalysts towards overall oxygen electrochemical reactions including oxygen reduction reaction (ORR) and oxygen evolution reaction (OER) [1-3]. The recent mechanistic understandings (e.g. active sites) of carbocatalysis on ORR and OER, however, are still rife with controversies because of the complicated surface species of catalysts and missing of effective investigation methods. In the present work, we demonstrate the potential roles of typical edge configurations, O and N species by using aromatic molecular catalysts with single species as model probes (Figure 1). Aromatic molecules with tunable π -conjugated system and edge configuration show a similar carbon network structure and thus represent a valuable model to understand the nature of carbocatalysts at a molecular level and to develop new applications. The real role of each species in ORR and OER can be directly revealed by comparing their TOF values and dynamic behaviors.

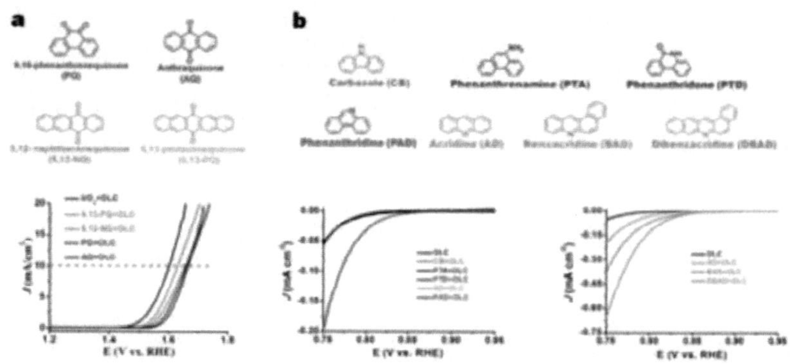

Figure 1 (a) Carbon model molecules with designated oxygen species and their performance towards OER process. (b) Carbon model molecules with designated nitrogen species and their performance towards ORR process.

References

[1] Yangming Lin, et. al. Angew. Chem. Int. Ed. 60 (2021) 3299-3306.
[2] Yangming Lin, et. al. J. Am. Chem. Soc. 140 (2018) 14717–14724.
[3] Yangming Lin, et. al. Angew. Chem. Int. Ed. 58 (2019) 8917 –8921.

Nanoconfined Carbonization Enabling High-Density Porous Carbon for Jointly Superior Gravimetric and Volumetric Zinc-Ion Storage

Yeru Liang and Jiacong Lu

Key Laboratory for Biobased Materials and Energy of Ministry of Education, College of Materials and Energy, South China Agricultural University, Guangzhou 510642, P. R. China

Carbon materials have shown promising potential for zinc ion storage. However, it is still a formidable challenge to fabricate porous yet dense carbon materials for simultaneously achieving both high gravimetric and high volumetric performances. Herein, we introduce a nanoconfined carbonization approach to elaborately engineer porous carbon cathode with a well-balanced porosity and density for realizing jointly superior gravimetric and volumetric zinc-ion storage. Key to this strategy is encapsulation of carbon sources (i.e., glucose and urea) into a silica network based on a sol-gel reaction. To this end, the Maillard reaction products are confined in the silica nanonetwork and in situ carbonized into a dense carbon substance. Such a nanoconfined carbonization is demonstrated to effectively avoid the possible formation of expanding carbonaceous char, and thus be crucial for realizing high density. Furthermore, the as-formed silica nanonetwork can be used as the template of pore, which then can be finely tuned by changing the amount of silica. By adapting the pore size to the zinc ions, the trade-off between density and porosity is optimized, allowing the obtained high-density porous carbon cathode to simultaneously reach high gravimetric capacitance of 452 F g-1 and high volumetric capacitance of 353 F cm-3.

Releasing Free Radicals in Precursor Triggers the Formation of Closed Pores in Hard Carbon for Sodium-ion Batteries

Yilin Wang[1,2], Fangyuan Su[1] and Cheng-Meng Chen[1]

[1]CAS Key Laboratory of Carbon Materials, Institute of Coal Chemistry, Chinese Academy of Sciences, Taiyuan 030001, China

[2]University of Chinese Academy of Sciences, Beijing 100049, China

Increasing closed pore volume in hard carbon is considered to be the most effective way to enhance the electrochemical performance in sodium-ion batteries. However, there lack of systematic insights into the formation mechanisms of closed pores at molecular level. In this study, a regulation strategy of closed pores via adjustment of the content of free radicals is reported. Sufficient free radicals are exposed by part delignification of bamboo, which is related to the formation of well-developed carbon layers and rich closed pores. In addition, excessive free radicals from nearly total delignification lead to more reactive sites during pyrolysis, which competes for limited precursor debris to form smaller microcrystals and therefore compact the material. The optimal sample delivers large closed pore volume of 0.203 cm3 g−1, which leads to high reversible capacity of 350 mAh g−1 at 20 mA g−1 and enhanced Na+ transfer kinetics. This work provides insights into the formation mechanisms of closed pores at molecular level, enabling rational design of hard carbon pore structures.

Hybrid carbon spheres derived from polystyrene and pitch for optimized sodium-ion batteries

Yonghong Ye [1,2], Taotao Guan [1,2], Jianlong Wang [1,2], and Kaixi Li [1,2]

[1] Institute of Coal Chemistry Chinese Academy of Sciences, Taiyuan, Shanxi, China
[2] University of Chinese Academy of Sciences, Beijing, China

Owning generous the Elemental Abundance in the Crust, sodium-ion batteries (SIBs), as a supplement to lithium-ion batteries (LIBs), alleviate the shortage and uneven distribution of lithium1. However, there are still many challenges in the practical application of SIBs' anodes derived from hard carbon (HC)2. In this study, the strategy to combined the advantages of soft carbon (SC) and HC is reported. We design and synthesize a series of anodes derived from hybrid carbon spheres (HCSs) with uniform size and stable structure, which effectively optimizes the specific capacity and initial coulombic efficiency of SIBs.

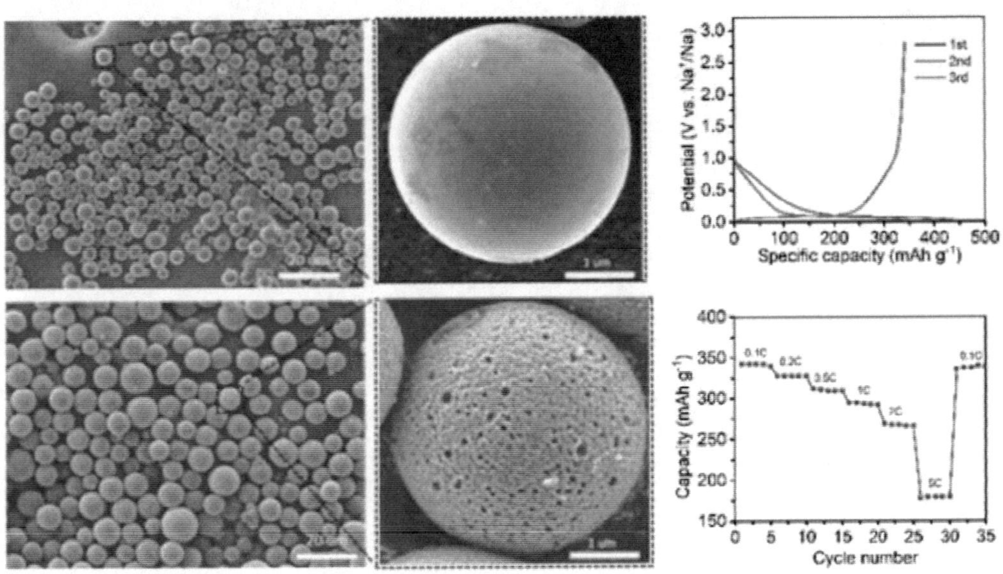

References

[1] Yin X, et al. Small 2021, 18 (5).
[2] Chen H, et al. Energy Storage Materials 2023, 56, 532-541.

Hard carbon promises a superior anode in sodium-ion batteries

Zhenyu Guo[1], and Magda Maria-Magdalena Titirici[1]

[1]Department of Chemical Engineering, Imperial College London, London, UK

Emerging sodium-ion batteries (NIBs) and potassium-ion batteries (KIBs) show promise in complementing lithium-ion battery (LIB) technology and diversifying the battery market. Hard carbon is a potential anode candidate for LIBs, NIBs and KIBs due to its high capacity, sustainability, wide availability, and stable physicochemical properties. Herein, a series of hard carbons are synthesized by hydrothermal carbonization and subsequent pyrolysis at different temperatures to finely tune their structural properties. When tested as anodes, the hard carbons exhibit differing ion storage trends for Li, Na and K, with NIBs achieving the highest reversible capacity. Extensive materials and electrochemical characterizations are carried out to study the correlation of structural features with electrochemical performance, and to explain the specific mechanisms of alkali-ion storage in hard carbons. In addition, the best-performing hard carbon is tested against a sodium cathode Na3V2(PO4)3 in a Na-ion pouch cell, displaying a high-power density of 2172 W kg-1 at an energy density of 181.5 Wh kg-1 (based on the total weight of active materials in both anode and cathode). The Na-ion pouch cell also shows stable ultralong-term cycling (9000 hours or 5142 cycles) and demonstrates the promising potential of such materials as sustainable, scalable anodes for beyond Li-batteries.

Highly uniform distribution of metallic lithium and carbon in nanoscale for highly stable carbon-based lithium metal anode

Zipeng Jiang [1,2], Huaihe Song [2*]

[1] Qinghai Provincial Key Laboratory of Advanced Materials and Applied Technology, Qinghai University, Xining, 810016, PR China.

[2] State Key Laboratory of Chemical Resource Engineering, Beijing Key Laboratory of Electrochemical Process and Technology for Materials, Beijing University of Chemical Technology, Beijing, 100029, PR China

*Corresponding Author: E-mail: songhh@mail.buct.edu.cn.(Huaihe Song)

Unfortunately, the next-generation lithium (Li) metal batteries present limitations for practical applications due to dendritic Li growth and infinite volume expansion of Li metal anode (LMA) during cycling. Lithium carbon composite exhibits great potential to address those issues owing to the promotion of Li ions diffusion by offering a stable Li host. Herein, we report the fabrication of a carbon-based lithium metal (GKBLi) anode through a simple ball milling approach by exploiting a mixture of passivated lithium metal powder and carbon nano materials (graphitic Ketjen carbon black, GKB). The in-situ mechanical chemistry lithiation reaction between GKB and Li metal forms graphite intercalation compounds that exhibit lithiophilicity and generate Li3N, which effectively promotes the transport of Li ions. The highly uniform distribution of Li and GKB in the anode matrix can enable the deposition and stripping of Li to occur homogeneously. The GKBLi symmetrical cell can stably cycle for more than 1200 h with low voltage hysteresis. Impressively, the full cell assembled with a GKBLi anode and commercial LiFePO4 (LFP) cathode achieves extraordinary cycling stability with a capacity retention of 90.9% and an extremely high Columbic efficiency of ~99.9% over 1000 cycles at a high current of 2 C. This highly uniform distribution of lithium in the electrodes by a simple dry preparation method provides new insight into high-performance LMA on an industrial scale.

Keywords

Passivated lithium powder, graphite intercalation compounds, lithium carbon composite, lithium metal anode

Acknowledgments

This work was supported by the National Natural Science Foundation of China (U1610252).

References

[1]Jiang, Z.; Meng, C.; Zhu, W.; Yuan, M.; Chen, G.; Hou, R.; Li, A.; Chen, X.; Song, H.*, Interfacial Anchored Sesame Ball-like Ag/C To Guide Lithium Even Plating and Stripping Behavior. ACS Appl Mater Interfaces 2023, 15 (1), 1934-1943.
[2]Jiang, Z.; Meng, C.; Chen, G.; Yuan, R.; Li, A.; Zhou, J.; Chen, X.; Song, H.*, Lithiophilic onion-like carbon spheres as lithium metal uniform deposition host. J Colloid Interface Sci 2022, 627, 783-792.
[3]Jiang, Z.; Li, A.; Meng, C.; Chen, X.; Song, H.*, Strategies and challenges of carbon materials in the practical applications of lithium metal anode: a review. Phys Chem Chem Phys 2022, 24 (43), 26356-26370.

Hard carbon promises a superior anode in sodium-ion batteries

Zhenyu Guo [1], and Magda Maria-Magdalena Titirici [1]

[1]Department of Chemical Engineering, Imperial College London, London, UK

Emerging sodium-ion batteries (NIBs) and potassium-ion batteries (KIBs) show promise in complementing lithium-ion battery (LIB) technology and diversifying the battery market. Hard carbon is a potential anode candidate for LIBs, NIBs and KIBs due to its high capacity, sustainability, wide availability, and stable physicochemical properties. Herein, a series of hard carbons are synthesized by hydrothermal carbonization and subsequent pyrolysis at different temperatures to finely tune their structural properties. When tested as anodes, the hard carbons exhibit differing ion storage trends for Li, Na and K, with NIBs achieving the highest reversible capacity. Extensive materials and electrochemical characterizations are carried out to study the correlation of structural features with electrochemical performance, and to explain the specific mechanisms of alkali-ion storage in hard carbons. In addition, the best-performing hard carbon is tested against a sodium cathode Na3V2(PO4)3 in a Na-ion pouch cell, displaying a high-power density of 2172 W kg-1 at an energy density of 181.5 Wh kg-1 (based on the total weight of active materials in both anode and cathode). The Na-ion pouch cell also shows stable ultralong-term cycling (9000 hours or 5142 cycles) and demonstrates the promising potential of such materials as sustainable, scalable anodes for beyond Li-batteries.

Prediction of Electrochemical Reaction Potentials in Energy Storage Devices Regarding Carbon Materials

Zonglin Yi[1], Fangyuan Su [1] and Cheng-Meng Chen [1]

[1]CAS Key Laboratory of Carbon Materials, Institute of Coal Chemistry, Chinese Academy of Sciences, 27th Taoyuan Rd., Taiyuan, China

Carbon materials are widely used in electrochemical energy storage area due to their high stability, high conductivity and low cost. The electrochemical inert nature of carbon enables its integrity during the practical charge-discharge processes in energy storage devices. However, lattice defects and heteroatom doping that inevitably exist in carbon materials increase their electrochemical reactivity, resulting in the change of reaction potentials in the energy storage devices. Therefore, accurate prediction of electrochemical reaction potentials is of vital importance for improving the electrochemical performances and safety of energy storage devices.

Numerous works have been done in order to understanding the reaction mechanisms of electrochemical reactions in energy storage devices regarding carbon materials. However, little attention has been paid on the theoretical prediction of reaction potentials. Herein, we used computational hydrogen electrode (CHE) that have already been widely used in electrocatalysis to develop an efficient method for reaction potential prediction. With the help of CHE, the reaction potentials of Li2S2 conversion reaction in lithium sulfur battery and the decomposition reaction of propylene carbonate (PC) as electrolyte solvent are predicted. Density functional theory (DFT) calculations and thermodynamics analysis unveil their varied reaction mechanisms with the presence of multiple kinds of metallic heteroatoms doping on carbon surface. These works provide new methods of accurate prediction on the electrochemical reaction potential in energy storage devices, along with theoretical guidance on the design of advanced carbon materials for electrochemical energy storage area.

Pretreatment Strategies for Saccharide-based Hard Carbon Anodes with Advanced Performance in Sodium-ion Batteries

Zoya V. Bobyleva[1], Anastasia M. Alekseeva[1], Sergey V. Ryazantsev[2], Oleg A. Drozhzhin[1], Evgeny V. Antipov[1,3]

[1] Lomonosov Moscow State University, 119991, Moscow, Russia
[2] Shenzhen MSU-BIT University, 518172, Shenzhen, China
[3] Skolkovo Institute of Science and Technology, 121205, Moscow, Russia

Sodium-ion batteries (SIBs) have found broad applications due to a lower cost of stored energy per kilowatt-hour and the availability of raw materials for the production of these batteries. Non-graphitizable carbons, or hard carbons, appear to be the material of choice for negative electrodes (anodes) in SIBs [1]. In this work, we utilized monosaccharides (glucose and fructose) and disaccharides (sucrose and lactose) to produce HCs by two-step synthesis, with pretreatment at variable temperatures (200–350 °C) followed by pyrolysis at 1300 °C. The pretreatment step conditions are shown to influence the properties of the hard carbon (defectiveness, inhomogeneity, surface functionalities and type of microporosity) and its electrochemical performance. The pretreatment temperature up to 250 °C was identified as optimal for most sugars to improve the initial Coulombic efficiency (ICE) by 40% (up to 89%) and achieve discharge capacities of 250–300 mAh g 1. Additionally, we measured the electrochemical performance of full cells and prototypes with three different cathode materials (Na3V2(PO4)3, β -NaVP2O7, O3-Na(Ni1/3Fe1/3Mn1/3)O2) and demonstrated the high ICE of 80–86% depending on the material used.

This work was supported by the Russian Science Foundation (grant No. 24-23-20052)

References

[1] E. N. Abramova, et. al. (2024). Uspekhi Khimii 93(2) (2024) 1-26

New Carbon Allotropes: Graphene-P-phenyl-Graphene

Huanxin Li[1,2*], Kostya S. Novoselov[2], Grant Ritchie[1], and Richard Compton[1]

[1]Physical & Theoretical Chemistry Laboratory (PTCL), Department of Chemistry, University of Oxford, South Parks Road, Oxford, OX1 3QZ, UK
[2]Chongqing Two-Dimensional (2D) Materials Institute, Chongqing 400714, China
Email: huanxin.li@chem.ox.ac.uk

Topic: Carbon for energy storage and conversion

Over the past three decades, carbon nanomaterials, including fullerenes [1,2], carbon nanotubes [3], and graphene [4], along with their derivatives and composites, have profoundly transformed our comprehension of nanomaterials. The discovery of various carbon allotropes has garnered significant recognition, as evidenced by two Nobel Prizes: the 1996 Nobel Prize in Chemistry for C60 and the 2010 Nobel Prize in Physics for graphene. Their enduring impact continues to unfold. Here, we present a novel carbon nanomaterial allotrope termed Graphene-P-phenyl-Graphene (GPG). GPG is synthesized by integrating π-π-conjugated p-phenyls into graphene layers, connecting them via C-C σ bonds (Figure 1). This carbon allotrope exhibits both thermal and dynamic stability, confirmed through Density Functional Theory (DFT) and Molecular Dynamics (MD) simulations. The incorporation of p-phenyls within the graphene structure expands the layer spacing of GPG from approximately 0.34 to 0.56 nm. This expansion diminishes van der Waals forces between graphene layers and expedites electron migration within GPG layers, resulting in a notably high Hall mobility (ranging from 10,000 to 13,000 cm2 V-1 s-1) for free-standing films. Furthermore, the increased layer spacing of GPG facilitates the storage and swift migration of large ions.

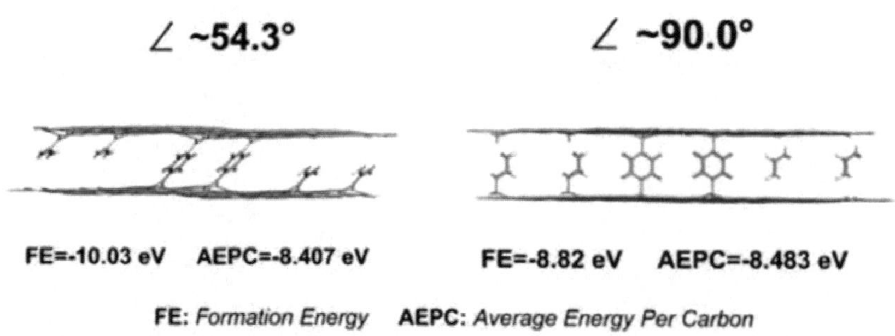

Figure 1: The Z-type (54.3 degrees) and H-type (90.0 degrees) models of GPG

References

[1] Nature, 1985, 318: 162-163. [2] Nature, 1993, 363:603-605. [3] Science, 2004, 306: 666-669. [4] Science, 2019, 365: 1299-1301.

Potential bifunctional carbon-supported Pd-based catalysts for the production and storage of H2 attained by formic acid-bicarbonate couple

Paula Riquelme-García, Miriam Navlani-García and Diego Cazorla-Amorós

Department of Inorganic Chemistry and Materials Institute, University of Alicante, Ap. 99, Alicante, E-03080, Spain

Formic acid (HCOOH) has emerged as a very interesting Liquid Organic Hydrogen Carrier (LOHC) because of its high volumetric hydrogen density, low toxicity and high stability. Moreover, it is involved in an H2 production and storage system based on an ideally carbon-neutral cycle, formed by the dehydrogenation of formic acid and the hydrogenation of CO2. However, this reaction is much more challenging than the dehydrogenation of HCOOH due to its unfavourable thermodynamics (CO2 (g) + H2 (g) → HCOOH (l), Δ G298K= 32.9 kJ·mol-1) [1]. The most straightforward approach to overcome the thermodynamic limitation is to conduct the reaction in the liquid phase using base additives such as bicarbonates, amines, and hydroxides) [2]. In this work, Pd-based catalysts, with different compositions of the metal nanoparticles and supported on activated carbon derived from biomass residues, were prepared and assessed in the H2-formate interconversion via formic acid dehydrogenation and bicarbonate hydrogenation reactions. The catalytic activity results show that the developed materials are potential bifunctional catalysts for H2 production and storage attained by a sustainable approach.

Acknowledgements:The autors would like to thank project PID2021-123079OB-I00 and project TED2021-131324B-C22 funded by MCIN/AEI/ 10.13039/501100011033 and by the "European Union NextGenerationEU/PRTR.

References

[1] Paula Riquelme-García, et al. Energies 17 (2024) 260.
[2] Priyanka Verma, et al. Journal of CO2 Utilization 54 (2021)101765.

Pt nanoparticles for improving the performance and durability of Fe Nx based materials towards ORR in ADMFCs

G. Alemany-Molina [1], V. Baglio [2], C. Lo Vecchio[2], S.C. Zignani [2], A.S. Aricò [2], E. Morallón [3], D. CazorlaAmorós [1].

[1]Department of Inorganic Chemistry and Materials Institute, University of Alicante, Ap. 99, Alicante, E-03080, Spain.

[2]Istituto di Tecnologie Avanzate per l' Energia "Nicola Giordano" ITAE, Consiglio Nazionale delle Ricerche CNR, Messina, 98126, Italy.

[3]Department of Physical Chemistry and Materials Institute, University of Alicante, Ap. 99, Alicante, E-03080, Spain.

Large-scale commercialization of alkaline direct methanol fuel cells (ADMFCs) is limited by the development of highly conductive and durable membranes and the preparation of cost-effective electrocatalysts for the methanol oxidation reaction (MOR) and the oxygen reduction reaction (ORR) [1]. For the ORR, besides requiring highly

active and durable electrocatalysts towards the 4-electron pathway, the cathode must present good methanol tolerance. Materials based on Fe-Nx sites have been reported to lose activity after long durability tests even in alkaline media. In this work, we have studied the preparation of Pt nanoparticles with very low Pt loading (2.5 wt. %) over two different supports based on highly dispersed Fe-Nx sites. The half-cell characterization revealed an enhanced stability after 1000 cycles tests and excellent methanol tolerance because of the Pt-Fe interaction. In an ADMFC, the durability of the Pt/Fe-Nx containing MEAs was studied for 100 h showing 5-10% higher maximumpower densities compared to the Fe-Nx containing MEAs, and vastly superior stability to the Pt/C MEA.

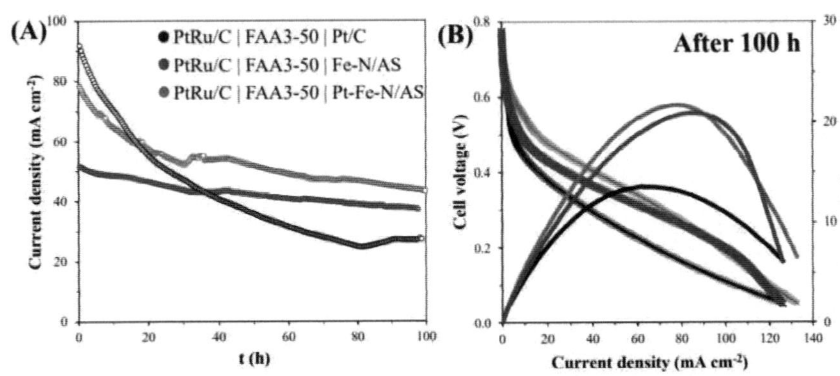

Figure 1: (A) Long-time potentiostatic curve at 0.3V and (B) polarization curves after 100 h stability (1 M KOH, 60°C).

Acknowledgements: The authors would like to thank PID2021-123079OB-I00 project funded by MCIN/AEI/10.13039/501100011033 and "ERDF A way of making Europe".

References

[1] C. Lo Vecchio, A. Carbone, I. Gatto, V. Baglio. Polymers 2023, 15, 1555.

Synthesis of Fe-Nx based electrocatalysts from different biomass residues by a single heat treatment and avoiding the washing step

G. Alemany-Molina[1], K.M. Pabón-Román[1],
E. Morallón[2], D. Cazorla-Amorós[1]

[1]Department of Inorganic Chemistry and Materials Institute, University of Alicante, Ap. 99, Alicante, E-03080, Spain.
[2]Department of Physical Chemistry and Materials Institute, University of Alicante, Ap. 99, Alicante, E-03080, Spain.

Materials prepared from biomass are potential candidates to substitute the Pt-based commercial electrocatalysts in alkaline fuel cells[1]. Although some investigations have reported activity and durability comparable to Pt/C using biomass-derived catalysts based on Fe-N-C sites, it remains a challenge to decrease the environmental impact of the production of these materials[2]. We have developed a straightforward procedure to obtain excellent ORR electrocatalysts from biomass residues which requires a singlehigh temperature treatment. Washing steps were avoided by optimizing the initial amount of Fe precursor and highly dispersed Fe sites were formed together with iron oxide nanoparticles. Graphitic carbon shell structures covering some of the nanoparticles were also observed. The best performance for almond shell-based materials was achieved for the sample containing 4.6 wt. % of Fe. The differences on the ORR performance depending on the biomass precursor were studied and the best

electrocatalyst was measured in a primary Zn-air battery obtaining a maximum power density of 59 mW cm-2and 755 mAh gZn-1 capacity.

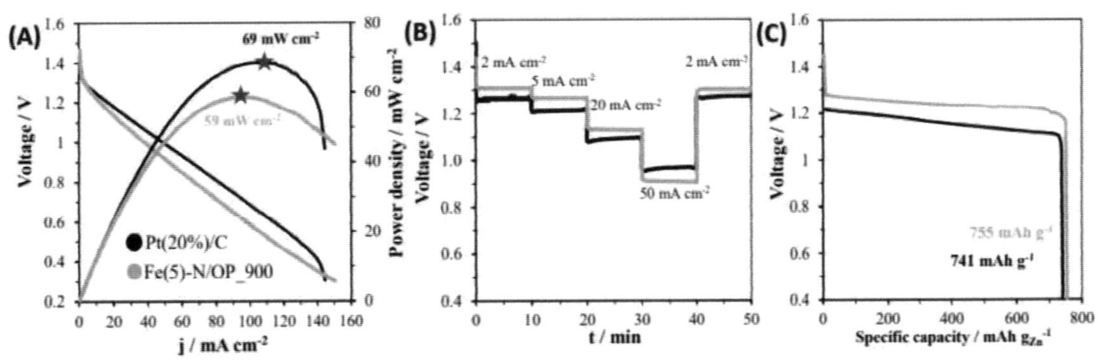

Figure 1: (A) ZAB polarization curves, (B) galvanostatic discharge and (C) long-time galvanostatic discharge curve at 5 mA cm-2(6 M KOH + 0.2 M Zn(O2CCH3)2 solution).

Acknowledgements: The authors would like to thank PID2021-123079OB-I00 project funded by MCIN/AEI/10.13039/501100011033 and “ERDF A way of making Europe” .

References

[1] M. Shao, Q. Chang, J.P. Dodelet, R. Chenitz, Chem. Rev. 116 (2016) 3594–3657.
[2] Y. Jiao, Y. Zheng, M. Jaroniec, S.Z. Qiao, Chem. Soc. Rev. 44 (2015) 2060–2086.

Coal-based Carbon Materials for Ion-batteries

Yaxin Chen*, Zhicheng Ju

School of Materials Science and Physics, China University of Mining and Technology, Xuzhou 221116, China

Carbon materials have attracted extensive attention as potassium-ion battery anode because of the merits of their low cost, high electric conductivity, and thermal and chemical stability, etc. By tuning the particle size, morphology, surface area and porosity, some carbon anodes exhibit capacities above the theoretical value predicted based on hybrid ion storage mechanisms of insertion and adsorption. However, in most cases, it is hard to simultaneously maintain the high specific reversible capacity and cyclic stability. In addition, in contrast to conventional observations of loss upon cycling, capacity increases are sometimes observed, especially porous carbons with high activity. Here, we will introduce our current understanding of the mechanism on storage activity and stability in coal-based carbon anodes for ion batteries. These works give some insights into the mechanism of reversible ion storage of carbon-based anodes.

References

[1]Yaxin Chen; Baojuan Xi; Man Huang; Liluo Shi; Shaozhuan Huang; Nannan Guo; Da Li; Zhicheng Ju*; Shenglin Xiong*; Defect-Selectivity and "Order-in-Disorder" Engineering in Carbon for Durable and Fast Potassium Storage, Advanced Materials, 2022, 34: 2108621 (Frontispiece)

[2]Zongfu Sun; Yaxin Chen*; Chao Geng; Ying Li; Weijia Guo; Jiangmin Jiang; Yongli Cui; Yueli Shi; Quanchao Zhuang; Zhicheng Ju*; Oxygen-driven bulk defect engineering in carbon to reduce voltage hysteresis for fast potassium storage at low voltage, Applied Catalysis B: Environmental and Energy, 2024, 343: 123473.

[3]Jianzhen Xiong; Zecheng Yang; Rui Zhou; Anyong Xiao; Xiangkai Kong; Jiangmin Jiang; Liang Dong; Quanchao Zhuang; Zhicheng Ju; Yaxin Chen*; In-Situ Defect Engineering in Carbon by Atomic Self-Activation to Boost the Accessible Low-Voltage Insertion for Advanced Potassium-Ion Full- Cells, Small, 2024, DOI: 10.1002/smll.202402037. (Cover)

[4]Zongfu Sun; Yaxin Chen*; Baojuan Xi; Chao Geng; Weijia Guo; Quanchao Zhuang; Xuguang An; Jie Liu; Zhicheng Ju; Shenglin Xiong; Edge-oxidation-induced densification towards hybrid bulk carbon for low-voltage, reversible and fast potassium storage, Energy Storage Materials, 2022, 53: 482-491.

[5]Yaxin Chen; Liluo Shi; Qiong Yuan; Ang Li; Shaozhuan Huang; Hui Ying Yang; Xiaohong Chen; Jisheng Zhou; Huaihe Song; Crystallization-Induced Morphological Tuning Toward Denim-like Graphene Nanosheets in a KCl-Copolymer Solution, ACS Nano, 2018, 12: 4019-4024.

Ultralight and Dual-conductive Carbon-Based Anode Current Collector for Anode-Free Lithium Metal Battery

Yidan Cao[1,*], Geng Zhong[1], Yufei Ouyang[1], Nanrui Li[1], Feiyu Kang[1]

[1] Institute of Materials Research, Tsinghua Shenzhen International Graduate School, Tsinghua University, Shenzhen 518055, China

The anode-free Li metal batteries (AFLMBs), by pairing Li-containing cathode with bare anode current collector (CC), have the potential to enable the maximum improvement (>70%) in energy density of lithium batteries.[1,2] Reasonable design of light-weight carbon-based current collector is a promising strategy to enable AFLMBs.[3] We have developed ultralight, dual-conductive all-carbon-fiber based material as anode CC, which takes advantage of the high electronic conductivity of carbon nanotube, good ionic conductivity of copper-coordinated cellulose nanofiber[4], and the three-dimensional porous structure by nanofibers. These enable the CC good mechanical strength, high electrical conductivity, low density and good wettability with electrolyte. The 3D microstructure effectively homogenizes the Li+ flux and buffers the volume changes during cycling. Besides, carbon-coated ZnO nanotube arrays with good lithiophilicity are constructed on carbon-fiber CC, which effectively promotes the formation of robust solid electrolyte interphase and mixed ionic-electronic conductive (MIEC) interface, ensuring rapid electron/ion transportation and dendrite-free Li deposition. As a result, the capacity retention for the anode-free full cell with carbonate-based electrolyte maintains 47.0% after 600 cycles. The anode-free cell also shows superiority in rate performance and energy density. The work provides a promising alternative for light-weight anode CC to enable high-energy-density AFLMBs.

References

[1] C. Shan, Z. Qin, Y. Xie, et al. Carbon, 2023, 204, 367.
[2] P. Molaiyan, M. Abdollahifar, B. Boz, et al. Advanced Functional Materials, 2023, 34(6), 2311301.
[3] H. Kwon, J.H. Lee, Y. Roh, et al. Nature Communications, 2021, 12(1), 5537.
[4] Yang C, Wu Q, Xie W, et al. Nature, 2021, 598, 590.

Carbonized polymer enhanced electrochemical performance of T_3C_2Tx Mxene

Fang Yi [1]

[1]Sun Yat-sen University, 132 Outer Ring East Road, University Town, Guangzhou, P. R. China

Supercapacitors, with merits including long cyclic life, are one of the most important energy storage devices in daily life. Electrode materials are essential for supercapacitors, and the rapid development of modern electronics, especially the burgeoning flexible/wearable electronics, demands supercapacitor electrodes with not only improved electrochemical performance but also good mechanical flexibility. In this presentation, I will talk about our recent works about carbonized polymer enhanced electrochemical performance of flexible Ti3C2Tx MXene supercapacitor electrodes and devices.

Graphene Materials: Green Synthesis and Robust Energy Conversion

Junzhong Wang [1], Junying Wang [2]

[1]Anhui Graphene Carbon Materials Research Center, Anhui University,
111 Jiulong Road, Hefei 230601 , China
[2]Institut of Coal Chemistry, Chinese Academy of Sciences, 27 Southroad Taoyuan, Taiyuan 030001, China

We have succeeded in producing graphene and graphene energy materials by green processes including electrochemical & mechanical exfoliation at large scale. We proved that our crystalline graphene is a good carbon platform for fabricating robust catalysts with superior stability and activity. Our fabricated graphene-based catalysts by different approaches exhibit high activity and stability for O2/CO2 reductions enabled for high-performance Zn-air battery, fuel cells or carbon dioxide conversion flow cells [1-4]. For example, The fuel cell loading with graphene/FeN5 catalytic sites tends to decay little after initial hours under H2-air measurement, remarkably inhibiting commonly fast decline tendency of M-N-C catalysts. Very recently, we thinned graphene layers along with building catalytic sites to obtain a bifunctional oxygen catalyst (overall overpotential of 0.64 V) through vapor intercalation for flexible and wearable cable-type Zn-air battery enabling smartphone charging [2]. The great potential of graphene energy materials will be discussed and commented.

References

[1] Z. Yang, et al, Chin. J. Catalysis, 2024, in press.
[2] Yu M. et al, Adv Funct Mater, 2024, revising.
[3] H. Zhang, et al, Angew. Chem., Int. Ed. 2019, 58: 14871.
[4] J.Y. Wang, et al, Energy Storage Mater., 2018, 12:1.

Aligned channel Gelatin@nanoGraphite aerogel supported form-stable phase change materials for solar-thermal energy conversion and storage

Xiaowen Wu*, Jian Feng, Xianjie Liu, Fankai Lin, Shengzhi Duan, Zhaohui Huang, Xin Min, Minghao Fang

Engineering Research Center of Ministry of Education for Geological Carbon Storage and Low Carbon Utilization of Resources, Beijing Key Laboratory of Materials Utilization of Nonmetallic Minerals and Solid Wastes,
National Laboratory of Mineral Materials, School of Materials Science and Technology,
China University of Geosciences (Beijing), 29 Xueyuan Rd, Haidian District, 100083, Beijing, China

To overcome the poor shape stability, low thermal conductivity, and weak photo response of phase change materials (PCMs), we designed a kind of Gelatin@nanoGraphite aerogel with aligned channel structure of graphite nanosheets and a new type of stable Paraffin/Gelatin@nanoGraphite composite PCMs was fabricated in this work. The graphite nanosheets were used as thermal conduction and photon absorption fillers. The microstructure, thermal conductivity, thermal stability, and solar-thermal conversion of the composite PCMs have been investigated. The results show that the rich porosity of the prepared Gelatin@nanoGraphite aerogel effectively loads a large amount of paraffin wax (>93.2 wt%). Meanwhile, the composite PCMs had anisotropic solar-thermal conversion and thermal conductivity (radial 3.75 and axial 2.59 W/(m·K)), high melting enthalpy (>146.4 J/g), and excellent phase change cycle and shape stability. This material has intensive application prospects in the fields of building energy saving, human body hyperthermia and solar energy collection system.

Multiple Dimensional Carbon Supported Single Atom Catalysts for Electrochemical Energy Application

Shiyong Zhao[1,2,3], Chang Liu [1], Zaiping Guo [2], Sanping Jiang [3]

[1] Institute of Metal Research, Chinese Academy of Sciences,
No.72 Wenhua Road, Shenhe District, Shenyang, 110016, China
[2] School of Chemical Engineering and Advanced Materials, The University of Adelaide,
Adelaide, SA 5005, Australia
[3] Department of Chemical Engineering, Curtin University, GPO Box U1987, Perth, WA 6845, Australia

The demand for energy is increasing dramatically at an alarming rate, resulting from rapid economic development and the ever-increasing requirements of energy-based appliances. Single-atom metal catalysts (SACs) comprising monodispersed metal atoms on appropriate substrates have a theoretical 100% atom utilization efficiency, and therefore have a much higher activity than conventional bulk metal and nanoparticle catalysts. Carbon Materials especially nano carbon owing the advantages of high surface area, sufficient defects sites and versatile bonding environmental, are ideal single atom supports. Nano carbon materials like carbon nanotubes, graphene and three dimensional integrated carbon are used to tune the single atom catalysts structure, including loading, varieties. In this talk, I will introduce our progress on utilizing the multiple dimensional carbon structure to design and synthesize the high loading SACs for a series of electrochemical applications, including Li-S batteries, carbon dioxide reduction and so on.

References

[1] Shiyong Zhao, et al. Energy Environ. Sci., 2023, 16, 4630–4640
[2] Zhiyuan Han, et al. Adv. Mater. 2021, 33, 2105947
[3]Shiyong Zhao, et al. Frontiers in Chemistry, 10.3389/fchem.2023.1150885
[4] Shiyong Zhao, et al. Adv. Funct. Mater. 2020, 30, 1906157
[5] Shiyong Zhao, et al. J. Colloid Interface Sci. 656 (2024) 58–67

Topologically Defective Carbon Anodes toward High-Rate and Large-Areal-Capacity Potassium-Ion Batteries

Shaohong Liu[1], Junlong Huang[1], Haozhen Yang[1], Ruowen Fu[1] and Dingcai Wu[1]

[1]School of Chemistry, Sun Yat-sen University, Guangzhou, China

Potassium-ion batteries (PIBs) have been regarded as one of the highly promising alternatives of lithium-ion batteries in terms of the abundant and widely distributed source of potassium (K) as well as the low redox potential of K/K+ couple. Carbonaceous materials, including graphite, hard carbon, and soft carbon, have been widely investigated as one of the most promising anodes for PIBs. Exploration of advanced preparation strategies, enabling facilitated kinetics of K+ de-solvation and diffusion across the SEI film as well as fast solid-phase K-diffusion, through simple structure optimization of carbon framework, holds significant importance for the development of high-rate PIBs. Herein, a simple temperature-programmed co-pyrolysis strategy is proposed for the efficient synthesis of topologically defective carbon anodes. We have demonstrated that the topological defects can effectively regulate the electronic structure of the resulted carbon anodes, which not only enables efficient potassium adsorption energy yet low potassium migration barrier, but also induce preferential decomposition of anions to produce high potassium ion conductive solid electrolyte interphase film with decreased potassium ion desolvation and transfer barrier. These synergistic structural advantages contribute to high potassium storage capacity, excellent rate capability, and long-term cycling stability of carbon anodes.[1-2]

References

[1] Junlong Huang, Yongqi Chen, Zongheng Cen, Tan Yi, Min Liang, Youlong Zhu, Ruliang Liu, Ruowen Fu, Shaohong Liu, Dingcai Wu. Advanced Materials, 2024, DOI: 10.1002/adma.202403033

[2] Haozhen Yang, Junlong Huang, Shaohong Liu, Yongqi Chen, Zongheng Cen, Chenguang Shi, Yuheng Lu, Ruowen Fu. Small, 2023, 19 (39), 2302537.

All Points Dynamic Monitor and Estimate Lithium-ion batteries Charge of State by Ultrasonic Imaging Technology

Yong Xiang[1], Zhiguo Zhang[1]

[1]Tsing Bosch (Zhuhai) Technology Co., Ltd, Zhuhai, China

Graphite anode lithium-ion batteries are extensively utilized across variousindustries, and their safe operation necessitates precise estimation of the state ofcharge (SOC). Ultrasound technology has been validated as an effective method forSOC estimation by tracking the acoustic changes that occur during the charging anddischarging. However, conventional methods rely on measurements from single ormultiple discrete points, which can lead to inaccuracies due to the non-uniformstructural changes within the battery during cycling. This paper introduces a novel approach that leverages ultrasonic imaging technology to capture data from all pointson the battery, thereby enhancing the accuracy of SOC estimation. Firstly, the paper dynamic monitors of the battery' s charging and discharging processes to characterize the amplitude changes at each point. Subsequently, the distribution of these amplitude points is found to correspond to a Gaussian distribution, with distinct shifts in the positions of Gaussian peaks observed at varying SOC. The paper concludes by discussing the potential of utilizing the distribution of amplitude points as a reliable indicator for evaluating battery SOC.

Keywords: Graphite, state of charge (SOC), ultrasonic imaging

MERCURY REMOVAL FROM WATER USING MXENE-MODIFIED CARBON AEROGELS FROM RICE HUSK LIGNIN

Armanbek Nursharip[1], Chingis Daulbayev[1,2], Jakpar Jandosov[3], Joseph C. Bear[4], Rosa Busquets[4], Vassilis J. Inglezakis[5] and Alzhan Baimenov[6,7*]

[1]National Laboratory Astana, Nazarbayev University, Astana, Kazakhstan
[2]Institute of Nuclear Physics, Almaty, Kazakhstan
[3]Institute of Combustion Problems, Almaty, Kazakhstan
[4]School of Life Sciences, Pharmacy and Chemistry, Kingston University, Greater London, UK
[5]Department of Chemical & Process Engineering, University of Strathclyde, Glasgow, UK
[6]Laboratory of Engineering Profile, Satbayev University, Almaty, Kazakhstan
[6]Institute of Physics and Technology, Satbayev University, Almaty, Kazakhstan

In the search of sustainable and effective solutions to eliminate heavy metal pollution in water, current research focuses on the development of innovative adsorbents to immobilize pollutants. The development of low-cost adsorbents from biological waste with high removal efficiency is a priority. Here, macroporous aerogel composites derived from lignin extracted from rice husk and modified by Ti3C2Tx MXene are explored as filtration media for removing heavy metals. Their preparation involved first synthetizing cryogels which had embedded MXenes. They were carbonized at 600 °C in inert atmosphere to obtain aerogels. Obtained materials had a developed 3D macroporous structure with open-end channels and micro-mesoporosity of chains, high specific surface area (up to 320 m2/g) and oxygen-containing functional groups of the MXene nanosheets and the rest of the carbonaceous structure. These showed remarkable potential for the targeted removal of mercury ions from aqueous solutions. Notably, the modification of lignin aerogel by 20 wt.% MXene increased the maximum adsorption capacity > 1.5 folds (135.8 mg Hg2+/g of composite) compared to the bare lignin aerogel (82.4 mg Hg2+/g). This is superior performance than commercially available adsorbents. The performance of this composite remained when filtering 10 mg/L of Hg from mercury-spiked tap water.

Life-cycle carbon emissions and decarbonization potentials of carbon fiber and carbon nanotube in China

Jiayi Xie[1,a], Zichen Liu[2,a], Shaojun Zhang[1] and Ye Wu[1]

[1]School of Environment, Tsinghua University, Beijing, China
[2]Tanwei College, Tsinghua University, Beijing, China

Carbon fiber and carbon nanotube are characterized by exceptional physicochemical properties that preferred by a spectrum of applications, such as wind turbine, aerospace, and lithium-ion batteries. However, their energy-intensive property in production may raise the concern of life-cycle carbon emissions in the expanded market. This study investigated plant-level life cycle inventories of carbon fiber and carbon nanotube as well as their precursors (e.g. acrylonitrile and polyacrylonitrile). By incorporating the plant-level input-output of material and energy, we assessed their life-cycle carbon emissions to be 46.5 t CO2/t and 7.5 t CO2/t for carbon fiber and carbon nanotube, respectively. Steam and electricity are major contributors with nearly 85% while the initial feedstocks (i.e., light hydrocarbon substances) account for no more than 15%. Hence, mitigation strategies should focus on energy saving, energy decarbonization, and material efficiency improvement, which are expected to result in 20% of reduction benefits. This study contributes to the refinement of the current knowledge of life-cycle impacts associated with carbon materials, and is expected to provide decision-making insights for carbon mitigation in the future application.

a J.X. and Z.L. contributed equally to this work

Molecularly Imprinted Adsorbent Based on β-COF/Graphene Aerogel for Phenol Adsorption and Removal

Yong Zhong[1], Lei Qin[1,2], Boxi Jing[1], Taoran Han[1], Yongzhen Yang[1,2] and Xuguang Liu[3]

[1]Key Laboratory of Interface Science and Engineering in Advanced Materials, Ministry of Education, Taiyuan University of Technology, No.79 West Yingze Street, Taiyuan 030024, China
[2]Shanxi-Zheda Institute of Advanced Materials and Chemical Engineering, No. 67 Zhengyang Street, Taiyuan 030600, China
[3]College of Materials Science and Engineering, Taiyuan University of Technology, No.79 West Yingze Street, Taiyuan 030024, China

Phenol pollutant in wastewater poses a serious threat to environment and public health, while phenol molecules are valuable raw chemicals. Graphene aerogel (GA) based surface molecularly imprinted composites that possess unique advantages in molecular recognition exhibit great potential for phenol separation and utilization. The combination of β-COF and GA to construct β-COF/GA composite as support for surface imprinting can rectify the inertness defects of GA, enhance adsorption performance of corresponding imprinting materials, and maintain high solid-liquid separation efficiency as a block material. In this study, β-COF nanoparticles are prepared and then assembled with graphene through hydrothermal processes to form β -COF/GA. Subsequently, surface molecularly imprinted adsorbent β-COF/GA@SMIP is synthesized by liquid-phase polymerization. Field emission scanning electron microscope, X-ray diffractometer, nitrogen adsorption specific surface area analyzer, etc. are adopted to characterize the morphology and structure of each product. Besides, UV-visible spectrophotometer is used to measure phenol adsorption capacities of the prepared adsorbents. The results show that the addition of β-COF enables the adsorption capacity of β -COF/GA@SMIP achieves 63.57 mg g-1, which is twice that of graphene based imprinted material. Moreover, β -COF/GA@SMIP also gets good adsorption selectivity and renewability. This work contributes to the promotion and application of high-performance phenol adsorbents.

Exploring the Effects Behind the Outstanding Catalytic Performance of PdAg Catalysts Supported on Almond Shell-Derived Activated Carbon Towards the Dehydrogenation of Formic Acid

Paula Riquelme-García, Jessica Chaparro-Garnica, Miriam Navlani-García and Diego Cazorla-Amorós

Department of Inorganic Chemistry and Materials Institute, University of Alicante, Ap. 99, Alicante, E-03080, Spain

Email: paula.riquelme@ua.es

In this work, highly efficient carbon-supported Pd-catalysts were synthesized by a straightforward wet impregnation-reduction method for H2 release via formic acid dehydrogenation [1]. The carbon support was obtained from a biomass residue (almond shell) prepared via H3PO4-assisted hydrothermal carbonization (HTC) and thermal activation. This support was doped with nitrogen groups to study the effect on the electronic properties and catalytic activity of the catalysts. Investigating the formation of PdAg alloys with varying Pd:Ag molar ratios resulted in catalysts exhibiting enhanced catalytic activity compared to monometallic Pd counterparts. Notably, the Pd1Ag0.5/NAS catalyst displayed outstanding catalytic performance, achieving an initial turnover frequency (TOF) of 1716 h-1 (calculated in the first 3 minutes of reaction and expressed per mole of Pd) and maintaining substantial activity over 6 consecutive reaction cycles. This work elucidates the successful synthesis of effective catalysts, emphasizing the influence of nitrogen doping and alloy composition on catalytic behaviour and stability.

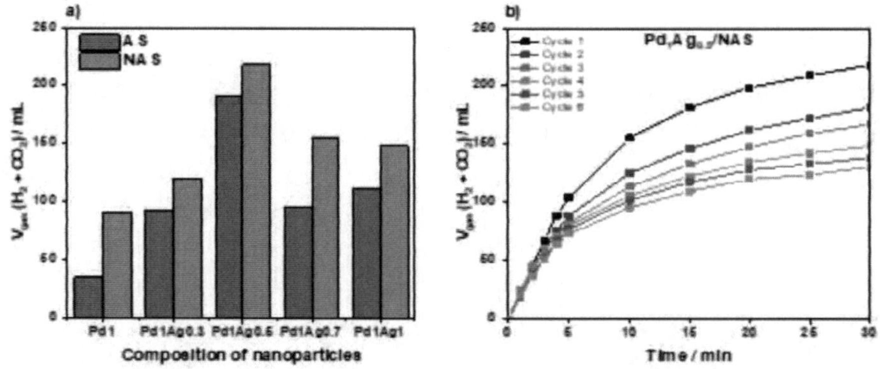

Figure 1. a) Total volumes of gas generated by the catalysts. B) Stability of Pd1Ag0.5/NAS.

Acknowledgements:The autors would like to thank project PID2021-123079OB-I00 and project TED2021-131324B-C22 funded by MCIN/AEI/ 10.13039/501100011033 and by the "European Union NextGenerationEU/PRTR.

References

[1] Jessica Chaparro-Garnica, et al. Fuel 320 (2022) 123900.

Thermocatalytic decomposition of methane to hydrogen and nano filamentous carbon over Ni-Ce bimetallic catalysts on porous carbon

C. Srinivasakannana[1], Adeel Ahmad Iftikhar[1], Iqra Reyaz Hamdani[1], Ahmed Al Shoaibi[1], Mohammad Mozahar Hossain[2]

[1]Chemical Engineering Department, Khalifa University, Abu Dhabi, UAE
[2]Department of Chemical Engineering, King Fahd University of Petroleum & Minerals, Saidi Arabia

Nickel and ceria-based bimetallic catalysts on a porous carbon template were synthesized using the incipient wetness impregnation approach and investigated for catalytic methane decomposition (CMD) reaction. Ceria was employed as a promoter varying the concentration ranging from 5 to 15wt. %. The effects of space velocity and methane feed concentration were evaluated using the best catalyst. A conversion close to the equilibrium at 850 C could be achieved with a feed concentration of 40% methane, accompanied by excellent stability of the catalyst. The low concentration of promoter ceria led to increased metal dispersion over the catalytic surface, resulting in suitable metal-support interaction. A key outcome of the research is that the combination of carbon support and promoter ceria was able to increase the thermal stability of the nickel catalyst up to 850°C, resulting in a high conversion rate and catalytic stability. The morphological study of the optimal catalyst indicates the development of compact carbon nanotubes through a tip-growth mechanism exhibiting a high aspect ratio, as opposed to the monometallic Ni catalyst, which promoted the development of amorphous carbon in addition to filamentous carbon nanotubes.

Rational Design of High-Efficiency Interfacial Solar Evaporator with Low Evaporation Enthalpy Based on Biomass-Derived Materials and Silica Aerogel

Linyi Wu[1], Hironori Ogata [2], Morinobu Endo [3], Zhipeng Wang [1]

[1]Engineering Laboratory for Functionalized Carbon Materials and Shenzhen Key Laboratory for Graphene-based Materials, Graduate School at Shenzhen, Tsinghua University, Shenzhen 518055, China
[1] Institute of Advanced Materials, College of Chemistry and Chemical Engineering, Jiangxi Normal University, 99 Ziyang Avenue, Nanchang 330022, China
Email: wangzhipeng@jxnu.edu.cn
[2]Department of Chemical Science and Technology, Faculty of Bioscience and Applied chemical and Graduate School of Engineering, Hosei University, 3-7-2, Kajino-cho, Koganei, Japan
[3] Research Initiative for Supra-Materials, Faculty of Engineering, Shinshu University,4-17-1 Wakasato, Nagano 380-8553, Japan

Solar-driven interfacial evaporation is a potential solution to address freshwater shortages and water pollution, but low evaporation rates limit its practical applications [1]. The proper design of evaporator structure and the selection of effective solar absorbers are the main ways to increase the evaporation rates. In this work, we have selected a self-floating carbonized natural wood (CNW) as a support for water transportation and a heat-absorbing SiC and carbon composite (SCC) as an absorber which is derived from biomass-natural wood and rice husk by high temperature annealing, respectively, as well as a thermal insulation layer of silica aerogel (SA) obtained from water glass by freeze-drying. All material components are assembled them into a trilayered solar evaporator, denoted as CAS. Due to the abundant hydrophilic groups, super wettability, and vertical array channels of CNW, ultralow thermal conductivity and porous structure of SA, and strong light absorption and excellent solar heat conversion of SCC, the trilayered CAS has a low evaporation enthalpy of 0.8118 MJ kg-1, and demonstrates a high evaporation rate (4.21 kg m-2 h-1), an excellent evaporation efficiency of 94.94%, as wells as good salt-resistance and superior long-term stability. In addition, the CAS can be utilized to purify water from a variety of contaminants with excellent performance. This work provides a sustainable method to prepare cost-effective evaporator components from biomass waste or cheap sources for interfacial solar evaporation, which is conducive to solve the global energy crisis, freshwater shortage issues and eliminate biomass waste.

References

[1] Wang J., Kong Y., Liu Z., Wang H. Solar-driven interfacial evaporation: design and application progress of structural evaporators and functional distillers. Nano Energy, 108, 108115 (2023).

Gas-generator compositions based on activated carbon from a gas mask for destruction of rocks

S. Tursynbek[1,2]*, V. E. Zarko[3], R.Q. Shen[4], Z.A. Mansurov[1,2]

[1]Kazakh National University. al-Farabi, Almaty, Kazakhstan.
[2]Institute of Combustion Problems, Almaty, Kazakhstan.
[3]Institute of Chemical Kinetics and Combustion named after. V.V. Voevodsky Siberian Branch of the Russian Academy of Sciences, Novosibirsk, Russia.
[4]Nanjing University of Science and Technology, Nanjing, China

In world practice, there are various methods for recycling activated carbon from a gas mask. Namely water purification, air purification, medicine, etc.

In this work, we studied the performance of a three-component mixture, activated carbon from a gas mask, sodium nitrate, magnesium and carbon obtained by carbonization of walnut shells. Laboratory and field tests of the studied compositions for the destruction of rocks during deflagration combustion in a gentle mode were carried out.

Key aspects are the study of the thermochemical properties of a mixture of 60% - NaNO3, 20% - Mg, 20% - C, reaction kinetics, the influence of physical conditions (especially high pressure) on the combustion process, as well as analysis of the influence of different types and concentrations of carbon powders, taking into account safety and environmental aspects associated with combustion products.

Carbon samples were crushed and dispersed onto sieves; in the experiments, the fraction 100÷200 μm was mainly used.

Preliminary experiments have shown that at a low carbon content, combustion of the mixture occurs at a high speed in a convective mode. Uniform flame propagation was recorded in the combustion wave of the composition 60/20/20. Accordingly, the working composition was chosen with a component ratio of 60% - NaNO3, 20% - Mg, 20% - C, characterized by fairly high values of gas productivity and the RT parameter, proportional to the performance of the mixture.

Table. Basic physico-chemical and operational characteristics of the gas-generating composition NaNO3-Mg-C (from a gas mask)

Efficiency of combustion products, kJ/kg	475÷615
Volume of gases released, l/kg	460÷480
Efficiency of combustion products, kJ/kg	550÷620
Bulk density in the cartridge, g/cm³	0.1÷0,17
Burning speed (P=0.1 MPa), mm/sec	3÷5

Based on the field tests carried out, a gas generator composition $NaNO_3$ – 60% was developed; Mg – 20%; C – 20% which can be used for directed destruction in the deflagration mode of hard rocks with a strength coefficient of 8. Thus, effective utilization of activated carbon from a gas mask is shown.

Single Molecule Capturing and Detection on Graphene Analogues

Yu Lei [1]

[1] Tsinghua Shenzhen International Graduate School,
Institute of Materials Research & Center of Double Helix

Extensive research efforts have been dedicated to the field of two-dimensional (2D) materials in a short period of time. From semi-metallic graphene, to semiconducting MoS2, to insulating hexagonal boron nitride (hBN), 2D materials provide a wide range of physicochemical properties that can be engineered for diverse applications. Recently, defect engineering has emerged as a primary approach to tailor the physiochemical properties and further extend the functionalities of 2D materials. Comprehensive understanding between the atomic-scale defects and the effects on the physio-chemical properties is challenging and essential for the materials community to design multi-functional 2D materials. It is also equally important to develop scalable approaches to control and tailor the defect types and concentration. With the urgency, I have dedicated in defect engineering in 2D materials: 1) Developed universal approaches to introduce defects in 2D materials; 2) Utilized the atomic-scale defects to activate the basal plane reactivities of 2D materials, and proposed a multi-scale strategy to understand the correlation between the reactivities and the atomic-scale defects; 3) Explored the application of defects in 2D materials for bio-sensing, including electrochemical sensing and surface enhanced Raman scattering. Overall, the work aims to utilize defects and surface potential tuning to reach single molecule detection on 2D platform.

References

[1]Sci. Adv., 2020, 6, eabc4250
[2]ACS Nano, 2017, 11, 5103-5112
[3]Small, 2021, 17, 2004827 (Cover)
[4]Nano Research Energy, 2023, 2, e9120043
[5]Mater. Today, 2023, 62, 168-189 (Cover)
[6]Adv. Mater, 2023, 35, 2300940
[7]Mater. Today, 2021, 51, 108-116
[8]ACS Nano, 2021, 15, 9796-9807
[9]ACS Nanosci. Au, 2022, 2, 450-485 (Editor`s Choice)
[10]ACS Appl. Energy Mater., 2019, 2, 8625-8632

Hard Carbon Anodes Derived from Crosslinked Waste PS and PVC and Their Sodium Ions Storage Performances

Jiahui Ding, Ying Zhou, Kening Xing, Qi Han, Yuhang Xie, Chunlei Wang*

Dalian University of Technology, Dalian, China

*Corresponding author e-mail: clwang-chem@dlut.edu.cn

Plastic has the advantages of durability, health, molding, etc., and is widely used in various fields such as building materials, electronic equipment, packaging materials, etc. However, the rapid growth of plastic production coupled with a short shelf life has led to the production of a large number of waste plastics, and the adoption of appropriate methods to recycle waste polymers will promote the solution of plastic waste pollution. Recently, converting waste plastics into carbon materials has attracted widespread attention, which achieving a win-win situation of recycling and high value-added utilization waste plastics. In this work, hard carbon anode was prepared through knitting of waste polystyrene (w-PS) and waste polyvinyl chloride (w-PVC) with FeCl3 as catalyst followed by carbonization at high temperature. The results show that hard carbon anode exhibit high reversible capacities of 301 mAh g-1 and initial coulombic efficiency (ICE) of 72.37%. After 200 cycles, the reversible capacity of 194mAh g-1 can be maintained.

Key words: Plastic recycling; Hard carbon; Sodium ion batteries; Anode materials

Acknowledgments: This work was supported by the National Natural Science Foundation of China (No. 22178053).

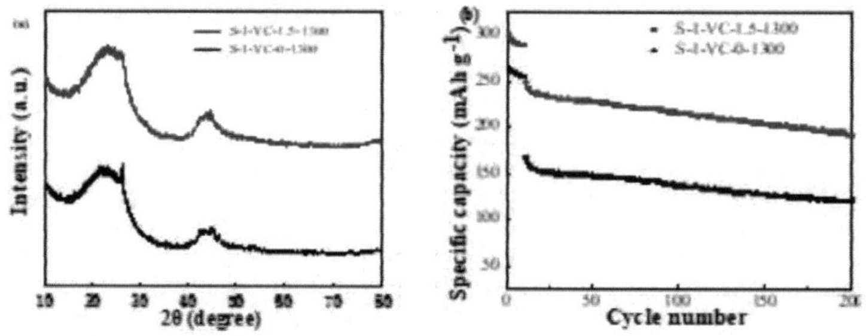

Fig.1 (a) XRD spectra of the sample. (b) Sample cycling performance at a current density of 0.03A g-1.

References

[1] K. Thileep Kumar et al. Nanoscale. 2022, 14, 4312–4323.

[2] Fonseca,W.S et al. ACS Sustainable Chem. Eng. 2015, 3, 2153– 2159.

Synergistic Effect of Indium Single Atoms and Pt-In Alloy Nanoparticles on Carbon Support for Selective Oxidation of Glycerol

Zhiyan He[1,2], Hao Yu[2]

[1]Hubei Key Laboratory of Biomass Fibers & Eco-Dyeing & Finishing, Wuhan Textile University, Wuhan, China

[2]State Key Laboratory of Pulp and Paper Engineering, South China University of Technology, Guangzhou, China

Developing single-atom catalysts (SACs) complexed with metal alloy nanoparticles (ANPs) on carbon materials has proven challenging, but also highly desirable. To date, the marriage of SACs and ANPs has not yet been reported in glycerol oxidation. In this study, we developed a one-step salt-modified method for co-loading In SACs and Pt-In ANPs on carbon nanotubes (CNTs). The optimized catalyst can efficiently catalyze the selection of glycerol to yield 1,3-dihydroxyacetone (DHA) with selectivity 51.73% at 20.40% glycerol conversion under mild conditions, which is around 4 times that of Pt/CNT. We proposed that the synergistic function of In SACs and Pt-In ANPs could enhance the adsorption activation of the secondary hydroxyl groups of glycerol and oxygen. The concept of synergistic SACs and ANPs cocatalysts can be extended to other systems, thus contributing to the development of more effective cocatalysts.

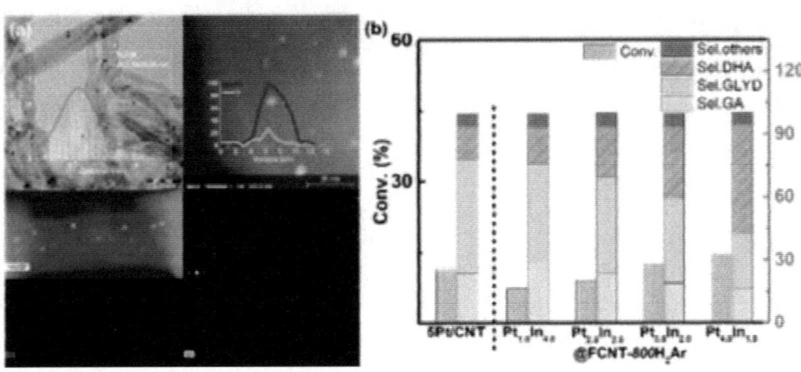

Fig.1 (a) TEM, HAADF-STEM images, and associated EDS spectrum of PtIn@(-F800)CNT-800H2Ar, (b) Catalytic performance evaluation for glycerol oxidation of PtxIn5-x@FCNT-800H2Ar catalysts.

References

[1] Wenli Wang, et al. Chem. Eng. J.(2022) 436: 131758

Biomass-derived 3D Porous Carbon Materials for Advanced Environmental Applications

Bei Zhang [1,2], Toshihiro Isobe [3], Manabu Fujii [4] and Baoliang Chen [1,2]

[1] Innovation Center of Yangtze River Delta, Zhejiang University, Hangzhou, Zhejiang 311400, China
[2] Department of Environmental Science, Zhejiang University, Hangzhou, Zhejiang 310058, China
[3] Department of Materials Science and Engineering, Tokyo Institute of Technology, Meguro-ku, Tokyo 152-8552, Japan
[4] Department of Civil and Environmental Engineering, Tokyo Institute of Technology, Meguro-ku, Tokyo 152-8552, Japan

The transformation of biomass to carbon materials has recently attracted increasing attention because of their potential cost-effective, large-scale, and quality-controllable production in an eco-friendly manner [1, 2]. Compared to 0D to 2D carbon materials (e.g., CQDs, CNT and graphene) derived from biomass [3], 3D porous carbon materials (e.g., carbon foam) were synthesized from modified lignin components of biomass in this study [4]. A novel hydrothermal treatment with a thermal foaming process was developed to synthesize biomass-derived 3D porous carbon materials. Comprehensive characterization results (e.g., 2D NMR and RRDE) revealed that 3D porous carbon materials possess a multi-porous structure and graphitic carbon, contributing to superior physicochemical and electrochemical properties. Accordingly, advanced environmental applications, including synergistic carbon sequestration [5] and electro-Fenton treatment [4], were validated utilizing 3D porous carbon materials as novel adsorbents and electrodes. Overall, our study demonstrates a sustainable strategy for upgrading biomass to 3D porous carbon materials for advanced environmental applications.

References

[1] Mingjie He, et al. Nature Reviews Earth & Environment 3 (2022) 444-460
[2] Bei Zhang, et al. Carbon (2019) 147 312-322
[3] Bei Zhang, et al. Chemical Reviews (2023) 123 7193-7294
[4] Bei Zhang, et al. Resources, Conservation and Recycling (2024) 203 107413
[5] Bei Zhang, et al. Resources, Conservation and Recycling (2022) 185 106453

Self-supporting spiral carbon forest clusters prepared by ethanol flame method and their performance in capacitive deionization

Die Gong [a], ChangKangLe Xu [a], FeiYu Zhou [a], QingShan Fu [a,b*]

[a] College of Material Science and Engineering, Sichuan Univer sity of Science and Engineering, Zigong 643000, China.

[b] Key Laboratory of Material Corrosion and Protection of Sichuan Province, Zigong 643000, China

*Corresponding author: sendysan@suse.edu.cn (Q Fu)

Capacitive deionization (CDI) is considered to be an alternative water purification technology because of its low cost and low driven energy. Generally, powder carbonaceous materials must be mixed with a polymer binder and an electrical conductor (such as carbon black) when they are used as electrodes in the application of capacitive deionization; however, binders increase internal resistance and block some pores in the carbon material, reducing the adsorption sites of the active substance, resulting in reduced adsorption capacity. Therefore, carbon electrode materials with self-supporting structures have become a hot topic. In this work, the ethanol flame method is used to prepare self-supporting carbon nanostructure on graphite paper without the need for additional carbon sources and complex reaction conditions; the ethanol flame can simultaneously provide the reactive heat and the required carbon sources for the growth of carbon nanomaterials under atmospheric condition. The prepared self-supporting carbon has a spiral forest cluster structure. The self-support carbon-modified graphite paper is used as the capacitive deionization electrode to investigate the desalination effect. At 25°C, the electrode has a salt adsorption capacity of 46.94 mg g-1 and an average salt adsorption rate of 0.78 mg g-1min-1 in 50 mg/L NaCl solution at 1.2 V.

Two-dimensional nanosheets with 1-nm-thick adlayers for Efficient and Near-instant Removal of PFAS

Dingxin Xu[1], Wenhui Ding[1], Yuxiong Huang[1], Yang Su[1,2], Hui-Ming Cheng[3,4,5]

[1] Tsinghua-Berkeley Shenzhen Institute, Tsinghua Shenzhen International Graduate School, Tsinghua University, Shenzhen 518055, P.R. China.

[2] Institute of Materials Research, Tsinghua Shenzhen International Graduate School, Tsinghua University, Shenzhen 518055, P. R. China

[3] Shenzhen Key Laboratory of Energy Materials for Carbon Neutrality, Institute of Technology for Carbon Neutrality, Shenzhen Institute of Advanced Technology, Chinese Academy of Sciences, Shenzhen 518055, P. R. China

[4] Faculty of Materials Science and Energy Engineering, Shenzhen Institute of Advanced Technology, Shenzhen 518055, P. R. China

[5] Shenyang National Laboratory for Materials Science, Institute of Metal Research, Chinese Academy of Sciences, Shenyang 110016, P. R. China

The environmental occurrence of anthropogenic chemicals, especially persistent micropollutants of per- and poly-fluoroalkyl substances (PFAS) raises pressing concerns for global drinking water safety. Adsorption is an effective technology for removing PFAS, but is limited by unsatisfactory adsorption capacity and efficiency. We report a strategy to attach polyamine adlayers to graphene oxide (GO) nanosheets, that produces highly-charged and monodispersed two-dimensional (2D) adsorbents of a GO nanosheet sandwiched between two 1-nm-thin polyamine adlayers. This adsorbent has a high adsorption capacity for PFAS of ~3070 mg/g, tens of times greater than that of GO. It also provides almost instant adsorption of a variety of PFAS and reaches 57-95% of its equilibrium capacity in a minute and removes ~100% of PFAS from contaminated water sources within a few minutes, transforming real-life PFAS-contaminated water into safe drinking water. Experiment and theory show that the planar nature of the 2D adsorbent combined with its abundant surface adsorption sites that electrostatically attract the polar groups of the PFAS, and hydrogen bonding and hydrophobic-hydrophobic interactions with their non-polar groups, account for its ultrahigh adsorption capacity and rapid removal efficiency. We also show that regeneration of the adsorbents removes the adsorbed PFAS and allows its subsequent destruction, demonstrating a closed-loop treatment solution for micropollutant contamination.

Hydrodynamic tearing of bacteria on carbon-coated nanotips for sustainable water disinfection

Yang Cheng[1], Wu Qianyuan[2], Hu Hongying[3]

[1] Institute of Materials Research, Tsinghua Shenzhen International Graduate School, Tsinghua University, Shenzhen, China
[2] Institute of Environment and Ecology, Tsinghua Shenzhen International Graduate School, Tsinghua University, Shenzhen, China
[3] School of Environment, Tsinghua University, Beijing, China

Water disinfection is often associated with a high carbon footprint and the formation of toxic byproducts. Here, we describe a nano-structured material that is highly effective at killing bacteria in water through a hydrodynamic mechanism. The material consists of carbon-coated, sharp Cu(OH)2 nanowires grown on a copper foam substrate. We show that mild water flow (e.g. driven from a storage tank) can efficiently tear up bacteria through a high dispersion force between the nanotip surface and the cell envelope. This mechanism produces rapid inactivation of bacteria in water, and achieved complete disinfection in a 30-day field test. Our approach exploits fluidic energy and does not require additional energy supply, thus offering an efficient and low-cost system that could potentially be incorporated in water treatment processes in wastewater facilities and rural communities.

References

[1] Peng, L.; Zhu, H.; Yang, C.; Wu, Q.; Hu, H. et al. Nat. Commun.; 2023, 14, 5734

Super-structured porous hydrogel composites and carbons for soft tissue repair

Rongkang Huang[1], Weiwen Liang[1], Pengwei Ma[2], Dongtian Miao[2], Fuxin Tang[1], Bingna Zheng[2], and Dingcai Wu[3]

[1] The Sixth Affiliated Hospital, Sun Yat-sen University, Guangzhou, P. R. China
[2] School of Chemistry, Sun Yat-sen University, Guangzhou, P. R. China
[3] The Eighth Affiliated Hospital, Sun Yat-sen University, Shenzhen, P. R. China

Annually, over 100 million soft tissue defects arise from surgical procedures, trauma, and tumor lesions, imposing a significant economic burden on healthcare. Most of these defects require biomaterials to protect the wound or promote tissue healing. However, current soft tissue repair materials exhibit limited pore structures and inadequate barrier and repair functions, failing to adequately address the intricate and varied clinical requirements. Here, a series of novel super-structured porous hydrogel composites, such as wet-adhesive patches and porous meshes, have been developed with hierarchical structures and multifunctional frameworks. We demonstrate that our super-structured hydrogel composites can integrate ultralow swelling, ultrahigh burst pressure tolerance and bio-adaptive mechanical support properties, matching the different requirements of soft tissue repair [1-2]. Furthermore, through the incorporation of porous carbons into the fabrication of multifunctional frameworks, a novel injectable hydrogel can effectively promote soft tissue healing with anti-inflammatory and drainage properties. The design and construction of our super-structured porous composites can provide a new idea for the development of high-performance materials for soft tissue repair.

References

[1] Pengwei Ma, Weiweng Liang, Rongkang Huang, et al. Adv Mater, 2024, 2305400.
[2] Fuxin Tang, Dongtian Miao, Rongkang Huang, et al. Adv Mater, 2024, 2307845.

Photoluminescent Properties Of Metal-containing Carbon Dot Nanostructures: Potential Applications in MRI Imaging

Guido Perrin [1], Cornel Stan [2], Adina Coroaba [3], Cristina Mariana Utiru [3,4], Conchi O. Ania [1,3] *

[1]CEMHTI, CNRS (UPR 3079), Université d' Orléans, 45071 Orléans, France.
[2]TUIasi - Gheorghe Asachi Technical University of Iasi, Romania
[3]Centre of Advanced Research in Bionanoconjugates and Biopolymers (IntelCentre), Petru Poni Institute of Macromolecular Chemistry, Iasi, Romania
[4]Advanced Center for Research & Development in Experimental Medicine "Prof. Ostin C. Mungiu", "Grigore T. Popa" Univ. Medicine and Pharmacy, Iasi, Romania
E-mail: conchi.ania@cnrs-orleans.fr

Carbon dot nanostructures have become interesting materials in many different application fields such as optoelectronic and sensing devices, photovoltaics, medical imaging owing to their photoluminescent properties in the UV and near-infrared regions [1]. Despite the origin of their radiative transitions is still unclear, the photoluminescence emission of carbon nanostructures can be quite easily modulated upon the choice of the precursor, the synthetic protocol and the post-functionalization methods. In this communication we report the preparation of metal-containing carbon dot nanostructures from various precursors, to explore the impact of adding metallic cations -Fe, Co, Mn- on their photoluminescence and physicochemical characteristics [2,3]. The samples displayed broad emission bands between 400-520 nm and absolute photoluminescence quantum yields of 3-35 %. The incorporation of the metals provoked a large shift in the emission at longer wavelengths. The Mn-containing carbon nanostructures were tested for their application in-vitro and in-vivo MRI imaging yielding similar results than commercial gadolinium-based compounds [2].

Acknowledgements: The authors thank the financial support of the LEAP-RE program (grant NANOSOLARCELLS) and the Romanian Ministry of Research, Innovation and Digitization (PNRR-III-C9-2022-I8-291, contract 760081/23.05.2023) within the National Recovery and Resilience Plan.

References

[1] N.A. Humaera et al. Lumin. 2021, 36, 1354-1364.
[2] A. Tiron et al. Pharmaceutics 2021, 13 1982.
[3] C.S. Stan et al. Nanomaterials 14 (2024) 83-97.

Targeting Carbon Dot Complex for Integration of Diagnosis and Treatment

Lin Chen[1], Yongzhen Yang[1*], Shiping Yu[2], Xuguang Liu[3]

[1]Key Laboratory of Interface Science and Engineering in Advanced Materials, Ministry of Education, Taiyuan University of Technology, Taiyuan 030024, China

[2]Department of Interventional Therapy/Shanxi Province Cancer Hospital, Shanxi Hospital Affiliated to Cancer Hospital/Chinese Academy of Medical Sciences/Cancer Hospital Affiliated to Shanxi Medical University, Taiyuan 030013, China

[3]College of Materials Science and Engineering, Taiyuan University of Technology, Taiyuan 030024, China

Carbon dots have excellent biocompatibility, optical stability, and low cytotoxicity, showing broad application prospects in the fields of biological imaging and targeted drug delivery [1]. Focusing on the safe and efficient diagnosis and treatment of tumors, the key issues for realizing the properties of carbon dots, such as organelle-targeting, fluorescent/phosphorescent imaging, and controllable delivery for drugs/genes were studied. The design strategy for nano carbon functional materials used for the integration of diagnosis and treatment was proposed, and systematic progress was made in the design, preparation, performance regulation, and efficient application in tumor therapy of carbon dot complex. The active/passive targeting strategy was adopted to synthesize organelle-targeting carbon dots based on the chiral structure and ligand receptor binding interaction, the precise regulation of its emission wavelength/targeting function was also realized, and they can achieve accurate tumor monitoring [2, 3]. By combining heteroatom doping with physical embedding strategy, long-lived phosphorescent carbon dot complex with tight cross-linking structure and rich hydrogen bonding network was synthesized, which can achieve high-precision afterglow imaging in vitro and in vivo [4]. Organelle-targeting carbon dot complexes with targeting delivery and precise controlled release properties were constructed by loading anti-tumor drugs/genes via electrostatic effect and chemical bonds, so as to achieve the integration of tumor diagnosis and treatment [5].

References

[1] Lin Chen, et al. Sensor and Actuators: B. Chemical 390 (2023) 133946
[2] Xin Zhang, et al. Dyes and Pigments 201 (2022) 110213
[3] Mingxiu Lei, et al. RSC Advances 12(2022) 27431–27441
[4] Yingying Wei, et al. Frontier Materials Science 15 (2021) 253–265
[5] Shicai Wang, et al. Materials Science & Engineering C 116(2020) 111233

Tailoring Pore Structure of Hollow Carbon Spheres for Blood Purification

Xiaohuan Zhao[1,2], Chong Wang[3], Yun Zhao[2,*], Yuehong Li[3,*], Qianming Gong[1,*]

[1]School of materials science and engineering, Tsinghua University, Beijing, China
[2]School of Chemistry and Chemical Engineering, Beijing Institute of Technology, Beijing, China
[3]Beijing Tsinghua Changgung Hospital, School of Clinical Medicine, Tsinghua University, Beijing, China

Human blood is a well-balanced complex system of plasma, erythrocytes, leukocytes and platelets. Among them, plasma is predominantly composed of water in addition to a variety of proteins, electrolytes, nutrients, waste products and other solutes. While accumulated and overloaded toxins or waste products in patients' blood would break the balance and cause severe diseases. In this case, blood purification is sometimes an irreplaceable treatment to remove toxins or pathogenic substances from the patients' blood via extracorporeal circulation equipment. Obviously, adsorbents are essential for the removal efficiency and selectivity of toxins with different sizes and physicochemical properties. Particularly, abnormally elevated middle-macromolecular toxins such as interleukin 6 (IL-6) and tumor necrosis factor alpha (TNF- α) in human blood are fatal for patients suffered from immune-related diseases. Herein, porous hollow carbon spheres (HCSs) with different size and pore structure have been successfully prepared. HCSs could remove 94.75 % and 98.33 % of parathyroid hormone (PTH) and β2-micro globulin (β2-MG) efficiently within 5–10 min. Moreover, HCSs with different pore structure exhibit distinct adsorption selectivity for IL-6 and TNF- α, which is of special significance for modifying the middle-macromolecular cytokine level in the complicated human blood environment.

References

[1] Yunhong Liu, et al. Materials Science & Engineering C 121 (2021) 111879
[2] Minjun Zhang, et al. Separation and Purification Technology 327 (2023) 124844
[3] Xiaohuan Zhao, et al. Chemical Engineering Journal. 454 (2023) 140213

The mechanical performance of the composite of the graphene and biofilm

Xiuyun Chuan[1], Mianshu Hu[1] Yiting Wang[2] and Jinsong Han[2]

[1]School of Earth and Space Sciences, Peking University, Haidian District, No. 5 Yiheyuan Road, 100871 Beijing, China

[2]Peking University Third Hospital, Haidian District, No. 49 North Garden Road, 100191 Beijing, China

As natural biological framework, biogenic patches preserved the three-dimensional framework structure of the extracellular matrix (ECM) of the source tissue, facilitating cell adhesion, proliferation, and differentiation in biomedical surgery. With good tissue compatibility of the biological patches, it can induce the anti-infectivity and lead the self-repaired ability of the damaged tissues, and was often used as medical materials. Due to its poor mechanical performance, insufficient strength, and fast degradation, it often induced serious problems for the patients, and lead the poor anatomical reduction effect at the implantation site and recurrence of pelvic organ prolapse. Therefore, there is an urgent need for a new type of biological patch materials with good mechanical properties, good biocompatibility, good chemical stability at the slow degradation rate.

As a typical biofilm of the biogenic patch, small intestinal submucosa (SIS) films were tried to be modified by graphene and the mechanical properties was increased here. With good hexagonal network structure, graphene-like layers based on graphite mineral were introduced into biological thin film materials in order to increase the mechanical properties and protect the film in biomedical state by the carbon coating. It will be a good way to increase the mechanical properties and chemical stability of current biological films. The formation mechanism could be claimed for the composites of the graphene like sheets/biological films on the base of characterization of the microstructure. It might be possible for new flexible biological film, with good effectiveness in clinical medical treatment in the future, and very important in mineralogy and biomaterials science, and beneficial in biomedical fields.

Probe the Interactions between the Graphene Analogues and Bio-molecules

Yu Lei[1]

[1]Tsinghua Shenzhen International Graduate School, Institute of Materials Research & Center of Double Helix

Extensive research efforts have been dedicated to the field of two-dimensional (2D) materials in a short period of time. From semi-metallic graphene, to semiconducting MoS2, to insulating hexagonal boron nitride (hBN), 2D materials provide a wide range of physicochemical properties that can be engineered for diverse applications. Recently, defect engineering has emerged as a primary approach to tailor the physiochemical properties and further extend the functionalities of 2D materials. Comprehensive understanding between the atomic-scale defects and the effects on the physio-chemical properties is challenging and essential for the materials community to design multi-functional 2D materials. It is also equally important to develop scalable approaches to control and tailor the defect types and concentration. With the urgency, I have dedicated in defect engineering in 2D materials: 1) Developed universal approaches to introduce defects in 2D materials; 2) Utilized the atomic-scale defects to activate the basal plane reactivities of 2D materials, and proposed a multi-scale strategy to understand the correlation between the reactivities and the atomic-scale defects; 3) Explored the application of defects in 2D materials for bio-sensing, DNA cleavage, and anti-bacterial coating. Overall, the work aims to provide a comprehensive understanding of the synthesis, characterization, properties, and application of 2D materials' defects in the context of bio-applications.

References

[1]Sci. Adv., 2020, 6, eabc4250
[2]ACS Nano, 2017, 11, 5103-5112
[3]Small, 2021, 17, 2004827 (Cover)
[4]Nano Research Energy, 2023, 2, e9120043
[5]Mater. Today, 2023, 62, 168-189 (Cover)
[6]Adv. Mater, 2023, 35, 2300940
[7]Mater. Today, 2021, 51, 108-116
[8]ACS Nano, 2021, 15, 9796-9807
[9]ACS Nanosci. Au, 2022, 2, 450-485 (Editor`s Choice)
[10]ACS Appl. Energy Mater., 2019, 2, 8625-8632Type:

Planted Graphene Quantum Dots for Long-term Visualization of Nanoparticles In Vivo

Hao Yan [1*], Feiyu Kang [1,2]

[1]Tsinghua Shenzhen International Graduate School, Tsinghua University, Shenzhen 518055, China.
[2]School of Materials Science and Engineering, Tsinghua University, Beijing 100084, China.

While photoluminescent graphene quantum dots (GQDs) have long been considered very suitable for bioimaging owing to their protein-like size, super-high photostability and in vivo long-term biosafety, their unique and crucial applications in fluorescence imaging in vivo remain unreachable. Herein, we present planted GQDs as an excellent tool for in vivo fluorescent and sustainable bioimaging for tumors in various scenarios. The GQDs were in situ seeded in the PEG layer of PEGylated nanoparticles via a bottom-up molecular approach to obtain the NPs-GQDs-PEG nanocomposite. The planted GQDs show more than four times prolonged blood circulation and 7~8 times increased tumor accumulation than typical GQDs in vivo. After accessible targeting modification in outer PEG, the planted GQDs enable long-term, real-time visualization of local pharmacokinetics of NPs in vivo, moreover, provide targeted, multimodal molecular imaging for various tumor models. Planting GQDs in PEGylated nanomedicine offers a new strategy for broad in vivo biomedical applications of GQDs.

References

[1] H. Yan, XD Sun*, FY Kang*, et.al. Adv. Mater., 2023: 2210809.
[2] Q. Wang#, Z. Zhang#, H. Yan*, et.al. Adv. Funct. Mater., 2024, 2400485.
[3] H. Yan, SH Yun*, et.al. Biomaterials, 2023: 122079.

Functional porous polymers and carbons towards efficient tissue repair

Bingna Zheng[1], Wenyi He[2], Weiwen Liang[3], Zifeng Yang[3], Shuyi Wu[4], Rongkang Huang[3] and Dingcai Wu[1,2]

[1] The Eighth Affiliated Hospital, Sun Yat-sen University, Shenzhen, P. R. China
[2] School of Chemistry, Sun Yat-sen University, Guangzhou, P. R. China
[3] The Sixth Affiliated Hospital, Sun Yat-sen University, Guangzhou, P. R. China
[4] Hospital of Stomatology, Sun Yat-sen University, Guangzhou, P. R. China

Porous materials have been widely used in the design of tissue repair materials. Since tissue repair is a complex physiological process, the design of functional porous tissue repair materials is one of the main trends to match the diverse or even contradictory requirements for materials' bioactive functions. However, existing porous tissue repair materials still suffer from relatively simple pore structures and poor integration of multiple functions. Here, inspired by natural hierarchical structures (e.g., peritoneum, leaf), a series of functional porous tissue repair materials (e.g., soft tissue patch, barrier membrane) have been developed with the well-organized pores and multifunctional frameworks. We demonstrate that with the asymmetric porous structures, our tissue repair materials can simultaneously realize anti-adhesion and proliferation functions, matching the different requirements at different locations in physiological envrionment. Moreover, with the integration of multifunctional porous carbons, which can ameliorate the undesirable microenvironment by scavenging reactive oxygen species, repairing bioelectrical environment, and killing bacteria, tissue defects can be efficiently repaired.[1-3] Our ingenious utilization of porous and multifunctional structures will contribute to the development of novel tissue repair materials.

References

[1] Wenyi He, et al. Adv Funct Mater (2024) 2316679.
[2] Weiwen Liang, et al. Adv Mater 34 (2024) 2108992.
[3] Zifeng Yang, et al. Small Sci 3 (2023) 2300138.

Nitrogen-containing carbon-based flexible devices for electrochemical detection of Malathion pesticide

Ashish Jaswal[1], Mamta Devi[1] and Swati Sharma[1]

[1]School of Mechanical and Material Engineering, Indian Institute of Technology Mandi, Kamand, Himachal Pradesh-175005, India

Electrochemical sensing with g-CN or other N-containing carbons(N-carbon) are limited to solution-based systems and face problems associated with long-time storage, resulting in loss of sensitivity. In this work, an attempt is made to fabricate a solid-state device of N-carbon via laser carbonization of graphitic carbon nitride quantum dots (g-CNQDs) [1] decorated on filter paper (FP). FP facilitates uniform decoration of g-CNQDs on its cellulose strands. The heat generated via laser results in coalescence of g-CNQDs to form g-CN particles and eventually to N-containing carbons (N-C) after the release of NH3 and water. Thus, a flexible device (N-Carbon/FP), comprising of N-containing particles and flakes, decorated on FP is fabricated which is utilized as an electrochemical sensor for malathion.

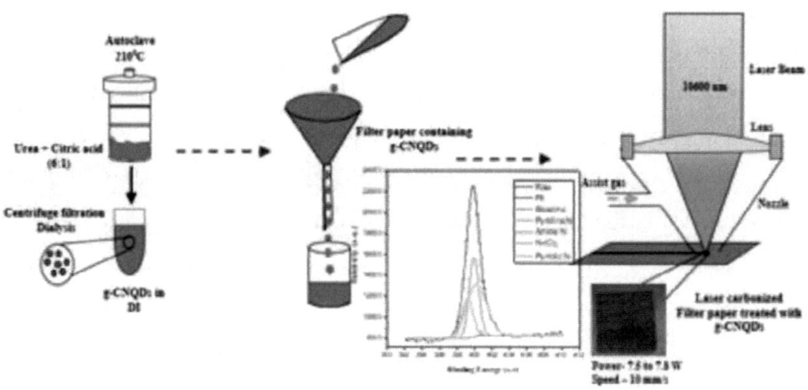

References

[1]Devi, M., Das, P., Boruah, P. K., Deka, M. J., Duarah, R., Gogoi, A., Neog, D., Dutta, H. S., and Das, M. R., 2021, "Fluorescent Graphitic Carbon Nitride and Graphene Oxide Quantum Dots as Efficient Nanozymes: Colorimetric Detection of Fluoride Ion in Water by Graphitic Carbon Nitride Quantum Dots," Journal of Environmental Chemical Engineering, 9(1), p. 104803.

Enhancing Performance of Carbon Quantum Dot-Based Memristors through Boron Doping for Artificial Synapse Application

Haotian Hao[1], Lingpeng Yan [1] and Yongzhen Yang [1*]

[1]Taiyuan University of Technology, No. 79, West Yingze Street, Taiyuan, China

Carbon quantum dots-based memristors (CQDMs) have emerged as a rising star in data storage and computing. However, a significant obstacle to their commercialization lies in memristance variability, primarily attributed to disordered conductive paths. Doping methodology can optimize electron and ion transport to help construct a stable conductive path for memristors. Herein, three kinds of comparable quantum dots were synthesized, including carbon quantum dots (CQDs), boron-doped CQDs (BCQDs), and boron quantum dots. The BCQDs-based memristor exhibits a ternary flash-type memory behavior with longer retention time and more controllable memristance stability. The doped B elements were proved to be able to directionally guide the migration of aluminum ions, resulting in ordered conductive paths and stable ternary memory behavior. Furthermore, the novel artificial synaptic device was constructed using a composite functional layer of triazine-based conjugated porous polymer and BCQDs. This device presents an analog-type resistive switching characteristic, enabling its reliable synaptic plasticity and good associative memory learning capabilities. This work establishes a theoretical foundation and template scheme for developing effective and stable CQDMs on artificial intelligence devices.

Corresponding author: Yongzhen Yang, yangyongzhen@tyut.edu.cn

Carbon Networks for Soft Electronics

Qingbin Zheng[1]

[1]School of Science and Engineering, The Chinese University of Hong Kong, Shenzhen, 518172, China
Email: zhengqingbin@cuhk.edu.cn

Wearable piezoresistive physical sensors have attracted tremendous attention due to their broad applications in human-machine interfaces, robotics, health-monitoring, personalized medicine and therapeutics[1-3]. Conventional metal- and semiconductor-based sensors are rigid, fragile, and opaque, restricting their applications in wearable electronics [4]. Among various nanostructured materials that have been explored to construct stretchable piezoresistive sensors, carbon-based networks show the potential of achieving improved sensing performance [5]. We present several novel carbon-based interconnected networks including expanded graphite, carbon nanofiber, graphene woven fabric, graphite nanoplatelet welded carbon nanotube networks, wrinkled and crack-bridging graphene film, unidirectional graphene aerogel, and graphene foam as highly sensitive, flexible, and stretchable physical sensors capable of detecting human motions [6-7]. By bridging the technological gap between signal collection, processing and transmission and allowing data transmission to a smartphone via a custom-developed application consisting of a user-friendly interface, several wearable applications including wireless wearable musical instrument prototype, human motion detection and switch controls are demonstrated [8]. The novel approaches reported here offer a wide range of practical applications, including medical diagnosis, health monitoring, early warning systems for structural failure, and wearable displays [9].

References

[1] Zheng Q. B., Lee J. H., Shen X., Chen X. D., Kim J. K.* Graphene-Based Wearable Piezoresistive Physical Sensors. Materials Today, 36, 158-179 (2020).
[2] Xue J., Liu D., Li D., Hong T. Z., Li C. B., Zhu Z. F., Sun Y. X., Gao X. B., Guo L., Shen X., Ma P. C., Zheng Q. B.*: New Carbon Materials for Multifunctional Soft Electronics. Advanced Materials, 2312596 (2024).
[3] Wu Y., An C., Guo Y. R., Zong Y. Y., Jiang N. S., Zheng Q. B.*, Yu Z. Z*. Highly Aligned Graphene Aerogels for Multifunctional Composites, Nano-Micro Letters, 16, 118 (2024).
[4] Zhang F., Sun Y. X., Guo L., Zhang Y. H., Liu D., Feng W.*, Shen X.*, Zheng Q. B.* Microstructural Welding Engineering of Carbon Nanotube/Polydimethylsiloxane Nanocomposites with Improved Interfacial Thermal Transport. Advanced Functional Materials, 2311906 (2023).
[5] Xie J. W., Zhou G., Sun Y. X., Zhang F., Kang F. Y., Li B. H., Zhao Y., Zhang Y. H. *, Lin L.*, Zheng Q. B.* Multifunctional Liquid Metal-Bridged Graphite Nanoplatelets/Aramid Nanofiber Film for Thermal Management. Small, 2305163 (2023).

[6] Xie J. W., Zhang Y. H., Dai J. M., Xie Z. X., Xue J., Dai K., Zhang F., Cheng J. Y., Kang F. Y., Li B. H., Zhao Y., Lin L.*, Zheng Q. B.* Multifunctional MoSe2@MXene Heterostructure-Decorated Cellulose Fabric for Wearable Thermal Therapy, Small, 19, 2205853 (2023).
[7] Cheng J. Y., Zhang H. B., Wang H. H., Huang Z. H., Raza H., Hou C. X., Zheng G. P., Zhang D. Q. *, Zheng Q. B.*, Che R. C. * Tailoring Self - Polarization of Bimetallic Organic Frameworks with Multiple Polar Units Toward High - Performance Consecutive Multi - Band Electromagnetic Wave Absorption at Gigahertz. Advanced Functional Materials, 32, 2201129 (2022).
[8] Cheng J. Y., Zhang H. B., Ning M. Q., Raza H., Hou C. X., Zhang D. Q.*, Zheng G. P., Zheng Q. B.*, Che R. C. * Emerging materials and designs for low- and multi-band electromagnetic wave absorbers: the search for dielectric and magnetic synergy? Advanced Functional Materials, 32, 2200123 (2022).
[9] Zhang F., Ren D. H., Huang L. Q., Zhang Y. H., Sun Y. X., Liu D., Zhang Q., Feng W.*, Zheng Q. B.* Three-Dimensional Interconnected Conductive Graphite Nanoplatelet Welded Carbon Nanotube Networks for Stretchable Conductors, Advanced Functional Materials, 31, 2107082 (2021).

Highly Stable Carbon Dots Enabled by Structural Regulation for White Laser Lighting Diode with High Color Rendering Index

Xinghua Liu[1], Jingxia Zheng[1], Lingpeng Yan[1], Yongzhen Yang[1*] and Xuguang Liu [1]

[1]Taiyuan University of Technology, No. 79, West Yingze Street, Taiyuan, China.

Carbon dots (CDs) phosphors in white laser diodes (WLD) have the advantages of low cost, low toxicity, and easy functionalization. However, the color rendering index (CRI) of single CDs phosphors based WLD is unsatisfactory (<80) owing to the uncontrollable synthesis process and low thermostability of CDs. Herein, CDs with adjustable optical and thermal properties were realized by modulating the graphitization degree and the surface structure. In order to obtain CDs with excellent optical stability and thermostability, various strategies including post-synthesis thermal annealing, heteroatom doping, and molecular vibration-induced heating were employed to regulate the structure of CDs, and resulting in the modulation of optical and thermal properties. Further, the influence of internal structures and surface composition on the optical and thermal properties of CDs were investigated. Finally, WLD devices were realized by assembling CDs based fluorescent film with 450 nm laser diode, and the highest CRI value is 87, reaching the advancing value for single phosphors-based WLD devices. This work will lay the theoretical and experimental foundation for the realization of the low-cost, high photothermal stability, high-color-rendering-index CDs-based laser lighting devices.

Corresponding author: Yongzhen Yang, yangyongzhen@tyut.edu.cn

Novel Hot-Carrier Transistors based on Low-dimensional Materials

Chi Liu[1,2], Dongming Sun[1,2] and Huiming Cheng[1,2,3]

[1]Shenyang National Laboratory for Materials Science, Institute of Metal Research, Chinese Academy of Sciences, 72 Wenhua Road, Shenyang, 110016, China.

[2]School of Materials Science and Engineering, University of Science and Technology of China, 72 Wenhua Road, Shenyang, 110016, China.

[3]Institute of Technology for Carbon Neutrality, Shenzhen Institute of Advanced Technology, Chinese Academy of Sciences, 1068 Xueyuan Avenue, Shenzhen, 518055, China.

Hot-carrier transistors use excess kinetic energy of carriers to enhance device speed and functionality, which are crucial for applications requiring rapid switching and high-frequency operations. However, traditional mechanisms for generating hot carriers limit device performance in terms of speed, power consumption, negative differential resistance (NDR) and photo gain. Devices based on low-dimensional materials can offer different mechanisms by utilizing rich energy band combinations. This report introduces novel hot-carrier transistors based on graphene and TMDCs: a) Using a Si membrane-graphene Schottky junction to emit hot carriers, the silicon-graphene-germanium transistor reduces delay time to less than one-thousandth of that reported for graphene-based transistors, providing a scientific alternative for developing THz transistors [1]. b) Using stimulated emission of heated carrier, the hot-emitter transistor achieves a subthreshold swing of less than 1 mV/dec and a peak-to-valley current ratio exceeding 100, providing a new method for achieving low-power and high-performance NDR devices [2]. c) Using the photo induced barrier lowering mechanism to inject hot carriers, the double heterojunction transistor achieves the highest detectivity among two-dimensional photodetectors, offering new insights for fabricating high-performance photodetectors [3]. These hot-carrier transistors provide innovative research avenues for constructing high-performance and multifunctional devices in the post-Moore era.

References

[1] Liu, C. et al. Nat. Commun. 10 (2019) 4873.
[2] Liu, C. et al. Nature, accepted in principle.
[3] Feng, S.#, Liu, C.# et al. Nat. Commun. 12 (2021) 4094.

Carbon Nanotube based Steep Slope Transistors and Tensor Processing Unit

Jia Si[1], Baizhe He[1] and Zhiyong Zhang[1]

[1]Key Laboratory for the Physics and Chemistry of Nanodevices and Center for Carbon-based Electronics, School of Electronics, Peking University, Beijing 100871, China.

The growth of data-intensive computing tasks requires processing units with higher performance and energy efficiency, but this is increasingly difficult to achieve with conventional semiconductor technology. One solution is introducing steep slope transistors with novel operation principles. We show that one-dimensional van Hove singularity of carbon nanotube can be leveraged as an efficient cold source of FET. Nevertheless, it provides easy process of scaling FET by integrating homogenous cold sources and high-performance one-dimensional channels. We experimentally construct this van Hove source FETs based on carbon nanotubes, achieving an SS of 50 mV/decade over two decades at room temperature. It shows comparable on-state current to 14-nanometer silicon FETs while operating at a reduced supply voltage of 0.5 volts.

Another potential solution is to combine advancements in devices with innovations in system architecture. Here, we report a tensor processing unit (TPU) that is based on 3,000 carbon nanotube field-effect transistors and can perform energy-efficient convolution operations and matrix multiplication. The TPU is constructed with a systolic array architecture, enabling parallel 2-bit integer multiply-accumulate operations. A five-layer convolutional neural network based on the TPU can perform MNIST image recognition with an accuracy of up to 88% and a power consumption of 295 μW.

Enhanced Efficiency and Durability of Blue Phosphorescent OLEDs Employing Iridium Carbene Pincer Emitters

Chengcheng Wu

Tsinghua-Berkeley Shenzhen Institute (TBSI), Tsinghua Shenzhen International Graduate School (SIGS), Tsinghua University, Shenzhen 518055, China

Blue phosphorescent OLEDs (Ph-OLEDs) have faced remarkable challenges due to their high-energy excitons, insufficient stability of materials, and large efficiency roll-off at high brightness. To address these issues, novel Ir(III) emitters featuring a carbazolyl (tBuCz) substituted tridentate pincer ligand were designed to improve steric encumbrance and minimize exciton accumulation. These complexes, tBuCz-m-CF3 and tBuCz-p-CF3, demonstrate notable improvements in both the horizontal transition dipole orientation and electrochemical stability. Emitting at approx. 485 nm, they achieve photoluminescence quantum yields (PLQY) of 98%, with observed decay rate constant up to 7.97 × 105 s-1 in doped thin films. Ph-OLEDs based on these emitters achieve maximum external quantum efficiencies (EQE) of over 31.62% and brightness up to 214,255 cd m−2. Notably, the EQE values remained at 20.58% and 20.87% for tBuCz-m-CF3 and tBuCz-p-CF3, respectively, even at a high brightness of 100,000 cd m−2. Furthermore, these devices exhibit exceptional stability, with the device stability (LT50) reaching 1237 and 1162 hours at 1000 cd m−2, setting new benchmarks for efficiency and durability of sky-blue Ir(III) based Ph-OLEDs.

Exceedingly Stable Luminescent Electron Transporting Material with doped Hole Blocking Layer for High-Performance Deep-Red OLEDs with LT95 over 7500 h at 1000 cd m−2

Kai-Ning Tong

Tsinghua Shenzhen International Graduate School (SIGS), Tsinghua University, Shenzhen 518055, China

Among variety colors of OLEDs, deep-red and near-infrared OLEDs have promising applications in optical communication, information storage, biological therapy, night-vision-readable displays etc. However, as the band gap decreases, the probability of non-radiative recombination increases, which could cause low device lifetime and high efficiency roll-off. Meanwhile, the electron mobility is lower than the hole mobility in the OLED devices, which seriously decreases the external quantum efficiency (EQE) of OLED. In this work, a new electron transporting layer (ETL) material ET01 based on ANT-BIZ has been synthesized. The electron mobility of ET01 is closed to the hole mobility in the OLED device. Due to the high HOMO energy level of ET01, the OLED devices suffer the serious efficiency roll-off when the luminance over 10000 cd m−2. To solve the device efficiency roll-off at high luminance, the doped hole blocking layer (HBL) has been applied. In the HBL, the ET01 is doped with another ETL material with the different concentration, which could effectively restrict the exciton recombination zone and keep the high electron mobility at the same time. The deep¬-red OLEDs based on doped HBL and ET01 shows the maximum EQE of 33.3% and low efficiency roll-off (32.2% and 25.8% at the luminance of 1000 and 10000 cd m−2, respectively) with the CIE coordinates of (0.685, 0.315). Furthermore, these devices exhibit exceedingly stability, with the device stability (LT95) over 7500 hours at 1000 cd m−2.

Highly Efficient and Selective Extraction of Gold by Reduced Graphene Oxide

Fei Li[1,#], Jiuyi Zhu[1,#], Xiaolong Zou[1], A. K. Geim[1,2*], Yang Su[1,*], Hui-Ming Cheng[3,4,*]

[1]Tsinghua-Berkeley Shenzhen Institute & Institute of Materials Research, Tsinghua Shenzhen International Graduate School, Tsinghua University, Shenzhen 518055, P. R. China
[2]School of Physics & Astronomy, University of Manchester, Manchester M13 9PL, United Kingdom
[3]Faculty of Materials Science and Engineering / Institute of Technology for Carbon Neutrality, Shenzhen Institute of Advanced Technology, Chinese Academy of Sciences, Shenzhen 518055, P. R. China
[4] Shenyang National Laboratory for Materials Sciences, Institute of Metal Research, Chinese Academy of Sciences, Shenyang 110016, P. R. China

Materials that are capable of extracting gold from complex sources, especially electronic waste (e-waste) with high efficiency are needed for gold resource sustainability and effective e-waste recycling. However, it remains challenging to achieve high extraction capacity to trace amount of gold, and precise selectivity to gold over a wide range of complex co-existing elements. Here we report a reduced graphene oxide (rGO) material that has an ultrahigh extraction capacity for trace amounts of gold (1,850 mg/g and 1,180 mg/g to 10 ppm and 1 ppm gold). The excellent gold extraction behavior is accounted to the graphene areas and oxidized regions of rGO. The graphene areas spontaneously reduce gold ions to metallic gold, and the oxidized regions provide a good dispersibility so that efficient adsorption and reduction of gold ions by the graphene area can be realized. The rGO is also highly selective to gold ions. By controlling the protonation process of the functional groups on the oxidized regions of rGO, it shows an exclusive gold extraction without adsorption of 14 co-existing elements seen in e-waste. These discoveries are further exploited in highly efficient, continuous gold recycling from e-waste with good scalability and economic viability, as exemplified by extracting gold from e-waste using a rGO membrane based flow-through process.

References

[1] Fei Li#, Jiuyi Zhu#, Pengzhan Sun, A. K. Geim*, Yang Su*, Hui-Ming Cheng*, et al. Highly efficient and selective extraction of gold by reduced graphene oxide. Nature Communications 13, 4472 (2022).

Structural regulation of polymer electrolyte for high performance solid-state Li battery

Ruogu Xu[1,2], Shengjun Xu[1,2], Shuo Bai[1,2], Zhenhua Sun*[1,2], Feng Li*[1,2]

[1] Shenyang National Laboratory for Materials Science, Institute of Metal Research, Chinese Academy of Sciences, Shenyang 110016, China.

[2] School of Materials Science and Engineering, University of Science and Technology of China, Hefei 230026, China

Solid-state polymer electrolyte (SPE) is among the most promising candidates for solid-state electrolytes in lithium metal batteries owing to its advantage of non-leakage, light-weight, and low cost. However, due to the low ionic conductivity, poor electrode/electolyte interfacial stability, solid-state lithium batteries usually need to work at high temperature, and face challenges such as low capacity and poor cycle stability, which seriously restricts its widely application. Herein, 1) we proposed to construct homogeneous and fast ion transport channels from the microscopic scale of ion transport, which is conducive to achieve high performance SPE at room temperature; 2) a general solidified localized high-concentration electrolyte (S-LHCE) with the decoupling of ion pairing and ion conduction was proposed, which improve the interfacial compatibility and stable lithium plating/stripping process and wider electrochemical stability window. Those works provide new strategy to fabricate high-performance solid-state lithium batteries with high electrochemical performance.

References

[1] Ruogu Xu, Zhenhua Sun, Feng Li et al., Small Structures, 2023, 4, 2200206.

[2] Shengjun Xu1, Ruogu Xu1, Zhenhua Sun, Feng Li et al., Energy & Environmental Science, 2022, 15, 3379.

Design of functionalized carbon-based materials and electrocatalytic conversion of CO_2

Wenbin Li[1], Chang Yu[1] and Jieshan Qiu[2]

[1]Dalian University of Technology, Dalian, China
[2]Beijing University of Chemical Technology, Beijing, China

Carbon materials have attracted much attention due to their high electronic conductivity, excellent stability and cost advantages, especially in the emerging electrocatalytic CO2 conversion field. In addition to improving intrinsic catalytic activity directly as electrocatalysts, functionalized carbon materials can be used as an important component of catalysts to construct interconnected molecular diffusion channels and conductive networks, which can promote the diffusion of reactants and rapid electron conduction within the catalysts, and enhance the structural stability of the catalysts under severe electrocatalytic conditions. CO and formic acid are the most common products and have the highest commercialization potential in the CO2 catalytic conversion. Herein, we design a cross-linked carbon coupled with active bismuth catalyst with a stable spatial structure and unique molecular sieving effects, achieving the formate production at an industrial large current. [1] In addition, we utilize highly conductive carbon nanotubes to activate the formation of a built-in electric field, which activates intrinsically electrochemical inert phthalocyanine molecule for efficient CO2 electroreduction to CO. [2] Moreover, based on the highly active and stable carbon-based materials, we have made attempts and efforts in the optimization and scaling-up of electrochemical devices to enhance device and system stability. [1, 3]

References

[1] Wenbin Li, et al. ACS Catalysis, 14 (2024) 8056-8061.
[2] Wenbin Li, et al. Chemistry of Materials, 36 (2024) 1602-1611.
[3] Jieshan Qiu, et al. Energy & Environmental Science, 14 (2021), 765.

High-purity rayon-based carbon fiber felt prepared by halogen gas purification for superior thermal insulation and oxidation resistance

Biao Li[1,2], Xiao Wu[1,2], Ke Shen[1,2]*, Jinshui Liu[1,2]*

[1]College of Materials Science and Engineering, Hunan University, Changsha, Hunan, China
[2]Hunan Province Key Laboratory for Advanced Carbon Materials and Applied Technology, Hunan University, Changsha, Hunan, China

High-purity rayon-based carbon fiber (CFRay) felt holds significant promise for applications in thermal protection systems and semiconductor crystal growth furnaces. This study employs halogen gas purification and high-temperature purification techniques in the fabrication of high-purity CFRay felt. Additionally, the influence of the purification process on the thermal insulation and oxidation resistance of CFRay felt is investigated. The glow discharge mass spectrometer analysis reveals that purification at a lower temperature of 2300 ℃ achieves a purity level of 6.9 ppm, surpassing both purified CFRay felt without halogen at 2300 ℃ (56.1 ppm) and high-temperature treated CFRay felt at 3000 ℃ (10.4 ppm). Thermal insulation and dynamic-static oxidation resistance tests show that the properties of CFRay felt are concurrently affected by the heat treatment temperature. Compared to high-temperature purification, halogen gas purification operates at lower temperatures, preventing the graphitization of CFRay, thereby ensuring high purity while enhancing thermal insulation and demonstrating relatively superior oxidation resistance. This study offers novel insights into the preparation and performance adjustment of high-purity CFRay felt.

"Disruption" vs. "repairing" of carbon materials by ball-milling

Renlu Yuan, Yue Dong, Ruoyang Hou, Lei Shang, Xiaoyu Xu, Ang Li, Xiaohong Chen, Su Zhang, Huaihe Song*

State Key Laboratory of Chemical Resources Engineering, Beijing Key Laboratory of Electrochemical Process and Technology for Materials, Beijing University of Chemical Technology, Beijing 100029, P. R. China

Mechanical ball-milling is a commonly used method to shattering crystals or introducing defects in carbon materials. In our works, we revealed that the ball-milling of highly crystalline expanded graphene is a "destructive" defect engineering to prepare defect-enriched graphene block (DGB) with controllable defect density and high packing density. The introduced "self-doping" defects could act as active sites for aqueous supercapacitors, and sodium-ion capacitors and Zn-ion capacitors, possessing robust capacitive behavior and greatly improved capacitance. What' s more, the efficient utilization of the active sites in DGB can be realized by rational functionalization or pore structure design. While in our other works, we further observed that during ball-milling, the disordered and porous structure of activated carbon (AC) is gradually "repaired" to order carbon layers along with the introduction of oxygen in the bulk phase. The well-developed pores are gradually eliminated, leading to obviously decreased surface area from 1364 to 156 m2 g-1 and improved compaction density from 0.573 to 1.423 g cm-3. Due to the continuous particle cleavage, re-assembly, and cold-welding, the as-formed carbon blocks show 2 times larger in size and 4 times higher bulk oxygen content. This simultaneous crystallinity enhancement and oxygen functionalization is barely observed in other methods, leading to significantly improved volumetric capacitances and high-rate performance for aqueous supercapacitors. Furthermore, the ball milling-induced order rearrangement of carbon layers promotes the development of the large graphenic crystallites toward graphite structure during the subsequent 2800 ℃ treatment. A similar structure repair effect also be observed for various hard carbons derived from polymers and biomass. Therefore, rational ball-milling may be a universal pre-treatment strategy for "destroying" high crystalline carbons while "repairing" the disordered structure of hard carbons.

Micro-silicon/carbon anodes for the next-generation lithium-ion batteries

Ziyun Zhao[1] , Shichao Wu[2] and Quan-Hong Yang [1,2]

[1]Nanoyang Group, State Key Laboratory of Chemical Engineering,
School of Chemical Engineering and Technology, Tianjin University, Tianjin 300350, China
[2]Joint School of National University of Singapore and Tianjin University,
International Campus of Tianjin University, Binhai New City, Fuzhou, 350207, China

Silicon micro-particle (SiMP) anode with low-cost, low specific surface area and high capacity is an alternative for next generation of lithium-ion batteries. The lithiation of crystalline Si (c-Si) occurs via a two-phase alloying mechanism. And the lithiation of c-Si is highly anisotropic, inducing significant stress concentration during drastic volume expansion (~300%) process. Consequently, a marked change of properties including particle cracking and pulverization, SEI formation and "dead Si" formation. That is structure fade and electrical isolation are the sources of above a series of issues, hindering the further development of Si anodes1.

We focus on both mechanical control and adaptive electrical connection strategies for stable cycling of low-cost SiMP/carbon anodes. First of all, a strategy of covalent coating is developed to establish a robust encapsulation structure2. The obtained covalent Si–C bonds enable an effectively dynamic connection between the electrochemically deforming Si and the sliding graphene layers, preventing the evolution of gaps between Si and the carbon shell and maintaining persistent electrical connections as well as mechanical toughness. If the Si particle is inevitably broken up and pulverized during cycling, liquid metal (LM) can act as an adaptive conducting continuum to cure pulverized Si fragments upon cycling3. Therefore, the high-areal anodes give highly stable and superior volumetric lithium storage.

References

[1] Ziyun Zhao#, Quan-Hong Yang*, et al. Advanced Energy Materials 13 (2023) 2300367.
[2] Zhenshen Li#, Ziyun Zhao#, Quan-Hong Yang*, et al. Advanced Energy Materials 13 (2023) 2300874.
[3] Ziyun Zhao#, Quan-Hong Yang*, et al. Advanced Energy Materials 12 (2022) 2103565.

One-Step Construction of Closed Pores Enabling High Plateau-Capacity Hard Carbon Anodes for Sodium-Ion Battery: Closed-Pore Formation and Energy Storage Mechanisms

Chuang Qiu, Huaihe Song*

State Key Laboratory of Chemical Resources Engineering, Beijing Key Laboratory of Electrochemical Process and Technology for Materials,
Beijing University of Chemical Technology, Beijing 100029, P. R. China.

Closed pores play a crucial role in improving the low-voltage (< 0.1 V) plateau capacity of hard carbon anodes for sodium-ion batteries (SIBs). However, the lack of simple and effective closed-pore construction strategies, as well as the unclear closed-pore formation mechanism, have severely hindered the development of high plateau-capacity hard carbon anodes. Herein, we present an effective closed-pore construction strategies by one-step pyrolysis of zinc gluconate and elucidate the corresponding mechanism of closed-pore formation. The closed-pore formation mechanism during the pyrolysis of ZG mainly involves i) the precipitation of ZnO nanoparticles and the ZnO etching on carbon under 1100 °C to generate open pores of 0.45-4 nm, and ii) the development of graphitic domains and the shrinkage of the partial open pores at 1100-1500 °C to convert the open pores to closed pores. Benefiting from the considerable closed-pore content and suitable microstructure, the optimized hard carbon achieves an ultra-high reversible specific capacity of 481.5 mAh g-1 and an extraordinary plateau capacity of 389 mAh g-1 for use as the anode of SIBs. Additionally, some in-situ and ex-situ characterizations demonstrate that the high-voltage slope capacity and the low-voltage plateau capacity stem from the adsorption of Na+ at the defect sites and Na-cluster formation in closed pores, respectively.

Rationally designing closed-pores hard carbon structures to boost the low-voltage plateau capacity of sodium-ion batteries

Xiaoqi Yu, Xiaoyu Xu, Zipeng Jiang, Renlu Yuan, Ang Li, Xiaohong Chen, Huaihe Song*

State Key Laboratory of Chemical Resource Engineering,
Beijing Key Laboratory of Electrochemical Process and Technology for Materials,
Beijing University of Chemical Technology, Beijing 100029, PR China

The theoretical capacity of hard carbon anodes comprises a high-pressure slope region and a low-pressure plateau region. The high low-voltage plateau capacity makes hard carbon one of the strongest candidate materials for sodium-ion batteries. However, the mechanism of sodium storage in hard carbon remains controversial. Establishing a link between microstructure and electrochemical properties provides strong evidence for elucidating how hard carbon stores sodium. Therefore, in this study, phenolic resin-derived hard carbon (PRHC) anodes with a rich closed-pores structure were synthesized, and it was demonstrated that the presence of a closed-pores structure significantly increased the sodium storage capacity. This hard carbon anodes exhibited a reversible capacity of 352.16mAh/g at 20mA/g, with the platform capacity accounting for 78.58% and being 276.73mAh/g. As predicted, as the number of closed-pores decreases, the platform capacity gradually diminishes. The platform capacity gradually decreased to 106.85mAh/g and accounting for 58.35% of the total capacity. It is evident that the closed-pores structure of hard carbon promotes platform capacity and provides a reasonable framework for understanding the mechanism of sodium storage in hard carbon anodes.

Keywords: Phenolic resin-derived hard carbon, Closed-pores structure, Low-voltage plateau capacity, Sodium storage mechanism

Deciphering the Triple Relationship: Hard carbon (HC) precursor-Closed Pore-Plateau Capacity for Sodium-ion Batteries

Xiaotian Li[1], Jiapeng Zhang[1], Renlu Yua[1], Boyang Zhao[1], Chuang Qiu[1], Dengke Wang[1], Man Yuan[1], Ang Li[1], Hiayan Liu[2], Xiaohong Chen[1*], Huaihe Song[1*]

[1]State Key Laboratory of Chemical Resource Engineering, Beijing Key Laboratory of Electrochemical Process and Technology, Beijing University of Chemical Technology, Beijing 100029, China
[2]National Engineering Research Center of Coal Gasification and Coal-Based Advanced Materials, Shandong Energy Group CO, LTD, Jinan, PR China)

Hard carbon (HC) anodes are pivotal for the future advancement of sodium-ion batteries. However, significant challenges persist in achieving HC with high plateau capacity (< 0.1 V) and clearly defining the closed pore structure corresponding to its plateau capacity. Here, we manipulate the closed pore structure of HC at molecular dimension by altering the HC precursor molecular structure of phenolic resin. Our findings indicate that an increase in the cross-linking degree of the HC precursor modulates the closed pore structure of HC by affecting the size and curvature of the graphitic domains. Furthermore, we establish a reliable conformational relationship among HC precursor molecular structure, HC structure, and performance. Since different sizes of closed pores have distinct sodium storage capacities, a rise in the total volume or average pore size of the closed pores does not lead to a plateau capacity enhancement. What' s more, this work also reveals the sodium storage behavior of closed pores with different sizes, with the optimal radius of the closed pore for sodium storage is approximately 1.2 nm. As a proof, the capacity of the optimized HC at 50 mA g-1 was improved from 194.8 to 351.1 mA h g-1, and the plateau capacity was improved from 123.0 to 256.3 mA h g-1.

Carbon-tuned thermal field management under microwave irradiation: Towards highly active CO2 reduction electrocatalyst

Yafang Zhang[1], Chang Yu [1] and Jieshan Qiu [2]

[1]Dalian University of Technology, Dalian, China
[2]Beijing University of Chemical Technology, Beijing, China

As a common microwave absorption medium, carbon nanomaterials with rich conjugate π-electrons exhibit rapid heat response and localized heat microenvironment under microwave radiation, offering a platform to design functional materials for energy and catalysis. Herein, we provide two paradigms to govern thermal field over nanocarbon via microwave irradiation, towards the construction of highly active CO2 reduction electrocatalyst. On the one hand, a local high temperature field enabled by the designable defects over carbon nanotube (CNT) is stimulated under microwave irradiation. This defect-mediated integration strategy is presented to promote the in-plane configuration instead of out-plane/on-plane spatial geometrical microstructure as well as strong complex-support interaction. On the other hand, with the aim of achieving ultra-high nitrogen (N) doping density yet ordered N distribution configuration in metal-free carbon materials. Microwave could induce a sub-minute residence time for precursors and relatively mild surface temperature on CNTs, resulting to a gradient N-doping along the radial direction of CNTs with ultra-high surface N content of 30.64 at%. Finally, the ideal microstructure endows these catalysts with an impressive activity for electrocatalytic CO2-to-CO conversion. Our work confirms the unique advantage of carbon-tuned thermal field management under microwave irradiation, which deserve ever-expanding applications in the future.

Carbon Surface: A Key to Improved Anode Coulombic Efficiency in Sodium-Ion Batteries

Qiaowei Lin[1,2], Wei Lv [3], Quan-Hong Yang [4] and Da-Wei Wang [1,2]

[1] Institute of Technology for Carbon Neutrality, Shenzhen Institute of Advanced Technology, Chinese Academy of Sciences, Shenzhen 518071, China

[2] Faculty of Materials Science and Energy Engineering, Shenzhen University of Advanced Technology, Shenzhen 518071, China

[3] Shenzhen Geim Graphene Center, Tsinghua Shenzhen International Graduate School, Tsinghua University, Shenzhen, 518055, China

[4] Nanoyang Group, State Key Laboratory of Chemical Engineering, School of Chemical Engineering and Technology, Tianjin University, Tianjin, 300072, China.

Non-graphitic carbons are the promising anode materials for sodium-ion batteries (SIBs) [1]. Non-graphitic carbons are rich in defective surface, which can serve as active sites for charge storage. However, they will inevitably cause irreversible side reactions between solvents molecules and defects, leading to a low initial coulombic efficiency (ICE) and poor cyclic stability. Here, we systematically investigated the surface modification method to shield the defective surface for less electrolyte decomposition, stable electrochemical interphases, and better reversibility. We reported the deactivating method of surface defect using inert Al2O3 nanoclusters, of which the precursors were selectively adsorbed by the strong adsorption potential of defects [2]. The chemical modification of carbon surface was studied by using the presodiation agent of sodium naphthalene, promoting the formation of artificial solid electrolyte interphase (SEI) on its surface [3]. Moreover, we found that surface functionalization of carbons can also lead to regulated SEI with high stability. In addition, tunning the pore structure of non-graphitic carbons can bury the defective surface into pores with ultra-small size gating, which separates the solvent from defects through desolvation [4]. This series work should offer valuable insights for various applications beyond energy storage and conversion devices, where carbon surface plays vital roles.

References

[1] Qiaowei Lin, et al. Small, 1902603 (2019)
[2] Qiaowei Lin, et al. Advanced Energy Materials, 9, 1803078 (2019)
[3] Ganyu Zheng and Qiaowei Lin, et al. Infomat, 3(12), 1445-1454 (2021)
[4] Qi Li, et al. National Science Review, 9 (8), nwac084 (2022)

Electrocatalytic Electrodes for High-Current-Density Electrochemical Synthesis and Devices

Yuting Luo [1]

Email: yuting.luo@utoronto.ca

[1]University of Toronto, 35 St. George St., Toronto, Ontario M5S 1A4, Canada.

In the context of carbon neutrality, developing electrochemical technologies driven by renewable electricity is becoming ever more important. One of the challenges is developing electrocatalytic electrodes that can be used under high current densities and show high selectivity to specific product. Due to high conductivity and tunable mass-transfer capability of carbons (or carbides), they are used in the catalyst or gas-diffusion layers of electrodes. In this talk, I will introduce our progress in use of carbon/carbides for high-current-density carbon dioxide upgrading[1,2] and water splitting[3-5].

References

[1] Luo, Y. et al., Steering electroreduction of carbon dioxide to valuable C3+ products. Science China Materials, 2024, https://doi.org/10.1007/s40843-024-2862-9.

[2]Luo, Y. et al., Selective electrochemical synthesis of urea from nitrate and CO2 via relay catalysis on hybrid catalysts. Nature Catalysis, 2023, 6, 939–948.

[3]Luo, Y. et al., Recent advances in design of electrocatalysts for high-current-density water Splitting. Advanced Materials, 2022, 34, 2108133.

[4]Zhang, C.+; Luo, Y.+; et al., High-throughput production of cheap mineral-based two-dimensional electrocatalysts for high-current-density hydrogen evolution. Nature Communications, 2020, 11, 3724.

[5]Luo, Y. et al., Morphology and surface chemistry engineering toward pH-universal catalysts for hydrogen evolution at high current density. Nature Communications, 2019, 10, 269.

Revisiting self-discharge of nanoporous electrodes: insights from the multilayered graphene membrane-based electrodes

Xiaoyang Du[1], Qinghua Liang[2,3], Wen-jie Jiang[1], Lianhai Zu[1], Desheng Feng[4], Mengran Li[1], Yang Cao[5], and Dan Li[1]

[1]Department of Chemical Engineering, The University of Melbourne, Parkville, Victoria, Australia
[2]Key Laboratory of Rare Earth, Chinese Academy of Sciences, Ganzhou, Jiangxi, P. R. China
[3]Ganzhou Innovation Academy, Chinese Academy of Sciences, Ganzhou, Jiangxi, P. R. China
[4]School of Chemical Engineering, The University of Queensland, Brisbane, Queensland, Australia
[5]Australian Nuclear Science and Technology Organization, Clayton, Australia

Self-discharge (SD) is not only a persistent challenge for the community of electric double-layer capacitors but also has plagued the development of other electrochemical energy systems due to the dissipation of both electronic and ionic charges at the electrode/electrolyte interface. However, the intricate interplay of diverse interfacial electrochemical processes complicates the elucidation of the SD mechanism of electric double-layer capacitors. The complex nanostructure and surface chemistry of commonly used carbon-based nanoporous materials further complicate the analysis of the mechanism. In this work, using a series of additive-free and highly conductive multilayered graphene-based membranes with tunable slit sizes and self-consistent surface chemistry as a platform, we revisit the SD of nanoporous electrodes. By leveraging a commonly used hybrid SD model, our fitting results suggest a predominant role for the kinetic process of activation-controlled Faradaic reactions during the SD of multilayered graphene-based membranes under various testing conditions, including varied slit sizes and aqueous electrolytes. However, subsequent experiments on charging voltage-dependent SD and chemical characterization exclude the commonly deemed activation-controlled reactions, i.e., carbon oxidation and water electrolysis, implying the presence of previously unidentified activation-controlled processes. Our findings also provide new insights into the SD at the electrode/electrolyte interface and underscore the need to revisit the existing hybrid SD model to accommodate these novel observations.

Graphene and MXene for High performance Micro-supercapacitors

Pratteek Das [1], Zhong-Shuai Wu [1]

[1] State Key Laboratory of Catalysis, Dalian Institute of Chemical Physics, Chinese Academy of Sciences, Dalian 116023, China

Two-dimensional materials (2DMs) by virtue of their unique morphology have risen to prominence in the area of energy storage[1]. Graphene and MXene (particularly, transition-metal carbides) are such 2DMs that have greatly accelerated the progress of micro-supercapacitors (MSCs), which are viable energy storage solutions for powering future micro-electronics. Graphene is renowned for its high conductivity and tunable morphology, while MXene offers high capacitance and customizable surface properties. In our group, we leverage these materials to address various challenges, paving the way for high-performance MSCs in diverse applications. We have developed functionalized graphene through electrochemical exfoliation with enhanced voltage window, capacitance, and energy density. They were further formulated into inks for 3D printing of MSCs with unique porosity [2-3]. Additionally, our 3D-printed MXene MSCs have achieved not only record voltage and capacitance but also stable operation at low temperatures[4]. We have also achieved densely-integrated MSCs using MXene, yielding high areal voltage[5]. Recognizing the importance of device architecture, we have enabled microelectrodes constructed from these 2DMs to be highly resilient to extreme deformation[6]. We believe that rapid fabrication of graphene and MXene-based materials coupled with unique device fabrication strategies will not only elevate the resulting MSC performance but also expand their application scope.

References

[1]L. Zhang, Z. Yang, S. Feng, Z. Guo, Q. Jia, H. Zeng, Y. Ding, P. Das, Z. Bi, J. Ma, Y. Fu, S. Wang, J. Mi, S. Zheng, M. Li, D.-M. Sun, N. Kang, Z.-S. Wu, H.-M. Cheng, Nature 2024, 628, 313.

[2]P. Das, L. Zhang, S. Zheng, X. Shi, Y. Li, Z.-S. Wu, Carbon 2022, 196, 203.

[3]L. Zhang, J. Qin, P. Das, S. Wang, T. Bai, F. Zhou, M. Wu, Z.-S. Wu, Advanced Materials 2024, 36, 2313930.

[4]Y. Zhu, Q. Zhang, J. Ma, P. Das, L. Zhang, H. Liu, S. Wang, H. Li, Z.-S. Wu, Carbon Energy 2024, e481.

[5]S. Wang, S. Zheng, X. Shi, P. Das, L. Li, Y. Zhu, Y. Lu, X. Feng, Z.-S. Wu, Nature Communications 2024, 15, 2850.

[6]P. Das, X. Shi, Q. Fu, Z.-S. Wu, Advanced Functional Materials 2020, 30, 1908758.

Preparation, structure regulation and application of bituminous-based hard carbon materials

Xue Li[1], Keke Li[1,2], Yanping Hu[1], Youyu Zhu[1,2], Peng Chang[1,2], Guoyang Liu[1,2], Yating Zhang[1,2*]

[1]College of Chemistry and Chemical Engineering, Xi'an University of Science and Technology, Xi'an, China

[2]Key Laboratory of Coal Resources Exploration and Comprehensive Utilization, Ministry of Natural Resources, Xi' an, China

Sodium-ion batteries have attracted much attention due to their low cost, high safety and excellent low-temperature performance. Anode materials play a crucial role in sodium-ion batteries, significantly impacting their cost. Hard carbon stands out as one of the most promising choices owing to its abundant raw material sources, low sodium storage potential, and excellent cycling stability, making it highly suitable for industrial applications. Coal, as a naturally carbon-rich material, holds substantial potential in carbon material synthesis. Particularly, using coal as a precursor for hard carbon offers notable structural, performance, and cost advantages. Beginning with coal' s organic macromolecular structure, the study fully exploits its rich oxygen-containing functional groups and high surface activity, employing strategies such as structural and surface modulation, interfacial configuration, and various modification techniques like optimized carbonization, pre-oxidation, elemental doping, cross-linking reactions, and binder enhancement. This approach yields a range of high-performance coal-based hard carbon materials. Furthermore, these coal-based hard carbon materials are investigated as anodes for sodium-ion batteries to explore their electrochemical performance and sodium storage mechanisms. This research introduces novel technologies and insights into the materialization and utilization of coal.

Efficient Resource Utilization of Discarded Polypropylene Masks for High-Value Carbon Anode Materials

Ning Sun, Bin Xu

Beijing University of Chemical Technology, 15 Beisanhuan East Road, Chaoyang District, Beijing, China

Polyolefin with high carbon content has emerged as a promising carbon precursor for functional carbon materials. However, their linear molecular structures tend to disintegrate into low molecular fragments rather than aromatization, resulting in complete decomposition with no residues left after heating treatment. Here, discarded polypropylene (PP) masks were used as precursors to prepare a series of carbon materials via sulfonation/vulcanization/oxidation crosslinking and catalytic pyrolysis strategies [1,2]. The obtained carbon anode materials demonstrated satisfactory Na-storage performance by regulating the interlayer spacing, pore architecture and elemental doping. In addition, the structure-activity relationship between carbon structure and sodium storage performance has been elucidated. This work provides novel insights into transforming the waste masks into value-added hard carbons with promising prospects for efficient Na ion storage.

References

[1] Shaohong Zhang, Ning Sun*, Bin Xu*, et al. Carbon 209 (2023) 118034
[2] Shaohong Zhang, Ning Sun*, Bin Xu*, et al. Energy Storage Mater. 66 (2024) 103183

noble metal nanoparticles @ graphite edges for efficient pH-universal hydrogen evolution reaction

Qiqi Li, Xuanke Li,* Yongting Chen,* and Qin Zhang*

Wuhan University of Science and Technology, Wuhan, China

The catalytic activity of metal nanoparticles is closely pertinent with the substrate, while the influence of the support on the catalytic activity was acquainted scarcely. Herein, we proposed that the edge selective decoration of noble metal nanoparticles into an edge-rich carbon foam (CF) matrix for efficient pH-universal hydrogen evolution (HER) reaction. Highly localized density of states of the graphene edges enables increased reactivity to noble metal nanoparticles and tune their electronic states. As expected, the Pt nanoparticles decorated on CF (Pt NPs@CF) exhibit low overpotentials (η 10) of only 35, 49, and 53 mV under acidic, alkaline, and neutral conditions, respectively. Moreover, it can deliver the ampere-level current density of 1000 mA cm−2 with an overpotential of 97 mV in acidic solution. Above all, the other noble metal NPs decorated on CF also possess efficient pH-universal HER performance. This work presents a new avenue for developing high-activity electrocatalysts via support engineering.

Catalytic conversion of Li2S for high-performance lithium-sulfur batteries

Chuannan Geng[1,2], Wei Lv[2] and Quan-Hong Yang[1]

[1] Nanoyang Group, State Key Laboratory of Chemical Engineering,
[2] School of Chemical Engineering and Technology, Tianjin University, Tianjin 300072, China
Shenzhen Geim Graphene Center, Engineering Laboratory for Functionalized Carbon Materials,
Tsinghua Shenzhen International Graduate School, Tsinghua University, Shenzhen, 518055, China
Email: gengcn@tju.edu.cn

The rechargeable lithium-sulfur (Li-S) batteries have become one of the most promising systems for next-generation high energy-density energy storage. However, the sluggish kinetics of the sulfur evolution reaction (SER) that occur because of the high oxidation barrier of Li2S causes low sulfur utilization and the poor rate performance of Li-S batteries. The design of the catalysts to solve this problem is still hard to achieve because it is difficult to precisely correlate the catalytic oxidation ability with the electronic structure. Here, a layer transition metal oxide, NaxTi0.5Co0.5O2, is used as a model catalyst to probe such a correlation because it has a tunable electronic structure and good stability in the working potential window of Li-S batteries. By removing Na+, a partial phase change gradually increases the concentration of Co active sites while decreasing the work function with an upshift of the Fermi level, accelerating charge transfer on the catalyst surface and therefore improving its catalytic oxidation activity of Li2S.

On this basis, we further developed the homogeneous catalyst trithiocyanuric acid trilithium salt (TTCA-Li) dissolved in the electrolyte to improve the catalytic efficiency. It can not only catalytically decrease the initial overpotential of Li2S, but also form an interfacial layer by in-situ polymerization during charging. The produced coating effectively confines the shuttling of lithium polysulfides, thus inhibiting the redistribution of sulfur species and active sulfur loss upon cycling. The prototype full cell where the coated Li2S cathode couples with the Li anode has an extremely high CE of over 99.5%.

References

[1] Qu W. J., Geng C. N., Lv W.*, Yang Q.H.*, et al., Advanced Energy Materials 2022, 2202232.
[2] Geng C. N., Lv W.*, Yang Q.H.*, et al., Advanced Energy Materials 2023, 2204246.

Catalytic conversion of Li2S for high-performance lithium-sulfur batteries

Wang Zhou[1], Rui Tang[1], and Jilei Liu[1]

[1] College of Materials Science and Engineering, Hunan University, Changsha 410082, China

The intricate microstructures of hard carbons together with the diverse physicochemical properties of alkali metal ions contribute to the enigmatic nature of Li/Na/K ions storage mechanisms[1-2]. The current state of research still lacks comprehensive theoretical models and structural design strategies, making it challenging to optimize the performance of hard carbons for Li/Na/K-ion batteries in a targeted manner. Herein, we quantitatively investigated the extrinsic defects, crystallite parameters, and nanopores of resin-derived hard carbons and associated these microstructural parameters with the electrochemical performance. Our findings demonstrate that enabling a sufficient large graphene interlayer spacing (> 3.9 Å) could facilitate Li migration into the closed pores for clustering, resulting in satisfactory reversible capacity (497 mAh/g). Additionally, fabricating abundant closed pores with suitable graphene interlayer spacing (≈ 3.7 Å) is a compelling pathway to achieve fascinating reversible capacity in Na/K-ion batteries (530 mAh/g for Na, 520 mAh/g for K). Our work highlights the importance of the combined effects of adsorption, intercalation, and nanoclustering for achieving a reversible capacity of approximately 500 mAh/g in Li/Na/K-ion batteries, and provides clear design guidelines for the purposeful tuning of the structural characteristics to optimize their electrochemical performance.

References

[1] Wang Zhou, et al. Adv. Energy Mater. 13 (2023) 2202874.
[2] Wang Zhou, et al. Adv. Funct. Mater. 34 (2024) 2312994.

Exudate Triggered Electrical Dressing for Complicated Wound Management

Yuanbin Wang[2], Qian Ma[3], Rongkang Huang[2], Hui Wang[2], Dingcai Wu[1]

[1] PCFM Lab, School of Chemistry, Sun Yat-sen University, Guangzhou 510006, P. R. China
[2] Department of General Surgery (Colorectal Surgery), Guangdong Institute of Gastroenterology, Biomedical Innovation Center, Guangdong Provincial Key Laboratory of Colorectal and Pelvic Floor Diseases, The Sixth Affiliated Hospital, Sun Yat-sen University, Guangzhou 510655, P. R. China
[3] Research Center of Medical Sciences, Guangdong Provincial People's Hospital (Guangdong Academy of Medical Sciences), Southern Medical University, Guangzhou 510080, P. R. China

Exudate management and cell viability enhancement are vital to complicated wound healing [1-2]. However, current exudate management dressings indiscriminately remove exudate, tending to against cell viability enhancement. Herein, we have developed a electroactive bilayer dressing (MO@PVA C) by preparation of manganese oxide nanoneedle clusters decorated commercial carbon cloth (MO), followed by in situ casting of polyvinyl alcohol (PVA) hydrogel and charging under a constant voltage. Benefitting from unique hierarchical nanoneedle cluster structure of MO, abundant active sites are sufficiently exposed, so that favorable electric filed (EF) is built. EF synergistic coupling with physical absorption (by PVA hydrogel) realizes extraordinary selective clearance of noxious substances from the exudate. Simultaneously, these noxious substances trigger long term electrical stimulation therapy (EST) to further enhance cell viability and antibacterial property. As a result, this multifunctional MO@PVA C achieves rapid repair performance of full thickness wounds in diabetic rats. The concept and results presented in this study offer an advanced strategy for the treatment of complicated wounds.

References

[1]V. Falanga, R. R. Isseroff, A. M. Soulika, M. Romanelli, D. Margolis, S. Kapp, M. Granick, K. Harding, Nat. Rev. Dis. Primers 2022, 8, 50.
[2]S. Matoori, A. Veves, D. J. Mooney, Sci. Transl. Med. 2021, 13, eabe4839.

Interfacial Mechanism of Carbon Anode for Sodium-ion Batteries

Yibo Zhang[1], Si-wei Zhang[1,2], Jun Zhang[1], Ying Tao[1], Quan-hong Yang[1]

[1]School of Chemical Engineering and Technology, Tianjin University, Tianjin, China
[2]Shenzhen International Graduate School, Tsinghua University, Shenzhen, China

Non-graphitic carbons are believed as the most promising anode candidates for industrializing sodium-ion batteries, while how their closed pores regulate the solvation structure and interfacial properties remain poorly explored and understood.[1, 2] Herein, we reveal that the evolution of solvation structure highly depends on the pore mouth size (PMS) and thereby redefine closed pore as one with a PMS of 0.35 nm. The precise and uniform control of the PMS is achieved by using carbon molecular sieves as a model. We show when the PMS decreases from 0.5 nm to 0.35 nm, that is, from being inaccessible to N2 to CO2 molecular probes, both free solvents and coordinated solvents are thoroughly sieved to enable a highly anion-aggregation solvation structure inside the pores. As a result, an organic-poor and inorganic (NaF)-rich SEI is produced, resulting in a high ICE of 91 % and a high capacity retention of 95 % after 200 cycles. Furthermore, a high capacity retention of 82 % is demonstrated at -40 °C compared to 25 °C. We also show that intrinsic defects inside the closed pores are conducive to increasing the plateau capacity. This work indicates new avenues for the rational structural design of high-performance porous electrodes from the perspective of solvation structure, not limited to sodium-ion batteries.

References

[1] S. Zhang, et al. Energy Storage Mater. 3 (2016) 18-23.
[2] Q. Li, et al. Natl. Sci. Rev. 9 (2022) nwac084.

Controlling surface chemistry of activated carbon fibers for high product sulfate selectivity of H2S at room temperature catalytic oxidation

Yue Liu[1], Jianxiao Yang [1], Xin Zhang [2]

[1]Hunan University, Changsha 410082, China
[2]University of Chinese Academy of Sciences, Beijing, 101408, China

In recent years, the issue of odor control has become increasingly significant due to the rise in civil complaints and concerns regarding environmental and human health. Wastewater treatment plants located in cities are a major source of malodorous gases such as H_2S, which can cause great nuisance to neighboring communities. Hence, this work investigated the catalytic oxidation properties of H_2S at room temperature on a series of modified activated carbon fibers (ACFs) prepared from carbon fibers through NH_3 activation and non-thermal plasma (NTP) treatment in various modification atmospheres (N_2, O_2, Air). Among them, NH3-ACF-A10 achieved a high sulfur capacity (0.21 gH_2S/g), the highest products sulfate selectivity (85.74%), and excellent performance in water scrubbing regeneration (82.98 %). During catalytic oxidation, oxygen-containing functional groups (OFGs) could react with HS^- and regenerate by O_2. $O_2\cdot^-$ required for the reaction was facilitated by N-6 and sp2 ring carbon defects. H_2O was responsible for transporting catalytic oxidation products and continuously exposing active sites. Lewis acid sites (COOH) were proved that were the specific sites affected products sulfate selectivity, rather than the strength of the acid sites. This work presented a novel methodology for the design of high-regeneration catalysts and provided a comprehensive and detailed theoretical framework for the further understanding of the catalytic oxidation effect of ACFs.

New insight into boron catalytic graphitization of graphene lateral heat spreader

Dong Liu[1,2,#], Zhe-Fan Wang[1,2,#], Hui Jia[1], Zong-lin Yi , Li-Jing Xie[1], Fang-yuan Su[1], Zechao Tao[1], Xi Yan[1], Dong Jiang[1,3], Kegong Fang[1], Qing-Qiang Kong[1], Cheng-Meng Chen[1,3,*]

[1]CAS Key Laboratory of Carbon Materials, Institute of Coal Chemistry, Chinese Academy of Sciences, Taiyuan, 030001, China

[2]University of Chinese Academy of Sciences, Beijing 100049, China

[3]Center of Materials Science and Optoelectronics Engineering, University of Chinese Academy of Sciences, Beijing 100049, China

The boron catalytic graphitization mechanism is significant for increasing graphite crystallite size of carbon-based materials in thermal management, which is urgently required for the consumer electronics and aerospace industries. However, due to the lack of an in-depth recognition of the boron catalytic graphitization, the optimization of intrinsic structure is hindered and the design of thermal management carbon materials is restricted. Herein, we systematically unveil of boron species, their geometrical and electronic structures on reactivity of graphitization. Evolution of boron species is traced in whole heat treatment stage, revealing its thermalchemical evolution pathway BCO2→BC2O→BC3→B4C. Among them, BC3 configuration enhances edge electron activity, and the transformation of BC3 to B4C promotes growth of crystallites, as evidenced by the X-ray absorption near edge structure (XANES) and density functional theory (DFT) calculations. Based on the mechanism, the prepared boron-assistant catalytic graphene films (GBF) finally possess higher thermal conductivity (1223 W m−1 K−1) than pure graphene film (GF) (979 W m−1 K−1). This work sheds light on the graphitization process in microscale and paves the way for the development of carbon-based functional materials.

Hierarchical Graphite/SiCw Matrix for phase change composites toward High-efficiency Thermal Harvesting

Hongwei Zhong [a,b,1], Zhou Deng [a,b,1], Ke Wang [a,b], Qin Zhang [a,b], Jianguang Guo [a,b], Baoliu Li [a,b], Hui Zhu [a,b,*] and Xuanke Li [a,b,*]

[a]Hubei Pilot Test Research Base of Coal Conversion and Advanced Carbon Materials, Wuhan University of Science and Technology, Wuhan, Hubei 430081, PR China
[b]State Key Laboratory of Refractory Materials and Metallurgy, Wuhan University of Science and Technology, Wuhan, Hubei 430081, PR China
[1]These authors contributed equally to this work and should be regarded as co-first authors.
E-mail: zhuhui@wust.edu.cn; lixuanke@wust.edu.cn

Phase change composites (PCCs) for the thermal management of high-energy density device are required to have large heat storage capacity and rapid heat conduction and dissipation ability. A hierarchical graphite foam (GF)/Silicon carbide whiskers (SiCw) network skeleton with abundant pore capacity for loading paraffin wax (PW) was designed and fabricated. The highly oriented carbon walls of GFs as the main channels for fast thermal transportation and SiCw formed in GF pores as the high efficiency heat conductive networks provide GF-SiCw/PW with highly efficient thermal energy harvesting and releasing ability. The prepared GF-SiCw-2 composite loaded with 48.99 wt% of PW exhibits an excellent thermal conductivity of 249.7 W m-1 K-1 and fast thermal response characters. No PW leakage, homogeneity and superior phase change behavior were observed in GF-SiCw-2/PW. The excellent thermal management of GF-SiCw/PW was also confirmed. This innovative structure design suggests an efficient route for the development of phase change thermal management system with high heat conductivity, and shows extensive application prospects.

Spontaneous Electrochemical Reconstruction of NiNb2O6@C for High-Rate Lithium-Ion Batteries

Lidan Tan[1], Nianjun Yang[2,3*], and Ye Cong[1*]

[1]Wuhan University of Science and Technology, Wuhan, China
[2]University of Siegen, Siegen, Germany
[3]Hasselt University, Diepenbeek, Belgium

Niobium oxides are widely active in the arena of lithium-ion batteries due to their distinctive structure and rapid charging capability, nevertheless facing the serious challenge from their inherently low electrical conductivity. Herein, a molten salt method is proposed to prepare nickel niobate (NiNb2O6) anode materials for lithium-ion batteries, and a carbon coating derived from mesophase pitch is fabricated to construct NiNb2O6@C composites. Importantly, NiNb2O6@C undergoes spontaneous electrochemical reconstruction to generate metallic Ni and amorphous Nb2O5 during the initial charge/discharge cycle. The presence of the outer carbon layer not only serves as a uniform conductive coating to improve the conductivity and thereby deliver superior rate performance, but also effectively inhibits the stripping of Ni0 to guarantee a great cycling stability. The synergistic effects of the dual conductive interfaces, consisting of the carbon layer and the Mott-Schottky heterojunction between amorphous Nb2O5 and metallic Ni, culminate in exceptional electrochemical performance. This work underscores the substantial potential of nickel niobate anode materials in advancing the development of next-generation lithium-ion batteries.

Carbon-based atom catalysts for high-performance lithium-sulfur batteries

Zhiyuan Han[1] and Guangmin Zhou[2]

[1] Tsinghua-Berkeley Shenzhen Institute (TBSI), Tsinghua University, Shenzhen 518055, China

Carbon-based atom catalysts have been used as sulfur cathode additives to promote battery performance, although the material selection and mechanism that govern the catalytic activity remain unclear. We ever proposed that d-p orbital hybridization between the single-atom metal and the sulfur species can be used as a descriptor for understanding the catalytic activity of single-atom metal catalysts in Li–S batteries. [1] However, atomic-level interactions among multiple atom sites—ensemble effect—remain elusive. The challenge arises from the lack of methods to trace activity origins in complex catalytic systems with numerous combinations and unclear activation mechanisms. Here, we propose a multi-view machine-learned framework to automatically evaluate catalyst features using limited datasets and uncover factors influencing ensemble effect in multi-site catalysts, such as corrected d orbital properties. We demonstrate that orbital coupling among sites induces shifts in band centers and alterations in spin state, thus influencing interactions with polysulfides and resulting in diverse Li-S bond breaking and lithium migration barriers. Ah-level Li-S pouch cells with ensemble effect achieved an energy density of 480 Wh kg-1 with high sulfur loading and lean electrolyte operation. This ensemble effect reveals the activity origins of carbon-based multi-site catalysts and offers a universal strategy for their design and practical applications.

References

[1] Zhiyuan Han, et al. Advanced Materials 33 (2021) 2105947

Exceptional thermal interfacial properties of highly resilient three-dimensional carbon foam/CNTs matrices

Zhuo Deng, Xuanke Li,* and Hui Zhu*

Wuhan University of Science and Technology, Wuhan, China

With the development of modern electronics industry, high performance thermal interface materials (TIM) are in urgent demand.

Here, we choose to prepare GF/CNTs thermal interface materials for high-temperature applications by using 3D spongy porous mesophase pitch-based carbon foam as the backbone, in-situ growth of CNTs by float-catalyzed chemical vapor deposition, and filling with silicone rubber. The thermal conductivity and resilience of the complex material were improved by decreasing the diameter of CNTs and increasing the content of CNTs. When the CNTs content is 25 wt% and the diameter is 70 nm, the thermal conductivity of the GF/CNTs material is 122.88 W m-1K-1 and the resilience is up to 90%. The chip thermal performance shows that the developed TIMs have stable thermal conductivity for long-term applications. This work realizes the large-scale preparation of high-performance interface materials.

Preparation and thermodynamic properties of mesophase pitch based carbon fiber mat reinforced 3D-C/C composites

Ke Wang[1], Xuanke Li,* and Hui Zhu*

Wuhan University of Science and Technology, Wuhan, China

Mesophase pitch based carbon fiber mat was prepared by mesophase pitch melt spray spinning process. The 3D-C/C composites were prepared by hot molding process using mesophase pitch based carbon fiber mat as raw material and mesophase pitch as binder. After one slow carbonization, five liquid pressure impregnation and carbonization, and one graphitization to give a densities of 1.86 g/cm3. The effects of the structure of the carbon fiber mat on both the thermal conductivity and mechanical properties of the 3D-C/C composites was investigated. Results indicate that in the carbon fiber mat, the fibers are arranged in a network and the fibers are interwoven in the Z direction, forming a 3D heat conduction network structure and improving the thermal conductivity of the material in the Z direction. The thermal conductivities of the 3D-C/C composites in X-Y plane and Z directions are respectively 355 W/(m·K), 350 W/(m·K), and 110 W/(m·K). The bending strength reached up to 130 MPa while compressive strength reached 101 MPa for this particular material composition. The thermodynamic properties primarily rely on mesophase pitch based carbon fibers within 3D-C/C composites. Compared to conventional methods for preparing 3D-C/C composites, process significantly reduces the preparation time and greatly improves the Z-direction thermal conductivity of the composites.

Synthesis and characterization of methylene-bridged pitches derived from naphthalene via chloromethylation-dechlorination reaction

Wei Jin[1], Xingheng Sun[1], Yongsheng Tian[1], Zhijun Dong[1*], Yanjun Li[1*], Guanming Yuan[1], Ye Cong[1], Baoliu Li[1], Jianguang Guo[1], Qin Zhang[1], Xuanke Li[1*]

[1]Hubei Province Key Laboratory of Coal Conversion & New Carbon Materials, Wuhan University of Science and Technology, Wuhan, Hubei 430081, P.R. China

A novel methylene-bridged polymeric pitch was successfully synthesized from naphthalene (NAP) using chlorine as a sacrificial agent for C-C bond construction via chloromethylation-dechlorination reaction. The primary chloromethylation products of NAP are predominantly 1-chloromethylnaphthalene, with minor amounts of 1,4-dichloromethylnaphthalene. Thermodynamic and kinetic calculations indicate that the alpha-hydrogen on the NAP ring is preferentially substituted by chloromethyl during chloromethylation, leading to the formation of 1-chloromethylnaphthalene and 1,4-dichloromethylnaphthalene. Following a thermal dechlorination polycondensation reaction conducted at 320-380 °C, isotropic carbonaceous pitches with a softening point range of 221-265 °C were successfully synthesized. The synthesized polymeric pitches, characterized by repeated monomers with a linear structure of methylene-bridged NAP rings, exhibited homogeneous isotropic phase, optimal viscosity, and excellent spinnability, serving as precursors for isotropic pitch-based carbon fibers via melt spinning. Furthermore, the carbonization study of the synthesized pitches revealed that the isotropic pitch derived from NAP at 380 °C exhibited a 57% carbon yield at 800 °C. Notably, the semi-cokes formed exhibited a fine isotropic texture, even after being carbonized at 550 °C. These findings suggest that these unique pitches could be excellent precursors for preparing various carbon materials, owing to their advantageous spinnability and carbonization characteristics.

Delayed pyrolysis of binder pitch in fine-grained isostatic carbon blocks

Xincheng Yao[1,2†], Jiang Tang[1,2†], Weiwei Chen[1,2†], Guolian Wang[3], Jinping Liu[4], Hui He[3*], Hui Chen[1,2*], Ke Shen[1,2*]

[1]College of Materials Science and Engineering, Hunan University, Changsha, Hunan, China
[2]Hunan province key laboratory for advanced carbon materials and applied technology, Hunan University, Changsha, Hunan, China
[3]School of Nuclear Science and Engineering, Shanghai Jiao Tong University, Shanghai, China
[4]Hunan Changyu Science and Technology Development Co., Ltd, Changsha, Hunan, China

Artificial graphite, a key material in numerous industries, is typically produced through processes involving coal-tar pitch and coke filler particles. The baking process, integral to its production, influences the final product's structural integrity. Despite extensive research on coal-tar pitch pyrolysis, there is a gap in understanding the carbonization behavior within the green body, as the coal-tar pitch is carbonized in confined space and shows location-dependent pyrolysis behavior. This study investigates the location-dependent pyrolysis behavior of green bodies by introducing a novel parameter, baking degree (η). Experimental and simulation studies reveal delayed pyrolysis behavior within the central part of the green body, confirmed through examination of porosity development. Such delayed pyrolysis is not caused by temperature gradient, but should be attributed to the kinetic factors associated with carbonization behavior in the confined space. Introducing the η helps to grasp more precisely the pyrolysis process of the binder in the green body. By optimizing process parameters to minimize the difference in η between the center and surface of the green bodies during baking process, the delayed carbonization of binders within the green bodies can effectively be reduced, thereby effectively enhancing the performance of the artificial graphite.

Carbon nanosphere molecular brushes for treating inflammatory bowel disease

Pengwei Ma[1], Guangyuan Chen[2], Rongkang Huang[2], Bingna Zheng[3] and Dingcai Wu[1]

[1]School of Chemistry, Sun Yat-sen University, Guangzhou, P. R. China
[2]The Sixth Affiliated Hospital, Sun Yat-sen University, Guangzhou, P. R. China
[3]The Eighth Affiliated Hospital, Sun Yat-sen University, Shenzhen, P. R. China

Nanocarbons have been widely used in energy storage[1], catalysis[2], sensing[3], biomedicine[4,5] and other fields due to their good conductivity, high mechanical strength, and excellent biocompatibility. However, the lack of functionalization greatly limits their application. Polymer modified nanocarbons can achieve multifunctionality and processability, which has attracted widespread attention in the field of biomaterials. Here, a kind of carbon nanosphere molecular brushes with efficient catalytic activity by the carbon cores and intestinal colonization ability by the surface-grafted wet-adhesive side-chains have been obtained and applied for the treatment of inflammatory bowel disease. Carbon nanosphere molecular brushes can quickly form wet-adhesion with tissues through chemical bonds when injected onto the inflammatory area of intestine through the anus, thereby achieving long-term colonization on intestine surface. The colonized carbon nanosphere molecular brushes can effectively reduce the level of reactive oxygen species in intestinal inflammation, thereby achieving sustained and efficient regulation of intestinal inflammation. In summary, the carbon nanosphere molecular brushes has high catalytic activity for reactive oxygen species and intestinal colonization ability, making it a highly promising therapeutic material for inflammatory bowel disease.

References

[1] Ji Qian, et al. Advanced Materials 33(2021) 2100810.
[2] SongLin Zhang, et al. Angewandte Chemie International Edition 60(2021) 19068-19073.
[3] Shuncheng Yao, et al. Advanced Materials 34(2022) 2109568.
[4] Lijun Ye, et al. Advanced Materials 33(2021) 2102981.
[5] Fangfang Cao, et al. Nature Nanotechnology 18(2023) 617-627.

Carbon Nanozyme-Loaded Star Polymer Hydrogel with Long-Lasting Retention and Shear - Thinning Properties Facilitates Durable Long-Term Tumor Therapy

Dongtian Miao[1], Rui Luo[2], Yue Zhang[1], Rongkang Huang[2], Hui Wang[2] and Dingcai Wu[1]

[1]School of Chemistry, Sun Yat-sen University, Guangzhou, P. R. China
[2]The Sixth Affiliated Hospital, Sun Yat-sen University, Guangzhou, P. R. China

Peritoneal metastasis of colorectal cancer is a challenging clinical issue. The vascular supply to peritoneal tumors in the peritoneal cavity is typically limited, hindering the delivery of antitumor drugs via intravenous injection. The development of novel injectable antitumor materials to increase the long-term local drug concentrations for peritoneal tumors is extremely desirable but remains a significant challenge. Here, a multi-arm star polymer hydrogel (denoted as β CP hydrogel) is successfully developed by grafting poly(oligo(ethylene glycol) methyl ether methacrylate) (PEGMA) side-chains from β -cyclodextrin (β -CD) via atom transfer radical polymerization (ATRP). The reversible polymer network structure endows the materials with good shear-thinning property, and can achieve local delivery of drugs and carbon nanozymes loaded with Fe sites (CFe). Once the injection shear is ceased, winding of polymer side-chains takes place, thereby establishing a long-lasting depot for drugs and carbon nanozymes that can prevent tumor recurrences. Compared with intravenous injection, βCP-CFe hydrogel can be effectively enriched in the tumor site after being injected subcutaneously in rats (63 d vs 24 h). These results indicate an appealing prospect for new application of β CP-CFe hydrogel as a good antitumor material in treatment of peritoneal carcinomatosis.

N/S Co-doped Carbon Nanosheets for Efficient Electrochemical Extraction of Uranium from Seawater

Tan Yi[1], Junlong Huang[2,*], Zongheng Cen[1], Yiwei Ji[1], and Shaohong Liu[1,*]

[1]School of Chemistry, Sun Yat-sen University, Guangzhou 510006, China
[2]Medical Research Institute, Guangdong Provincial People's Hospital, Southern Medical University, Guangzhou 510080, China

At present, land uranium resources can no longer support the rapid expansion of future nuclear power scales[1]. Uranium extraction from seawater is a crucial approach to ensure the sustainability of nuclear energy[2]. Carbon-based catalysts are one of the most promising materials for electrochemical extraction of uranium from seawater[3,4]. However, the practical application of these materials is often limited by high costs and low catalytic activity. Herein, to achieve efficient uranium extraction from seawater, a class of nitrogen and sulfur co-doped carbon nanosheets (CNSs) are developed by simple heat treatment of low-cost commercial resin and melamine. Benefiting from the high conductivity and high catalytic activity brought by heteroatomic doping, the CNSs based electrodes can catalyze uranyl ions in seawater into easily recoverable Na2O(UO3·H2O)x precipitates at -2 V, and achieve an extraction capacity of up to 3923.7 mg g-1 with a uranium removal rate of 98.1% in uranium-spiked seawater. Even in natural seawater, the electrode can still recover 72.7% of the uranium, demonstrating excellent application prospects. We hope this work can provide new insights for designing low-cost, metal-free electrocatalysts for efficient uranium extraction from natural seawater.

References

[1]Abney, C. W.; Mayes, R. T.; Saito, T.; Dai, S., Materials for the recovery of uranium from seawater. Chemical Reviews. 2017, 117, 13935-14013.
[2]Xie, Y.; Liu, Z.; Geng, Y.; Li, H.; Wang, N.; Song, Y.; Wang, X.; Chen, J.; Wang, J.; Ma, S.; Ye, G., Uranium extraction from seawater: Material design, emerging technologies and marine engineering. Chemical Society Reviews. 2023, 52, 97-162.
[3]Guo, H.; Mei, P.; Xiao, J.; Huang, X.; Ishag, A.; Sun, Y., Carbon materials for extraction of uranium from seawater. Chemosphere. 2021, 278, 130411.
[4]Yang, H.; Liu, X.; Hao, M.; Xie, Y.; Wang, X.; Tian, H.; Waterhouse, G. I. N.; Kruger, P. E.; Telfer, S. G.; Ma, S., Functionalized iron–nitrogen–carbon electrocatalyst provides a reversible electron transfer platform for efficient uranium extraction from seawater. Advanced Materials. 2021, 33, 2106621.

Improved self-sintering mechanical performance of carbon blocks by regulating composition of green petroleum coke

Jiao Tan, Pei Gong, Xueli Wu, Run Li * and Chuanjun Tu *

College of Materials Science and Engineering, Hunan University, Hunan Province, Changsha, China

* Corresponding authors

Self-sintering of green petroleum coke (GPC) provides an important method to synthesis high-strength graphite blocks due to its relatively short production cycle. However, the volatilization of small molecular components in GPC inevitably results in the generation of pores and microcracks which seriously deteriorates the mechanical performance of carbon blocks. Herein, we report a strategy for regulating composition of GPC by using solvents, i.e. dichloromethane and toluene, to remove light molecular-weight compounds [1]. Owing to the removal of light molecular-weight compounds, the flexural and compressive strength of DCMI-C are 1.7 and 1.9 times that of GPC-C, respectively, indicating that DCM soluble compounds are the mainly reason for microcracks and pores generation. Furthermore, to further purify the GPC, an organic solvent with much higher solvency power should be employed. We propose a facile quinoline-assisted GPC composition regulation approach to prepare high mechanical strength carbon blocks [2]. Quinoline has three major functions, e.g. removing the unideal organic volatile, promoting the polymerization of aryl compounds and enhancing the interactions between different graphite layers. The carbon blocks obtained from quinoline-modified GPC illuminate a flexural strength of 39.61 MPa and compressive strength of 136.98 MPa, which is 3.39 and 4.53 times that of GPC counterparts, respectively.

References

[1]Jiao Tan, et al. Fuel 343 (2023) 127912

[2]Jiao Tan, et al. Fuel Processing Technology (2024), accepted.

Preparation and thermal conductivity of 2.5D Carbon/Carbon composites

He Huang [1,2], Zhijun Dong [1,2], Baoliu Li [1,2]

[1]The State Key Laboratory of Refractories and Metallurgy,
Wuhan University of Science and Technology, Wuhan, Hubei 430081, P. R. China
[2]Hubei Province Key Laboratory of Coal Conversion & New Carbon Materials,
Wuhan University of Science and Technology, Wuhan, Hubei 430081, P. R. China

In the preparation of 2.5D Carbon/Carbon composite, three kinds of reinforcement with different T700 fiber content were densified in different densificati on modes((Mesophase asphalt impregnation/Resin impregnation/Chemical vapor impregnation/CVI+MPI/CVI+RI)+1000℃ HT). Test results show that Fiber content plays a leading role in heat conduction. The mesophase carbon has the best performance which XY direction thermal conductivity is 251.2 W·m·K-1, Z direction thermal conductivity is 121.4 W·m·K-1. Resin carbon is second. This is because most of the components of pyrolytic carbon are smooth layers or even isotropic, and thermal stress occurs between resin carbon and fiber, resulting in additional graphitization after heat treatment at 3000℃, e.g. [1].

References

[1] Luz A P, et al. Ceramics International 43 (2017) 8171-8182

Design and preparation of multifunctional two-dimensional layered film barrier for lithium sulfur batteries

Xiang Long, Zhi-Hong Luo, and Jiao-Jing Shao

School of Materials and Metallurgy, Guizhou University, Guiyang, China

Two dimensional materials have structural characteristics such as anisotropy and large aspect ratios, which give them good film-forming properties. Some two-dimensional materials have unique properties, for example, two-dimensional clay has excellent lithium ion transport ability, and modified two-dimensional clay can further regulate the distribution of lithium ions; Two dimensional graphene has excellent electron mobility and high mechanical strength; The high conductivity and catalytic activity of two-dimensional MXenes have been widely confirmed. The above properties make two-dimensional materials have great potential for application in the field of electrochemical energy storage. The authors have been engaged in the research of preparation and controllable assembly of two-dimensional materials in liquid phase environment, as well as their application in electrochemical energy storage devices. In recent years, our team has mainly focused on the issue of the shuttle effect in high-energy density lithium sulfur batteries. Starting from the functional modification of the separator, a multifunctional two-dimensional layered film has been designed and prepared as the barrier for lithium sulfur batteries. Via surface modification and size control of two-dimensional nanochannels, on the basis of promoting efficient adsorption and catalytic conversion kinetics of polysulfides by the barrier layer, the selective transport of lithium ions by the barrier layer is further improved, and the effective suppression of polysulfide shuttle effect is achieved, with the goal of obtaining high energy density, high safety, and long life lithium sulfur batteries.

References

[1] Min Zheng, Zhihong Luo, Jie Wu, Jiao-Jing Shao*, Guangmin Zhou, et al., Composites Part B-Engineering, 2023, 110898, 1359-8368.

[2] Xiang Long, Zhi-Hong Luo, Zhi-Yuan Han, Guang-Min Zhou, Wei Lv, Jiao-Jing Shao*, et al., Energy Storage Materials, 2022, 52, 120–129.

Preparation of nitrogen-doped carbon points and their fluorescence characteristics by one-step electrolysis of graphite based on ionic liquid

Wei Zhao [a], Zhenming Lu [a,b*], Qin Zhang [a,b], Tian Gao [a,b]

[a] School of Chemistry and Chemical Engineering, Wuhan University of Science and Technology, Wuhan, Hubei 430081, PR China

[b] Hubei Province Key Laboratory of Coal Conversion and New Carbon Materials, Wuhan University of Science and Technology, Wuhan, Hubei 430081, PR China

A one-step electrochemical method for the preparation of carbon dots (CDs) by stripping graphite rods using ionic liquid (IL) as electrolyte is described. Here, through comparative experiments, CDs with sizes of 2.98, 3.43, and 5.92 nm were obtained using graphite rods and Pt as electrolytic material for anode and cathode. The method has the advantages of high reaction speed, low preparation cost, safety and controllability. The fluorescence properties of CDs prepared with graphite rod as anode and cathode are similar to those prepared with graphite rod as cathode and Pt as anode. Their optimal excitation/emission wavelength is (390/475 nm), which is much redshifted compared with the optimal excitation/emission wavelength (320/425 nm) obtained by using graphite rod as anode and Pt as cathode. Density functional theory calculation reveals that graphite nitrogen ratios of CDs have an effect on bandgap reduction, resulting in a redshift of the emission, which is in good agreement with the results of graphite nitrogen content of CDs obtained by XPS. These CDs are applied in light emitting diode (LED).

Poster Presentations

Efficient preparation of horizontal semiconducting carbon nanotube arrays by pre-hydrogenation and selective oxidation

Jia-Yang Zhang [1,2], Lingtong Ding [3,4], Mengke Zou [1,2], Xin Li [1,2], Zhong-Hai Ji [1,2], Lili Zhang* [1,2], Xiao Wang* [3,4], Chang Liu* [1,2]

[1]Shenyang National Laboratory for Materials Science, Institute of Metal Research, Chinese Academy of Sciences, Shenyang 110016, P.R. China

[2]School of Materials Science and Engineering, University of Science and Technology of China, Hefei 230026, P.R. China

[3]Institute of Technology for Carbon Neutrality, Shenzhen Institute of Advanced Technology, Chinese Academy of Sciences, Shenzhen, 518055, Guangdong, China.

[4]Guangdong Provisional Key Laboratory of Functional Oxide Materials and Devices, Southern University of Science and Technology, Shenzhen 518055, China.

Horizontal semiconducting single-wall carbon nanotube (s-SWCNTs) arrays are ideal materials for the next-generation integrated circuits due to their high carrier mobility and unique one-dimensional tubular structures. However, it is still of a great challenge to prepare high-purity horizontal s-SWCNT arrays. Herein, we predicted that hydrogenated SWCNTs might dramatically increase the difference in etching energy barriers between different conductivities by DFT calculations, then we prepared high-purity horizontal s-SWCNT arrays by pre-hydrogenating and a subsequent oxidative etching. We characterized the hydrogen chemisorption and dehydrogenation process of individual SWCNTs using quasi-in-situ Raman spectroscopy. It was found that the hydrogenation of SWCNTs is structure-dependent, i.e. smaller-diameter and metallic SWCNTs are more easily to chemisorb hydrogen. Most metallic SWCNTs were then removed during the oxidation process. Finally, we experimentally obtained horizontal s-SWCNT arrays with a high purity up to 99.0 %.

Influence of Precursor Structure on the Formation of Active Carbon Deposition Sites and Helical Graphite Cones Growth Mechanism

Zihan Chen[1], Yuhong Liu[2], Xingwei Zhang[1], Yan He[1], and Zhaokun Ma*[1]

[1]Beijing University of Chemical Technology, Beijing, P. R. China.
[2]GRIMAT Engineering Institute CO, LTD, Beijing, P. R. China.

Graphite cones (GCs) have large aspect ratios and high mechanical properties, which have attracted much attention due to their similarity to carbon nanotubes and vapor grown carbon fibers. But the growth conditions of GCs have not been explored from the perspective of the molecular structure of the precursor. Therefore, this paper is based on the in-situ preparation of helical graphite cones (HGCs) using different precursors of polyimide (PI) graphite fibers (GFs) doped with polycyclic aromatic hydrocarbons (PAH). By varying the type of precursor of PI fibers, it can be found that HGCs appeared on GFs with pyromellitic dianhydride/4, 4′-oxydiamine (PMDA/ODA) and biphenyl tetracarboxylic dianhydride/4, 4′-oxydiamine (BPDA/ODA) as precursors, but there is no growth of HGCs on GFs with biphenyl tetracarboxylic dianhydride/p-phenylenediamine (BPDA/PDA) as precursors. Based on this phenomenon it can be further discovered that the growth of HGCs is related to the ether-oxygen bonds in ODA. The ether-oxygen bonds broken during heat treatment provide active carbon deposition sites, which is a key factor for the growth of HGCs, thus producing a large number of HGCs. The new discovery at the molecular structure level has a significant impact on the future preparation of HGCs and the improvement of its growth mechanisms.

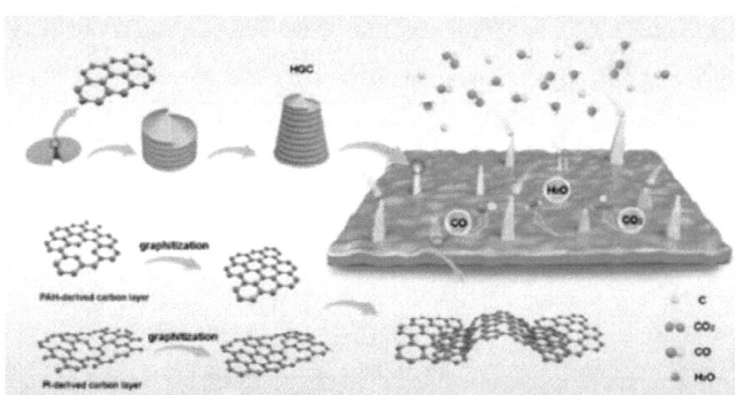

Figure. 1 Diagram of the grow mechanism of HGCs at the molecular level.

PCB or IC Substrate Microvia Metallization Based on Conductive Carbon

Qing-Sheng Zhu[1], Chang Liu[1], and Hui-Ming Cheng[1]

[1]Shenyang National Laboratory for Materials Science, Institute of Metal Research, Chinese Academy of Sciences, Shenyang, 110016, China

Electroless-plate of copper is the fundamental approach to metallization of microvia sidewalls in printed circuit board (PCB) or integrated circuit (IC) manufacture. However, the use of a large amount of formaldehyde and chelating agents in this process bring a terrible hazard to the environment and human health. Also, this process requires high-cost palladium as catalyst. We aim to develop a green, environment friendly, and low-cost metalization technology using conductive carbon. In this technology, the conductive film on sidewall of microvia is prepared by immersing the microvia into a nano-carbon aqueous dispersion and then drying. The following Cu electroplate can be directly carried out on the conductive carbon film instead of electroless plated copper film. The conductive carbon materials such as carbon black, CNTs and nano-graphite can be applied in this process. It requires special demands in carbon particles size, morphology and surface chemical groups. We have developed special nano-carbon materials for the copper direct electroplating process with high performances in stability, adsorption and copper deposition rate, which supported the advanced process such as "high-speed" and "one-step". This metallization technology based on conductive carbon will have a great application potential in PCB or IC substrate manufacture.

Enhancing Field Emission Properties of Cold Cathode Materials through Microstrcture-engineered Carbon Nanotube Fiber

Y.B Kim, S.H Gihm and K Jeong*

aweXome Ray Inc., Anyang-si, Gyeonggi-do 14056, Republic of Korea

Carbon nanotubes (CNTs) have garnered substantial interest as field electron sources for diverse applications, including X-ray tube cathodes, electron microscopy, and discharge tubes.[1,2] The notable advantages of CNT emitters—such as low work function, consistent electron emission, and superior thermal, chemical, and mechanical stability—make them ideal candidates for cold cathodes. Various types of CNT emitters have been investigated, including those grown via the chemical vapor deposition method, tailored bucky paper, and paste configurations. Despite numerous studies addressing performance variances among these emitters, the relationship between the microstructural characteristics of CNT emitters and their field emission properties has not been thoroughly explored. This study examines how the microstructure, including the orientation and integration of CNTs and their micro-aggregates, impacts the field emission properties of fibrous CNT materials. Utilizing the direct spinning method, we fabricate CNT fibers with controlled microstructures. The field emission characteristics of CNT fibres, including field enhancement factor, emission current density, and stability, are extensively evaluated. The results demonstrate that a high degree of CNT orientation and integration within the microstructure significantly improves the field emission stability and overall performance. These findings are expected to contribute valuable insights toward the development of advanced CNT-based cold cathodes.

References

[1] JEONG, et al. Applied Physics Letters, 102.2 (2013)
[2] ROSEN, et al. Applied Physics Letters 76.13 (2000)

Cu-containing carbon materials as catalysts for electrochemical CO2 reduction

Javier Quílez-Bermejo[1], Juan Herranz [1]

[1]Electrochemistry Laboratory, Paul Scherrer Institut (PSI), 5232 Villigen, Switzerland

Copper (Cu) stands out as the sole metal element known for the efficient electrochemical conversion of CO2 into hydrocarbons, with high catalytic performance and faradaic efficiencies [1]. This characteristic has generated considerable interest among researchers in the development of Cu-containing electrocatalysts. Nevertheless, recent works have revealed that seemingly similar Cu active sites show massive variations in selectivity and catalytic performance. These inconsistencies could stem from the catalytic support, often based on diverse types of carbon materials (e.g., with different pore size distributions, graphitization extents and/or surface functionalities).

Specifically, the features of the carbon support can exert a significant influence on the electronic properties of Cu active sites, thereby modifying the catalytic performance for CO2 reduction of Cu-containing carbon-based electrocatalysts. Therefore, in this contribution we have evaluated the effect of factors such as the Cu:C ratio, Cu nanoparticle size and carbon support on the CO2-reduction activity and selectivity of a series of Cu-containing carbon materials. These span from traditional nanoparticulate samples to single atom catalysts in which the active sites consist of isolated Cu-ions coordinated by N-functionalities hosted in the C-support (e.g., in the form of Cu-N4 sites alike those found in porphyrins and phthalocyanines). These catalysts' library was first submitted to electrochemical measurements (e.g., Pb underpotential deposition) to quantify their surface site density, along with complementary ex situ tests to determine their surface composition and the structure of the C-matrix. Subsequent electrochemical tests in a parallel plate cell with CO2-containing electrolyte served to establish relations among these catalytic features and those of their Cu- and C-sites.

References

[1] Lin.W, et al. Adv. Mater. (2024) 2313028.

Construction of L-Ascorbic Acid 2-Glucoside-assisted reduced graphene oxide foam for simultaneous electromagnetic interference shielding and thermal protection

Hailiang Huo[1,2] , Zechao Tao[2] and Shengliang Hu[1]

[1]North University of China, 3 College Road, Taiyuan, China

[2]Institute of Coal Chemistry, Chinese Academy of Sciences, 27 Taoyuan South Road, Taiyuan, China

Electromagnetic interference shielding materials with only a single shielding property cannot adapt to complex application environments. To address this demand, we have constructed an ultra-lightweight reduced graphene oxide foam with a remarkably low density of 0.06 g cm-3 with the assistance of L-Ascorbic Acid 2-Glucoside. In addition to giving an outstanding electromagnetic interference shielding effectiveness of 82.11 dB in the X-band, this foam also offered extraordinary thermal insulation performance with a low value of 0.025 W m-1 K-1. The exceptional properties of the foam were attributed to the formation of a continuous three-dimensional cross-linked highly electronic conductive network with uniformly distributed pore size. Not only did the network continually reflect, scatter, and eventually dissipate or absorb incoming electromagnetic energy, but multi-interfacial scattering due to cross-linked pore walls severely impeded heat transfer.

Influence of nano-sized carbon on chromium based electrochemical coatings

G. Yar-Mukhamedova, A. Kemelzhanova and G. Mussabek

Al-Farabi Kazakh National University, Al-Farabi av., 71, Almaty, Kazakhstan

When develop new nano-structured composition electrochemical coatings (nano-CEC) carbon is one of the most perspective of components for improving functional properties. Electrolytic chromium, in comparison with other electrodepositable metals [1-3], is the most corrosion-resistant, wear-resistant and has high hardness. Despite this, it is not protected against corrosion due to micro-porosity. Deposition of composite electrolytic coatings (CEC) using appropriate dispersed particles can improve the anti-corrosion properties of electrolytic chromium. The protective ability of the coating depends on the amount of the dispersed phase in the metal matrix. We present the results of experiments on introduction of nano-size particles of carbon C (11-100 nm) into chromium electrolytic coating and investigation of corrosion resistance in 3% NaCl solution. Nano-composition electrolytic coatings were obtained at the various current density: 2-7 kA/m2, temperature 303-343 K and nano-dispersion phase components' quantity content. For testing of corrosion resistance of chromium-carbon nano-composition electrochemical coatings the experimental device for corrosion resistance determining by a weight method adjusted for coatings working conditions was developed, assembled and completed. A test bench for corrosion resistance has been developed and created, which allows: conduct tests in various corrosive environments at temperatures from room temperature to 303-343 K. Quantitative assessment of the corrosion rate was carried out by a gravimetric method. The results showed that nano-composition electrolytic coatings Cr-C obtained from both electrolyte increase the corrosion resistance of steel St 3 in NaCl solutions up to 45-69 times. This research has been funded by the Science Committee of Kazakhstan' s Ministry of Science and Higher Education (BR21882188).

References

[1] Narivskyi O.E.et al. Materials Science (2021) 291-297.
[2] Yar-Mukhamedova, G.Sh. Materials Science (2023) 598–600.
[3] Aldabergenova T.M. et.al. Advanced Surface Science (2022) 1783-1788.

Fabrication of single-walled carbon nanotubes @ hexagonal boron nitride thin films with interconnected junctions

Changping Yu[1,2], Feng Zhang[1,2], and Chang Liu[1,2]

[1]Shenyang National Laboratory for Materials Science, Institute of Metal Research, Chinese Academy of Sciences, 72 Wenhua Road, Shenyang, China

[2]School of Materials Science and Engineering, University of Science and Technology of China, 96 Jinzhai Road, Hefei, China

Single-walled carbon nanotubes (SWCNTs) have great potential for thermal management materials due to their superior thermal conductivity. However, thermal resistance at the intertube joint leads to a serious degradation of thermal conductivity in SWCNTs macrostructures. Owing to the coherent lattice structures with SWCNT, hexagonal boron nitride (h-BN) can be integrated on SWCNT by epitaxial growth method [1], which could introduce more pathways for thermal conduction along the tubes. Herein, we propose a strategy to interconnect the junctions of SWCNT and introduce hexagonal boron nitride (h-BN) encapsulating layers on the surface of SWCNT. The strategy was achieved by adjusting the partial pressure of BN precursors in a two-step atmospheric pressure chemical vapor deposition method. Amorphous nanoparticles were firstly anchored at SWCNT junctions to interconnect the SWCNT networks. Crystallize h-BN was subsequently coated on the surface of SWCNTs and amorphous nanoparticles. The interconnected SWCNT@h-BN (I-SWCNT@h-BN) film exhibits significant enhancement (from 15.3 W/m·K to 67.9 W/m·K) in in-plane thermal conductivity after BN interconnection and encapsulation.

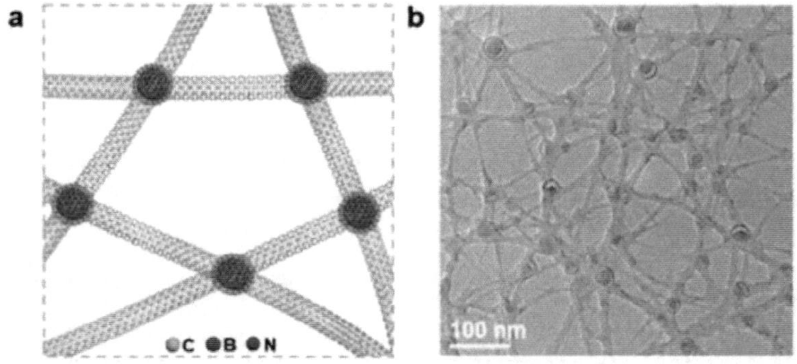

(a) Schematic of the typical network structure of I-SWCNT@h-BN, (b) Transmission electron microscopy image of I-SWCNT@h-BN.

References

[1] Yu, C, et al. Materials, 2023, 16: 1864.

High Performance Double-Wall Carbon Nanotube Fibers Prepared by Dry-Jet Wet Spinning

Hao-Zike Wang[1,2] , Peng-Xiang Hou[1,2*] and Chang Liu [1,2*]

[1]Shenyang National Laboratory for Materials Science, Institute of Metal Research,
Chinese Academy of Sciences, Shenyang 110016, China

[2]School of Materials Science and Engineering, University of Science and Technology of China,
Hefei 230026, China

*Corresponding Authors: pxhou@imr.ac.cn (P.X. Hou) or cliu@imr.ac.cn (C. Liu).

Carbon nanotube fibers (CNTFs) are considered an ideal candidate as next generation conducting wires due to their high conductivity and ampacity, low density, high strength and toughness. However, the orientation and compaction of the carbon nanotubes (CNTs) in a CNTF need to be further improved to achieve even higher conductivity and ampacity. Herein, we report the fabrication of double-wall CNTFs (DWCNTFs) by a dry-jet wet spinning method, which significantly improves the orientation and compaction of the DWCNTs. At the same time, we found that long DWCNTs with high molecular weight could be successfully spun under a stable and long air gap. As a result, the obtained DWCNTFs had high density (1.96 g/cm3) and orientation (0.994), showing a record high electrical conductivity of 1.1×107 S/m and ampacity of 8.0×108 A/m2, as well as a high tensile strength of 1.65 GPa and a toughness of 130.9 MJ/m3, among the highest for wet-spun CNTFs. Our CNTFs preserved their integrity and conductivity after more than 5000 bending cycles. Theoretical calculations indicate that in a 2 mm air gap under gravity effect could significantly promote the orientation of the CNTs along the axial direction of the fiber.

Isopropyl Alcohol-Assisted FCCVD Synthesis of Semiconducting-Enriched SWCNTs for High-Performance Photodetectors

Muhammad Naeem Nizam[1,2], Peng-Xiang Hou[1,2,*], Chang Liu[1,2,*], and Hui-Ming Cheng[1,2,*]

[1]Shenyang National Laboratory for Materials Science, Institute of Metal Research, Chinese Academy of Sciences, Shenyang, 110016, China

[2]University of Science and Technology of China, No.96, JinZhai Road Baohe District, Hefei, Anhui, 230026, P.R.China.

The selective synthesis of semiconducting single-walled carbon nanotubes (s-SWCNTs) has great significance in many high-end applications due to high carrier mobility, tunable band-gap, exceptional flexibility, and good stability. However, it remains a big challenge to directly grow high purity, high-quality s-SWCNTs in large scale. Here, we developed a general approach to grow s-SWCNTs utilizing the floating catalyst chemical vapor deposition (FCCVD) method with isopropyl alcohol (IPA) as the carbon source. IPA facilitated the preferential growth of s-SWCNTs, offering a promising route for tailored material synthesis with enhanced electronic properties. Our findings demonstrate that isopropyl alcohol played a crucial role in promoting the growth of s-SWCNTs within the FCCVD process. The resulting s-SWCNTs exhibit a high degree of enrichment, significantly improving their suitability for high-performance photodetector applications. We successfully implemented these s-SWCNTs in photodetector fabrication and observed a substantial enhancement in their performance. This research highlights the potential of FCCVD with isopropyl alcohol as a cost-effective and efficient approach for fabricating s-SWCNTs for high-sensitivity photodetectors.

Preparation of Small Diameter Carbon Nanotube Fibers with High Current Carrying Capacity

Chun-Yang Sun[1], Peng-Xiang Hou*, Chang Liu*, and Hui-Ming Cheng*

[1]Shenyang National Laboratory for Materials Science, Institute of Metal Research, Chinese Academy of Sciences, Shenyang, 110016, China

[2]University of Science and Technology of China, No.96, JinZhai Road Baohe District, Hefei, Anhui, 230026, P.R.China.

Carbon nanotubes (CNT) have been considered as one of the most promising materials with its excellent high strength, electrical conductivity, ampacity, and low density. However, assembling CNTs into macroscopic fibers (CNTFs) always greatly decrease the electrical properties, especially ampacity due to the local heat concentration from both the large current and low thermal conductivity of CNTs interface layer[1]. Herein, we optimized the method of wet spinning to fabricate CNTFs with small-diameter by replacing the toxic and volatile acetone coagulation bath in the traditional wet spinning process. The obtained CNTFs have a small diameter distribution of (3±0.5) μm, which shows an excellent current carrying capacity of 1×109 A/m2 and a tensile strength of 2 GPa.

References

[1]H. Li, Z. Yong, D. Liu, K. Wu, L. Guo, X. Wang, H. Jin, Q. Li, Carbon 2024, 219.

Preparation of Small Diameter Carbon Nanotube Fibers with High Current Carrying Capacity

Chun-Yang Sun[1], Peng-Xiang Hou*, Chang Liu*, and Hui-Ming Cheng*

[1]Shenyang National Laboratory for Materials Science, Institute of Metal Research, Chinese Academy of Sciences, Shenyang, 110016, China

[2]University of Science and Technology of China, No.96, JinZhai Road Baohe District, Hefei, Anhui, 230026, P.R.China.

Carbon nanotubes (CNT) have been considered as one of the most promising materials with its excellent high strength, electrical conductivity, ampacity, and low density. However, assembling CNTs into macroscopic fibers (CNTFs) always greatly decrease the electrical properties, especially ampacity due to the local heat concentration from both the large current and low thermal conductivity of CNTs interface layer[1]. Herein, we optimized the method of wet spinning to fabricate CNTFs with small-diameter by replacing the toxic and volatile acetone coagulation bath in the traditional wet spinning process. The obtained CNTFs have a small diameter distribution of (3±0.5) μm, which shows an excellent current carrying capacity of 1×109 A/m2 and a tensile strength of 2 GPa.

References

[1]H. Li, Z. Yong, D. Liu, K. Wu, L. Guo, X. Wang, H. Jin, Q. Li, Carbon 2024, 219.

Synthesis of single-wall carbon nanotube fiber non-woven fabrics as electrothermal heating element

Yu-ke Liu[1,2], Xin-yu Jiao[1,2], Xin-yang Sun[1,2], Peng-Xiang Hou[1,2,*], Chang Liu[1,2,*] and Hui-Ming Cheng[1,3]

[1]Shenyang National Laboratory for Material Science, Institute of Metal Research, Chinese Academy of Science, 110016, Shenyang, China

[2]School of Material Science and Engineering, University of Science and Technology of China, 230026, Hefei, China

[3]Faculty of Material Science and Engineering/Institute of Technology for Carbon Neutrality, Shenzhen Institute of Advanced Technology, Chinese Academy of Science, 518055, Shenzhen, China

Fabrics woven from carbon nanotube (CNT) fibers are highly desired in wearable electronics, smart textiles, and electrical heater. However, the weak interaction at fiber joints significantly limits electron and heat transfer and decreases their electrical and thermal conductivities. In this work, a simple pressure-fusing method to fabricate single-wall CNT (SWCNT) fiber non-woven fabrics (NWFs) is reported, these flexible NWFs are composed of interconnected SWCNT fibers with fused joints, with a density of 0.46 g/cm3, a high electrical conductivity of 3.7×105 S/m, and a record high specific electrical conductivity of 803 (S·m2)/kg. Furthermore, these NWFs have excellent electro-heating ability, which can be rapidly heated to about 160 ℃ at the voltage of 2 V[1]. Because of their low density, SWCNT fiber NWFs are prospective to be used as a heating unit of low temperature battery protection and de-ice devices.

References

[1] Xinyu Jiao, Lele Xu, Xinyang Sun, et al. Single-wall carbon nanotube fiber non-woven fabrics with a high electrothermal heating response. Nano Research 17(2024) 5621-5628.

Preparation of High-Quality SWCNT fiber Tows

You-Quan Zhang [1], Hao-Zike Wang [2] Xin-Yu Jiao [3], Peng-Xiang Hou*, Chang Liu*

[1]Shenyang National Laboratory for Materials Science, Institute of Metal Research (IMR), Chinese Academy of Sciences, 72 Wenhua Road, Shenyang, 110016, China.

[2]University of Science and Technology of China, No.96, JinZhai Road Baohe District, Hefei, Anhui, 230026, P.R.China.

Carbon nanotubes fibers (CNTFs) spun by wet spinning method have attracted much attention as the next generation conductor, because of their excellent electrical, mechanical, and thermal properties[1]. However, it is necessary to obtain CNTF tows in order to realize the practical applications. In this work, we use high quality employed single-wall (SW) CNTs with high aspect ratio as raw materials to prepare SWCNTF tows (SWCNTFTs) by a common wet spinning method. For the first time, we obtained SWCNTFTs composed of numerous highly-oriented, densely-packed SWCNTs which shows an excellent conductivity of 2 MS/m and ampacity of 7.1×10^3A/cm2, as well as high tensile strength.

References

[1] Jiao X , Shi C , Zhao Y ,et al.Efficient Fabrication of High-Quality Single-Walled Carbon Nanotubes and Their Macroscopic Conductive Fibers[J].ACS nano, 2022, 16(12):20263-20271.

Preparation and performance of nanoparticle polyANI compo-site-expanded graphite bipolar plate

Kai Fang [1], Kai Cheng [1], Shuyin Liu [1], Hongxin Liu [1], Haodong Tan [1], Qingting Liu [1,*]

[1]Key Laboratory of Green Light Industry Materials of Hubei Province, Hubei Univer-sity of Technology, Wuhan, China

Composite bipolar plates, while mitigating the shortcomings of metallic counterparts such as low mechanical strength, poor processability, and susceptibility to corrosion, offer advantages including cost-effectiveness and reduced weight. However, achieving a balance between electrical and mechanical performance remains challenging due to suboptimal compatibility between conductive fillers and the resin matrix. Li et al [1] blended polypropylene (PP), graphite, and epoxy resin to fabricate composite bipolar plates. The resultant material demonstrated a bending strength of 55 MPa and an elec-trical conductivity of 55.24 S/cm.

In this study, uniform growth of polyaniline (PANI) nanosheets on expanded graphite surfaces was achieved through in situ electrochemical polymerization, facilitating the development of composite bipolar plates with enhanced electrical conductivity and me-chanical properties. The electrochemical method employed allows for precise control over the polymerization rate and process, yielding polymers that are highly uniform and well-dispersed. Serving as an interface compatibility agent, the PANI nanosheets en-hance the interaction between the expanded graphite and resin, thus improving the elec-trical conductivity and mechanical robustness of the composite. The resulting bipolar plates exhibit an electrical conductivity of 266 S/cm and a bending strength of 46.2 MPa, alongside superior corrosion resistance and thermal conductivity. These attributes underscore the potential of PANI nanolayer-treated bipolar plates as a promising avenue for the advancement of composite bipolar plate technology.

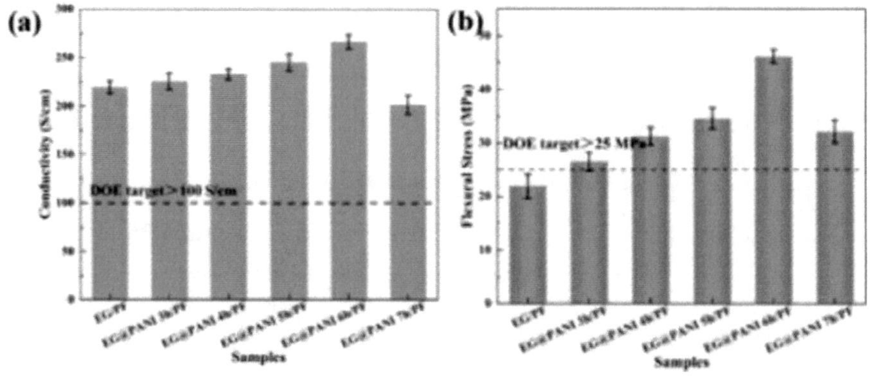

Figure 1 Conductivity (a) and flexural strength (b).

References

[1] Li Tuanfeng, et al Journal of Materials Engineering 522 (2024).

Study on molding process of expanded graphitephenolic resin composite bipolar plates

Kai Cheng[1], Kai Fang[1], Shuyin Liu[1], Hongxin Liu[1], Haodong Tan[1], Qingting Liu[1,*]

[1]Key Laboratory of Green Light Industry Materials of Hubei Province,
Hubei University of Technology, Wuhan, China

Composite bipolar plate has good electrical conductivity, mechanical strength, corrosion resistance, and good processing performance, low cost, but it has the problem of imbalance between electrical conductivity and mechanical properties. Hu et al. [1] fabricated bipolar plates composed of a dual conductive synergistic structure, consisting of Gr/MWCNTs with low filler contents of 5 wt% MWCNTs and 35 wt% Gr. This structure, characterized by its segregated bias configuration, exhibited an electrical conductivity of 161.57 S/cm and a bending strength of 43 MPa.

In this study, composite bipolar plates for fuel cells were fabricated using expanded graphite as the conductive substrate and phenolic resin as the impregnant, through processes of vacuum impregnation and hot press molding. The influence of vacuum impregnation duration, molding temperature, and hold time on the plates' performance was systematically evaluated. Vacuum impregnation facilitates the rapid, uniform, and thorough penetration of the impregnant into the fine pores of the substrate, thereby enhancing its overall properties. Optimal conductivity (261 S/cm) and flexural strength (60.62 MPa) were achieved at a molding temperature of 130°C and a hold time of 60 minutes. The superior electrical conductivity and mechanical properties of the expanded graphite/phenolic resin composite bipolar plates indicate significant potential for their application in fuel cells.

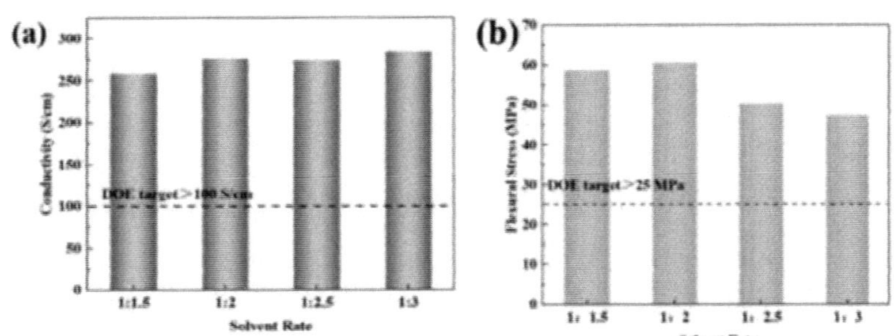

Figure 1 Conductivity (a) and flexural strength (b).

References

[1] Hu B, et al. International Journal of Hydrogen Energy, 2021, 46(50): 25666-25676.

Preparation of porous carbon with adjustable structure by pyrolysis of hypercrosslinked polymers

Zhengyang Liu[1,2], Tao Yang[1], Yan Song*[1,2], Xiaodong Tian

[1]Key Laboratory of Carbon Materials, Institute of Coal Chemistry,
Chinese Academy of Sciences, Taiyuan 030001, China
[2]Center of Materials Science and Optoelectronics Engineering,
University of Chinese Academy of Sciences, Beijing 100049, China
*Corresponding author: yansong1026@126.com

Achieving an adjustable pore structure of porous carbon through direct pyrolysis of polymers remains challenging. In this study, a series of hypercrosslinked polymers (HCPs) with porous structures were synthesized by using naphthalene as the raw material via Friedel-Crafts alkylation reactions in various dichloro-substituted solvents through the external crosslinking method. Subsequently, porous carbon was prepared by direct pyrolysis of HCPs (Figure 1). Characterization results indicated the involvement of both solvent and crosslinker in polymer construction. The presence of the crosslinker resulted in a microporous structure. The meso/macroporous structures of HCPs depended on the methylene chain lengths of the dichloride-substituted solvents. Additionally, HCPs-derived porous carbon exhibited a micro/mesoporous and macroporous structure. Adjusting the pyrolysis temperature enabled the specific surface area of the porous carbon to approach that of the precursor (approximately 1000 m^2 g-1). Due to the high carbon yield, adjustable pore structure and cost-effectiveness, HCPs-derived porous carbons showed significant potential in the field of energy storage.

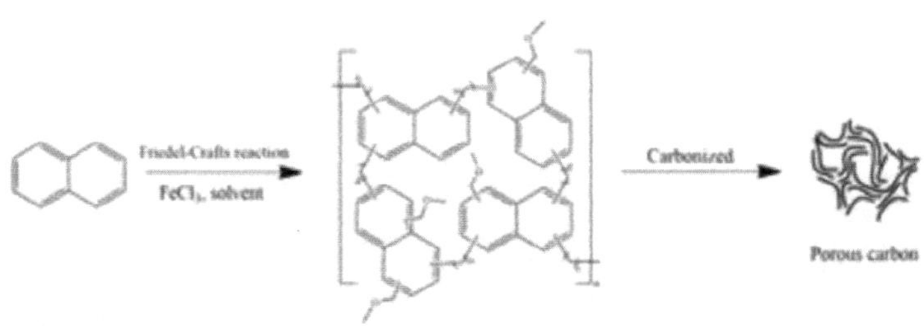

Figure 1. Schematic illustration of the synthesis for HCPs-derived porous carbons.

Reuse of Waste Road Asphalt: Fe/Fe3C nanoparticles encapsulated in N-doped porous carbon framework for rechargeable liquid and flexible all-solid-state zinc-air batteries

Juan Zhang[1]

[1]Institute of Coal Chemistry, Chinese Academy of Sciences, 27 Taoyuan South Road, Taiyuan, Shanxi 030001, P. R. China.

The conversion of the waste road asphalt into valuable carbon composites as efficient oxygen electrocatalysts has a significant potential for a clean and sustainable world. Herein, a facile emulsification-carbonization strategy is developed to prepare Fe and Fe3C nanocrystals encapsulated in N-decorated microcellular porous carbon from waste road asphalt. The mesopore-rich hierarchically porous structure accelerates the mass transport after the introduction of non-ionic emulsifier F-127, meanwhile, the powerful synergy between Fe-C and Fe-Nx bond serve as active sites. Simultaneously, the resultant catalyst shows a decent catalytic activity that is comparable to Pt/C toward the oxygen reduction reaction and an outstanding oxygen evolution reaction activity as well as excellent conductivity in alkaline mediums. It also delivers a considerable power density of 164.1 mW cm-2, a low charging-discharging voltage gap of 1.20 V and long-term durability of 200h at 5mA cm-2. This work opens new insights into improving electrocatalytic activity in terms of mass transport kinetics and synergy mechanism of the active sites.

Improved through-plane thermal and electrical conductivity of expanded graphite composites via in-situ deposited silver nanoparticles

Yuyuan Fan[1,2], Xueyao Feng[1,2], Zeyu Wang[1,2], Junfen Li[3], Zechao Tao[1,4], Xi Yan[1,4,*], Zhanjun Liu[1,4,*]

[1]CAS Key Laboratory of Carbon Materials, Institute of Coal Chemistry, Chinese Academy of Sciences, 030001, Taiyuan, China

[2]University of Chinese Academy of Sciences, 100049, Beijing, China

[3]State Key Laboratory of Coal Conversion, Institute of Coal Chemistry, Chinese Academy of Sciences, 030001, Taiyuan China

[4]Center of Materials Science and Optoelectronics Engineering, University of Chinese Academy of Sciences, 100049, Beijing, China

Expanded graphite (EG) has exhibited an increasingly prominent potential as a heat dissipation material (HDM) owing to its exceptional thermal conductivity and chemical stability. However, the inferior through-plane thermal conductivity stemming from the absence of transmission channels between graphite layers impedes its further practical applications. Herein, silver nanoparticles (AgNPs) were in-situ deposited on the surface and interlayers of EG using a one-step hydrothermal reduction method, and the expanded graphite/AgNPs composite (EGAs) was prepared followed by compaction. As AgNPs can serve as thermally conductive bridges to connect adjacent graphite layers, continuously three-dimensional heat transfer pathways were successfully established in the EG. The obtained composite achieved an enhanced through-plane thermal conductivity of 7.06 W m-1 K-1 and a through-plane electrical conductivity of 5915 S cm-1, respectively at an optimized Ag loading of 22.39%. This research provides an effective strategy for the design and fabrication of HDMs with upgraded through-plane thermal conductivity.

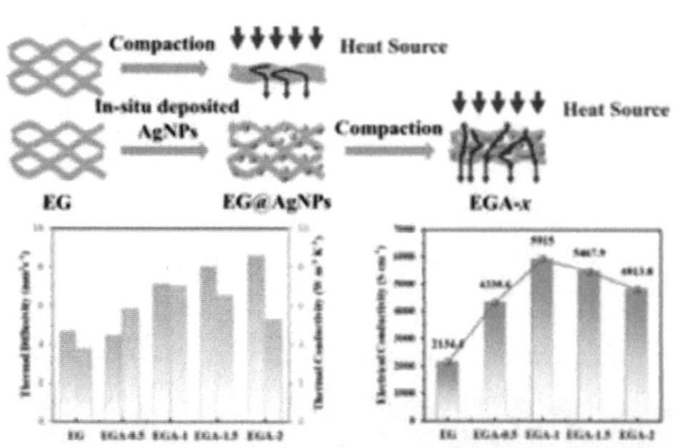

Promotion effect of trimeric acid on the oxidation of coal tar pitch

Yiting He[1,2], Yan Song*[1,2], Tao Yang[1,2], Xiaodong Tian[1,2]

[1]CAS Key Laboratory of Carbon Materials, Institute of Coal Chemistry,
Chinese Academy of Sciences, Taiyuan, China
[2]University of the Chinese Academy of Sciences,
Center of Materials Science and Optoelectronics Engineering, Beijing, China
*corresponding author: yansong1026@126.com

In this work, the effect of trimeric acid on air oxidation of coal tar pitch (CTP) were investigated. The results showed that the cross-linked oxygen-containing functional groups in the as-prepared samples increased to 5.86% and the softening point increased to 310 ℃, which were much higher than those without trimeric acid addition. Thus the addition of trimeric acid promote the efficiency of air oxidation of CTP. More crosslinked oxygen-containing functional groups improved the average molecular weight, softening point and thermal stability of CTP.

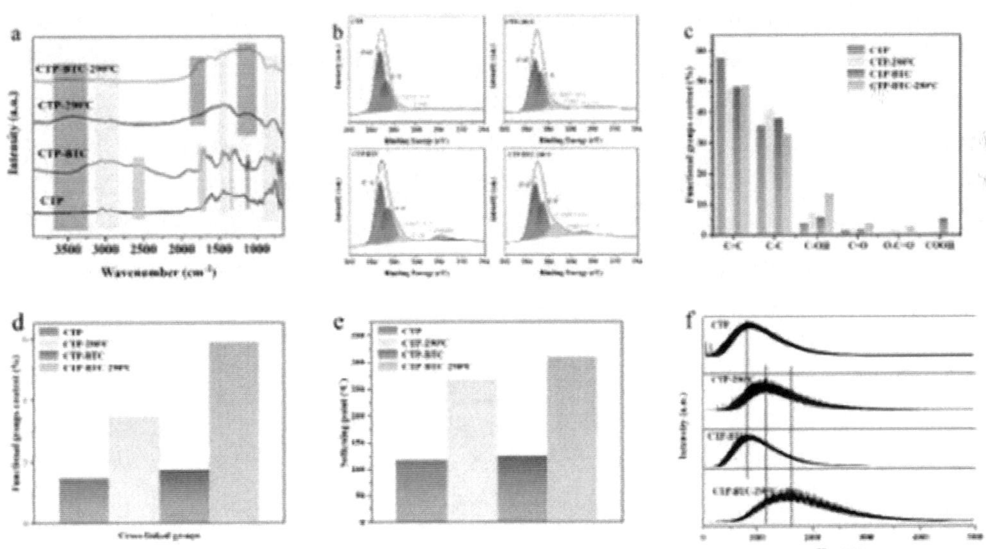

Fig. a. The FT-IR; b. The C1s XPS spectra; c. The content of different groups; d. The content of cross-linked functional groups; e. The softening point; f. The MOLDI-TOF spectra of the samples.

Breaking the "dead Li" Barrier: A Cross-Stacked Dual-function Framework by SWNTs in Graphite-Li Hybrid Anodes

Yan Liu[1], Hao Zhang[1]

[1]Research Institute of Chemical Defense, No.35 Huayuan North Road, Beijing 100191, China

The accumulation of "dead Li" between lithiated graphite particles leads to interfacial failure, rapid capacity degradation, and reduced longevity in graphite-Li hybrid anodes. To address this issue, we developed a cross-stacked dual-function framework incorporating single-walled carbon nanotubes (SWNTs) that facilitates efficient electron/ion transmission and creates a quasi-three-dimensional porous structure. This framework establishes unobstructed pathways for electron and ion transport between graphite particles, overcoming transmission barriers caused by "dead Li" accumulation and side reactions. Additionally, the quasi-three-dimensional porous structure accommodates more "dead Li" minimizing its accumulation on graphite particles and preserving the activity of graphite intercalation compounds (GICs: LiC12, LiC6, etc.). The dual-function framework ensures excellent cycling performance, achieving 345 cycles at a lithiation level of 500 mAh·g-1 with an average Coulombic efficiency of 98.7%. When paired with lithium iron phosphate (LFP) cathodes, the hybrid anode demonstrates remarkable capacity retention of 80% over 300 cycles. This work presents an effective strategy for enhancing the performance of hybrid anodes through rational optimization of mesoscale interfacial transmission and electrode structure.

Effect of heat treatment temperature on the structure and properties of mesocarbon microbeads based carbon blocks

Xueyao Feng[1,2] , Lei Zheng[1], Hongchao Zhao[1,2] and Quangui Guo[1,2]

[1]CAS Key Laboratory of Carbon Materials, Institute of Coal Chemistry, Chinese Academy of Sciences, Taiyuan, 030001, China.

[2]University of Chinese Academy of Sciences, Beijing, 100049, China

This study aimed to determine the effect of heat treatment temperature on the structure and properties of mesocarbon microbeads (MCMBs) based carbon blocks. It was found that 500-900°C was a crucial stage of the carbonization process, during which small gas components such as CH4, H2, and H2O escaped leading to significant volume shrinkage and mass loss, resulting in increased density of the carbon blocks. With the increase in heat treatment temperature, the internal structure of the carbon blocks tended to be orderly, the degree of graphitization gradually increased, and the microcrystalline defects decreased. The mechanical properties reached their maximum value during the carbonization stage of 1400-1800°C. At a heat treatment temperature of 2400 °C, the material density reached its maximum value, accompanied by the highest thermal conductivity and lowest electrical resistivity, while the dense structure restricted the reduction of the coefficient of thermal expansion.

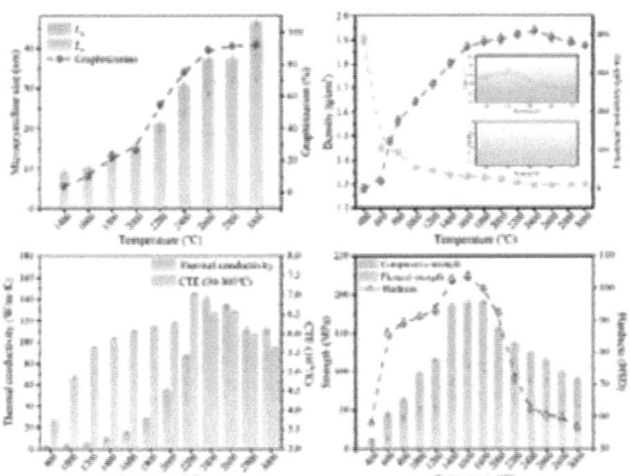

Effect of isopropyl chloride addition in AlCl3 catalytic system on the formation of coal tar pitch-based mesophases

Wensheng Chen[1], Chunfeng Duan[1], Zhipeng Yin, Mohammad Tabish, Zhaokun Ma, Xiaohong Chen, Huaihe Song*

State Key Laboratory of Chemical Resource Engineering,
Beijing Key Laboratory of Electrochemical Process and Technology for Materials,
Beijing University of Chemical Technology, Beijing 100029, PR China

Mesophase pitch (MP) is a high-quality precursor for the synthesis of high-performance pitch-based carbon fibers. The mesophase content, texture and softening point are the most important factors for the structure, properties and spinnability of mesophase pitch. Herein, we report a novel approach for the preparation of mesophase pitch with high anisotropic content and low softening point by AlCl3 catalytic pyrolysis in the presence of isopropyl chloride (PrCl). The variation of anisotropic content and mesophase texture under different PrCl additions and the effect of the introduction of isopropyl on the molecular stacking of aromatic layer molecules were investigated. The results confirmed that, except for their role in improving catalytic activity, carbocation co-catalysts can improve the structure and molecular assembly of aromatic molecules through the introduction of alkyl groups. The reaction of PrCl as a co-catalyst catalyzed pyrolysis of medium-temperature coal pitch at 370 ℃ for 8 h produced the 100% bulk mesophase. In addition, the introduction of isopropyl group significantly increases the stacking spacing of the mesophase layer molecules, which improves the mobility of the layer molecules causing the mesophase refinement. This work provides feasible ideas for the preparation of high-performance coal tar based mesophase pitch.

Influence of initial pore size of carbonization blocks on physical property and microstructural variation during impregnation process

U Sang Youn[1] and Jae Seung Roh[1]

[1]Kumoh National Institute of Technology, Deahak-ro 61, Gumi, Gyeongbuk, Korea 39177
(*jsroh@kumoh.ac.kr)

During the graphite block manufacturing process, the carbonization process causes internal pores to form due to pyrolysis or volatilization of the binder material, which degrades its properties[1]. The formed pores can be filled by an impregnation process, which can be repeated[2].

In this study, we observed the change in properties according to the number of impregnations during the impregnation of carbonization blocks with different initial pore sizes. The raw materials used in the study were needle coke, coal-based binder pitch (softening point 110 °C, QI 4-8 %), and impregnation pitch (softening point 91 °C, QI 2 %). After the needle coke was classified, it was kneading with a binder pitch and unidirectional pressing. It was carbonization at 1000 ℃. Impregnation and re-carbonization three times. The particle size-specific carbonization blocks are named Carbonization Block (CB)-D(50) using D(50), the average particle size of the needle coke used in the manufacture.

Before impregnation, CB-413 has the highest physical properties. However, after three impregnations, CB-52.8 has the highest physical properties. It is suggested that CB-52.8 has smaller pores than CB-413, which allows for deeper penetration of the impregnation pitch, resulting in better physical properties after impregnation.

References

[1] Sang Min Lee, et al. Carbon Letters 16 (2015) 135-146
[2] S. Ragan, et al. Journal of materials science 18 (1983) 3161-3176

Effect of air oxidation on the structure of toluene soluble fraction of coal liquefication pitch and derived carbon materials

Suxia Qi[1,2], Tao Yang[1], Yan Song*[1,2], Jinru Wu[1,2], Junqing Liu[3]

[1]CAS Key Laboratory of Carbon Materials, Institute of Coal Chemistry, Chinese Academy of Sciences, Taiyuan, 030001, China
[2]University of the Chinese Academy of Sciences, Center of Materials Science and Optoelectronics Engineering, Beijing, 100049, China
[3]National Institute of Clean-and-Low-Carbon Energy, Beijing, 102211, China
*corresponding author: yansong1026@126.com

The toluene soluble (TS) fraction of pitch has abundant aliphatic side chains which exhibits higher oxidative reactivity. In this work, the TS fraction of coal liquefication pitch was oxidized by air at different temperatures followed by carbonization to prepare pitch-based carbon materials. The effects of oxidation temperature on the molecular structure of TS fraction and the microstructure of subsequent carbonized samples were investigated. The results revealed that the cross-linking reaction occurred during oxidation, which led to in the transformation of TS fraction from thermoplastic to thermosetting in the carbonization stage to form a more disordered microcrystalline structure. With the increasing of oxidation temperature, more cross-linked structures were formed. At the oxidation temperature of 350 ℃, the (002) crystal plane of carbonized sample had the widest diffraction peak and the largest layer spacing, which was more in line with the hard carbon.

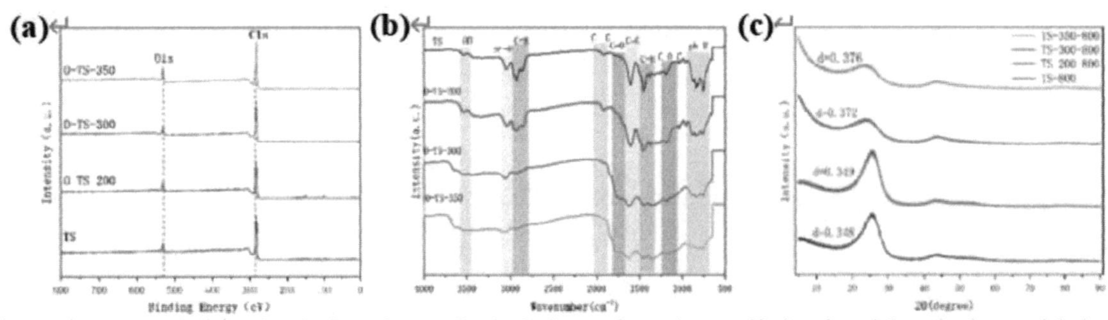

Fig.1. (a and b) XPS and FT-IR spectra of oxidation samples. (c) XRD patterns of carbonized sample at 800 ℃

Microstructural Representations and Reaction Mechanisms during Multiple High-Temperature Processes of Complex Amorphous Carbon Materials

Qifan Zhong [1,2], Zihan You [1], Liuzhou Zhou[1] and Yu Zhou

[1]School of Metallurgy and Environment, Central South University, Changsha 410083, China;
[2]National Engineering Research Center of Low-carbon Nonferrous Metallurgy, Changsha 410083, China;

The complex amorphous carbon materials (CAC) containing petroleum coke, calcined petroleum coke, and coal tar pitch are crucial feed stocks for artificial carbon and graphite products. Lack of exploration of microstructural transformation and reaction mechanisms of CAC during multiple high temperature processes have hindered the development of preparations and applications of graphite products. Here, the microstructural features and functional groups were investigated based on high-resolution transmission electron micrographs (HRTEM) analysis. The atomistic representations of CAC incorporating experimental parameters were firstly constructed for comprehensive observation and simulation via the self-developed automated modeling strategy [1]. The reaction mechanisms of CAC during pyrolysis, combustion, gasification, coking, calcination, graphitization, and molten electrolysis were explored via combining reactive force field molecular dynamics (ReaxFF MD), density functional theory (DFT) calculation, and advance detection methods. The molecular structural evolution, as well as the removal and catalytic mechanism pathways of the impurities (S, N, Na, V, Fe, and etc.) in CAC were ascertained [2].

References

[1] Qifan Zhong, et al. Carbon 129 (2018) 790-802.
[2] Lei Tang, et al. Carbon 188 (2022) 336-348.

Physical Property Variation of Graphite Blocks According to Graphitization Heating Rate in the Puffing Temperature Range

Min-Hyeok Son[1], U-Sang Yoon[1] and Jae-Seung Roh[1]

[1]Kumoh National Institute of Technology, Daehak-ro 61, Gumi-si, Gyeongsangbuk-do, Republic of Korea

The puffing that occurs in the temperature range of 1600 ~ 2300℃ during heat treatment is a phenomenon that causes pores and cracks due to the release of N and S gases[1]. It is essential to control the appropriate heating rate in the puffing zone.

In this study, the volume change of extrusions was measured in the temperature range between 1400℃~2200℃ to identify the start and end of puffing. The puffing temperature range was selected and the properties of anisotropic graphite blocks were measured after graphitization process up to 2800℃ with different heating rates of 1℃/min, 2℃/min, and 5℃/min.

At 1400℃, the volume change rate was 0.06% and the density change rate was -0.21%. At 2000℃, it was the highest at 3.72% and -4.46%, respectively, and decreased to 1.37% and -2.55% at 2200℃, respectively. From this, the puffing temperature range was selected from 1400℃ to 2000℃. At 1℃/min, the bulk density was 1.67 g/cm3 and the porosity was 17.8%. At 2℃/min, 1.66 g/cm3 and 16.7% were shown respectively. At 5℃/min 1.65 g/cm3 and 17.0% were shown respectively. This study confirmed that condition A was superior as a graphitization process condition.

References

[1] Lee, S. H, et al. Carbon 208 (2023) 443-451

Effect of molecular structure of pitch on microstructure and electrochemical performance of subsequent carbonized products

Jinru Wu[1,2], Tao Yanga, Yan Song* [1,2]

[1]CAS Key Laboratory of Carbon Materials, Institute of Coal Chemistry, Chinese Academy of Sciences, Taiyuan 030001, China

[2] University of the Chinese Academy of Sciences, Center of Materials Science and Optoelectronics Engineering, Beijing 100049, China

*corresponding author: yansong1026@126.com

To understand the reaction mechanism of molecular configuration on the microstructure and the performance of subsequent carbonized products, four coal tar pitches with different molecular configurations as raw materials were prepared. The results showed that long alkyl side chains provide a large number of free radicals during pyrolysis, which promotes the parallel arrangement of aromatic molecules, leading to an improvement in electrochemical performance for lithium-ion batteries. Carbonization products with a large amount of edges carbon can be obtained when aromatic molecules follow linear and nonlinear simultaneous addition patterns. Lithium ions are induced to be stored at the edges and surfaces of the carbon materials. The presence of cyclopenta-fused rings and oxygenated aromatics reduces the orientation of carbon stacks and the flatness of the microcrystalline, providing a large number of microspaces for lithium storage. This paper provides a framework for the targeted preparation of pitch-based carbon materials at the molecular level to satisfy the better electrochemical performance of lithium-ion batteries.

Enhanced self-sintering mechanical performance of carbon blocks via optimizing composition of green petroleum coke

Jiao Tan [1], Pei Gong [1], Ping Liu [1], Yanli Liu [1], Run Li [1,*] and Chuanjun Tu [1,*]

[1]College of Materials Science and Engineering, Hunan University, Hunan Province, Changsha, China

* Corresponding authors

Self-sintering of green petroleum coke (GPC) offers a critical approach to prepare high performance carbon blocks owing to its relatively short production cycle. However, the volatilization of small molecular chemicals in GPC inevitably results in the formation of pores and microcracks, thus severely limiting its commercial application. Herein, we report a strategy for optimizing composition of GPC by sequential extraction with dichloromethane and toluene, respectively [1]. Thanks to the removal of light molecular-weight compounds, dichloromethane-soluble (560.76 m/z, DCMS) fraction with long alkyl side chains, when the solvent purified GPC are employed as precursor, the obtained carbon blocks DCMI-C demonstrates flexural strength of 17.78 MPa and compressive strength of 54.67 MPa, which is 1.7 and 1.9 times that of pristine GPC-C counterparts, respectively, indicating that the DCM soluble compounds are the mainly reason for pores and microcracks generation. Moreover, the dichloromethane-insoluble-toluene-soluble (DCMI-TS, 624.39 m/z) fraction contains more naphthenic, oxygen-containing compounds and plasticizer species e.g. dibutyl phthalate compared with DCMS, which is favorable to carbonization due to its contribution to the improvement of the reactivity and mobility of GPC. As a result, a much more compacted pore structure is also observed in DCMI-C implying its remarkable mechanical strength.

References

[1] Jiao Tan, et al. Fuel 343 (2023) 127912

Mechanical and Electrical Properties Variation of Isotropic Carbonized Blocks with Kneading Holding Time

Eun Jeong Kim [1], U-Sang Yoon[1] and Jae Seung Roh[1]

[1]Kumoh National Institute of Technology, Daehak-ro 61, Gumi-si, Gyeongsangbuk-do, Republic of Korea

Graphite blocks are manufactured by mixing coke and binder pitch in a certain ratio and kneading at a temperature above the softening point of the binder pitch to produce green bodies by uniaxial pressure molding, etc[1]. Among them, the kneading process has influenced on the properties of graphite blocks, and enables the binder to uniformly coat the surface of the filler[2].

In this study, kneading at 170℃ in 1-hour increments from 0 to 2hours, the block was manufactured by uniaxial compression molding and carbonization. Subsequently, the bulk density, electrical resistivity, flexural strength, and electrical resistivity were quantified to ascertain the mechanical and electrical properties.

Fig. 1(a) shows bulk density and porosity variation with kneading holding time. At 0hr, the bulk density and porosity were 1.457 g/cm3 and 25.6%, respectively, and the bulk density increased and the porosity decreased as the kneading time increased, until at 2hr, the bulk density and porosity were 1.573 g/cm3 and 20.2%, respectively. Fig. 1(b), (c) are microstructures of the carbonized blocks after 0 and 2hr of kneading holding. It can be seen that the pore size decreased due to the increased connectivity of the filler particles as the kneading holding time increased.

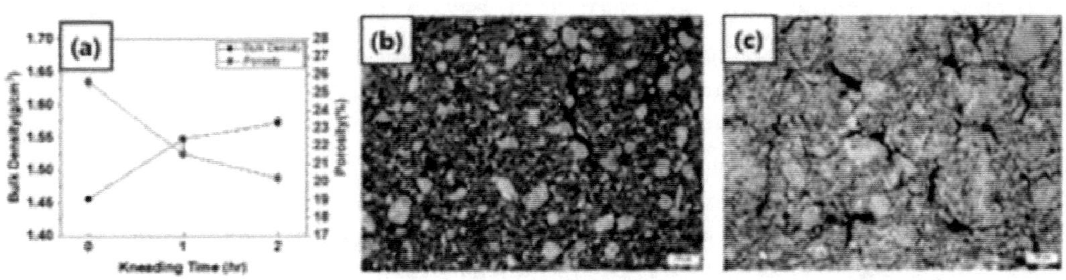

Fig 1.(a) bulk and porosity variation (b) microstructure of 0 hour (c) microstructure of 2 hours

References

[1] Jiao Tan, et al. Fuel 343 (2023) 127912
[1] K.H Kim, et al. Applied Chemistry for Engineering 29 (2018) 677-681
[2] D.H An, et al. Carbon Letters 31 (2021) 1357-1364

A control of beta resin of coal tar ptich through solvent extraction for binder and impregnation pitch

Seungjoo Park[1,2], Seon Ho Lee[1], Song Mi Lee[1], Doo-Hwan Jung*[1,2,]

[1]Hydrogen Convergence Materials Laboratory, Korea Institute of Energy Research (KIER), Daejeon 34129, Republic of Korea

[2]Energy Engineering, University of Science and Technology (UST), Daejeon 34113, Republic of Korea

(*doohwan@kier.re.kr)

C/C composites have excellent mechanical and thermal properties even in extreme environments. Thus, they are used in various fields such as aircraft, automobiles, rockets and etc. The characteristics of the raw materials are very important to reduce the swelling or pores that occur during the carbonization. In this study, the properties of coal tar pitch were controlled, one of the raw materials for C/C composites, using three solvents (acetone, toluene, and tetrahydrofuran (THF)). Using the solvent extraction method, the insoluble (CTP-AceI, CTP-TI, CTP-THFI) and soluble (CTP-AceS, CTP-TS, CTP-THFS) are separated and used as a binder and impregnation pitch, respectively. As a result, when CTP-TI is used as a binder, density is 1.75 g/cm3, which is the highest. And the highest density appeared in the order of CTP-TI, CTP-THFI, CTP-AceI, and CTP-Raw. This coincided with the order in which β -resin, which is an important parameter for binders. Additionally, the volatile matter that causes swelling could be reduced from 22.33wt% to 9.30wt%. Thus, the swelling was suppressed efficiently. When CTP-AceS was used as the impregnation pitch, the density was 1.89 g/cm3, the impregnation rate was 60%, and the porosity was 2.5%. The impregnation rates were higher in the order of CTP-AceS, CTP-TS and CTP-THFS. This is because QI reduces fluidity and interrupt penetration into pores.

References

[1] Park SJ, Lee SH, Lee SM, Jung DH, Effect of β-resin of binderless coke on mechanical properties of high-density carbon blocks at high molding pressure, Scientific Reports, 12, 12287 (2022)

[2] Lee SM, Lee SH, Park SJ, Jung DH, Preparation of mesoporous activated carbon by preliminary oxidation of petroleum coke with hydrogen peroxide and its application in capacitive deionization, Desalination, 539, 115901 (2022)

[3] Lee SH, Lee SM, Im US, Kim SD, Yoon SH, Lee BR, Peck DH, Shul YG, Jung DH. Preparation and characterization of high-spinability isotropic pitch from 1-methylnaphthalene-extracted low-rank coal by co-carbonization with petroleum residue, Carbon, 155, 186 (2019).

[4] Lee SH, Kim JY, Yoon SH, Peck DH, Kim SK, Jung DH. Preparation of chestnut-like carbon and its application for electrodes with high specific capacitance, Applied Catalysis B: Environmental, 158, 308 (2014).
[5] Park SJ, Lee SH, Lee SM, Park JW, Kim SS, Jung DH, The Effect of Oxygen Content in Binderless Cokes for High-Density Carbon Blocks from Coal Tar Pitch, Materials, 14, 1832 (2021).
[6] Lee SM, Lee SH, Jung DH, Surface oxidation of petroleum pitch to improve mesopore ratio and specific surface area of activated carbon, scientific reports, 11, 1460 (2021).

Influence of co-catalysts on preparation of coal tar based mesophase pitch

Hui Li[1,2], Tao Yang[1], Yan Song*[1,2], Zihui Ma[1,2]

[1]CAS Key Laboratory of Carbon Materials, Institute of Coal Chemistry, Chinese Academy of Sciences, Taiyuan, 030001, China

[2]University of the Chinese Academy of Sciences, Center of Materials Science and Optoelectronics Engineering, Beijing, 100049, China

*corresponding author: yansong1026@126.com

In this work, a series of mesophase pitch with different polycondensation degrees were produced through one-step catalytic polycondensation at an initial pressure of 1 MPa (self-generated) by employing medium-temperature coal tar pitch as raw material with AlCl3 and terephthalic acid (PTA) serving as the catalyst and co-catalyst, respectively. The impact of polymerization temperatures and co-catalyst contends on mesophase pitch preparation were investigated. The results showed that the anisotropy content and polycondensation degree of the pitch products gradually increased with the rise of polymerization temperature. With the PTA content increasing at a constant temperature of 350 °C, the melting behavior of the mesophase spheres showed more evident, indicating that PTA, as a co-catalyst, enhanced the catalytic activity of AlCl3 and accelerated the formation of mesophase pitch with domain optical texture. Further study on the distribution of functional groups and crystal parameters demonstrated that the -COOH group in the PTA structure could establish a stable complex with AlCl3, thereby decreasing the hydrolysis and deactivation in the catalytic system.

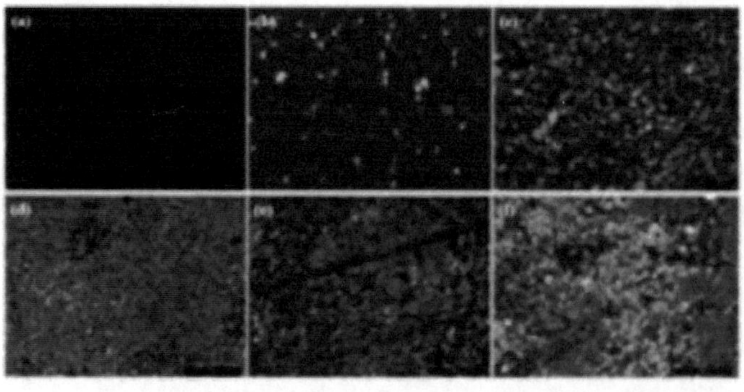

Fig.1. Polarizing microscopy images of mesophase pitch prepared at different polymerization temperatures(a-c) and different PTA contents (d-f)

PRESSURE DYNAMICS IN FLAME: IMPACT ON SOOT AND CARBON NANOMATERIALS FORMATION

Z. A. Mansurov [1,2]

[1]Institute of Combustion Problems, Almaty, Kazakhstan
[2]al-Farabi Kazakh National University, Almaty, Kazakhstan

It is known that the formation and synthesis of fullerenes in the traditional method of arc evaporation of graphite and in flames are carried out at pressures below 40 Torr [1]. In [2] experiments were carried out at 40 Torr. The formation of fullerenes occurs at low pressures, and here it is important to consider the steric factor. It should be taken into account that the formation of such a graceful molecule as C60 requires the necessary spatial orientation of two C30 molecules. There are various models of C60 fullerene formation, one of which is realized by the zip-mechanism [3]. The necessary condition for such a mechanism is low pressures. With increasing pressure, i.e. with the transition to atmospheric and higher pressure, where triple collisions prevail, which complicates the formation of fullerenes and single-layer graphenes, the coagulation of PCAH with the formation of soot clusters occurs [4, 5].

Earlier, Bockhorn in 1994 [6] proposed a phenomenological scheme for the reaction of soot formation in homogeneous mixtures of premixed flames, where the intermediate products are PCAH. The sequence of formation of combustion products in fuel-rich flames was shown: CO, H2, H2O, aliphatic hydrocarbons, aromatics with the increasing number of rings, up to coronene, then nucleation of soot particles and their growth to the size of 50 nm.

Taking into account the new data PCAH can be considered as the basis for the formation of fullerenes and graphenes in the flame as precursors to the formation of soot particles. We developed a complete scheme of soot formation [4, 5].

This model not only delineates the stages of soot particle formation but also incorporates the pathways for the generation of fullerenes and graphenes. Specifically, we have observed that at lower pressures, single-layered graphenes are predominantly formed, whereas at atmospheric pressure, the formation of multilayered graphenes is favored.

Furthermore, we have developed a scheme detailing the synthesis of fullerenes, graphenes, and carbon black within hydrocarbon-rich flames, with due consideration to varying pressure conditions. This synthesis process presents a promising alternative to the conventional methods of production utilizing electric graphite arcs.

In summary, our research underscores the pivotal role of PCAH in facilitating the formation of fullerenes and graphenes within the flame environment, offering insights into the potential applications of flame-based synthesis as an alternative technology in carbon material production.

References

1.J.B. Howard, Symposium (Int.) on Combustion 24 (1992) 933–946. DOI: 10.1016/S0082-0784(06)80111-6
2.C. Jager, F. Huisken, J.I. Lamas, Th. Henning. Astrophysical J. 696 (2009) 706–712. DOI: 10.1088/0004-637X/696/1/706
3.J. Ahrens, M. Bachmann, Th. Baum, J. Griesheimer, R. Kovacs, P. Weilmünster, K.-H. Homann, Int. J. Mass Spectrom. 138 (1994) 133–148. DOI: 10.1016/0168-1176(94)04036-2
4.Z.A. Mansurov, Soot formation. Almaty: alFarabi Kazakh National University, 2015, 167 p.
5.Z. Mansurov, Eurasian Chem.-Technol. J. 20 (2018) 277–281. DOI: 10.18321/ectj760
6.H. Bockhorn, Soot formation in combustion. Berlin–Heidelberg: Springer, 1994, 596 p. DOI: 10.1007/978-3-642-85167-4

Epitaxial growth of carbonaceous mesophase on coke surface

Xinlei Cao[1,2], Shen Lv[1,2], Hongbo Liu[1,2], Hui Chen[1,2], Ke Shen[1,2]*

[1]College of Materials Science and Engineering, Hunan University, Changsha, China

[2]Hunan Province Key Laboratory for Advanced Carbon Materials and Applied Technology, Hunan University, Changsha, China

Surface anchoring occurs when planar molecules of carbonaceous mesophase interact with a foreign substrate. This interaction can result in either face-on or edge-on surface anchoring, depending on the substrate's nature. However, our study delves beyond conventional surface anchoring phenomena, revealing a remarkable epitaxial growth mechanism between carbonaceous mesophase and needle coke (NC) surfaces. Here, we observe the replication of planar liquid crystal molecules onto the basal planes of coke, leading to epitaxial growth. Following carbonization, this growth results in a fused NC/binder interface. Our investigation extends throughout graphitization stages. Optical texture analyses demonstrate coherent orientation between the mesophase-derived binder and the coke substrate, maintained up to 2500 °C. Moreover, observations of intragranular crack morphology validate the strength of the fused filler/binder interface. Such findings underscore the pivotal role of epitaxial growth in forming interface structure and mechanical robustness in graphite. This study elucidates the fundamental role of epitaxial growth in shaping the interfacial structure of carbonaceous mesophase on coke surfaces, offering insights into graphite's mechanical properties and paving the way for advanced carbon material design.

INFLUENCE OF CARBON ADDITIVES ON STRUCTURE FORMATION OF THERMAL INSULATION MATERIALS

A.E. Bakkara [1,2], B.S. Sadykov [1], A.E. Maten [1,2], Artykbayeva A.B. [1,2], Zhapekova A.O. [1,2]

[1]The Institute of Combustion problems, Bogenbay batyr st., 192, Almaty, Kazakhstan
[2]Farabi University, al-Farabi ave., 71, Almaty, Kazakhstan
Email: Bakkara.Ayagoz@kaznu.kz

The most important task of modern construction is to increase the efficiency, quality, environmental safety, reliability and durability of structures and structures, taking into account reducing material consumption and capital costs. A significant group of building materials in terms of volume and significance are thermal insulation materials. When producing high-temperature heat insulators, the range of initial mineral compounds used expands. The starting components can be both natural (diatomaceous earth chips) and man-made raw materials (ash and slag and fly ash).

Modification of mineral powder particles directly during the grinding process is one of the areas of mechanochemical processing of inorganic materials. This process allows you to simultaneously bring the material to the required degree of dispersion, modify the surface of the dispersed particles and obtain plastic, good masses to be melted. To modify the surface and activate the crushed particles, carbon was introduced into the mill drum along with the original powders asa modifier.

Experimental work was carried out using natural mineral raw materials - diatomite crumbs of burnt grade "A" and technogenic raw materials (ash and slag from thermal power plants). The initial components were subjected to grinding in a planetary centrifugal mill for 10, 20, 30 and 40 minutes at different ratios of powder mass (Mp) and ball mass (Mb), namely at Mp/ Mb = 1/2, 1/3 and 1 /4.

As a result of the experiments, patterns of changes in the structure and morphology of diatomite particles and technogenic waste were established. The role of modifiers in changing the structural characteristics of particles, in activating the combustion process and the patterns of formation of the phase composition of synthesis products is shown. The most effective modifying additives and activation modes for each of the materials under study were determined. The obtained patterns make it possible to purposefully regulate the thermokinetic characteristics of the combustion process and phase formation when obtaining heat-insulating materials with increased thermal and mechanical stability and different densities.

Mechanochemical treatment of aluminum powder with graphite for energy intensive systems

B.S. Sadykov[1], A. Ye. Bakkara[1,2], A.B. Artykbayeva[1,2], A.O. Zhapekova[1,2], A.S. Khairullina

[1]The Institute of combustion problems, Almaty, Kazakhstan
[2]Farabi University, Almaty, Kazakhstan

The mechanochemical treatment (MCT) process in general can be represented as triaxial loading and shear deformation at the contacts between particles of a solid substance. It should be noted that during mechanical processing the impact on particles is of a pulsed nature, that is, chemical and physical processes do not occur during the entire stay in the mechanical reactor, but only at the moment of impact and for some time during the period of relaxation of the stress field [1, 2].In this work, a study was carried out of the effect of mechanochemical treatment on aluminum powders with the addition of graphite. To determine the optimal conditions for obtaining metal carbon, the time of mechanochemical treatment and the amount of graphite in the treated mixture were varied. It was found that when the Al-C mixture contains more than 20% graphite and the treatment time is more than 20 minutes, the activated mixture ignites in air. To determine the activity and optimal time of MCM of aluminum, an evaluation was carried out using the volumetric method in the Al-C composite.

Thus, the optimal MCM modes and aluminum grade, as well as the mass fraction of graphite in the Al-C composite, were selected for further work. Al grade PA4 after 20 minutes of MCT with a graphite content of 20% has high activity and resistance to external influences during long term storage.

References

[1] Soler, L., et al. Journal of Power Sources, Vol. 169, No. 1, (2007) 144-149.
[2] Streletsky A.N., et al. Colloid Journal. Vol. T. 66, No. 6 (2004) 811-818.

Acknowledgment
This work has been supported by the Ministry of Science and Higher Education of the Republic of Kazakhstan (grant number: Ap19680387).

Correlation between the composition and softening point of high-temperature coal tar pitch and its impact on needle coke preparation

Yiting Zhao, Jingdong Yang, Gang Li, Long Gao, Haiyong Zhang, Yonggang Wang*

School of Chemical & Environmental Engineering, China University of Mining and Technology (Beijing), D11 Xueyuan Road, Haidian District, Beijing, P.R. China.

High-temperature coal tar pitch was utilized as the raw material in this study, employing the solvent sedimentation method to eliminate quinoline insoluble substances so that the QI content below to 0.005%. Subsequently, the evaporation degree of lighter components of pitch was regulated by a vacuum distillation device to control parameters such as softening point, component composition, and molecular weight of refined pitch. As a result, a series of refined pitches with softening points ranging from 43.9 to 98 °C were obtained.

The correlation between the softening point and the constituents of processed pitch was as follows:

$$SP=\sqrt{110299.7567-1091.5978\times H-1016.0024\times T-1983.0332\times Q},\ \delta=3.22\%.$$

The variables H, T, and Q represent the percentages of HS, HI-TS, and TI-QS content respectively.

As the HS content in pitch decreases continuously, the HI-TS and TI-QS content increases, leading to an increase in network structure nodes within the pitch system. And this transition from sol type to gel type will increasing the softening point and molecular weight of the pitch. The refined pitches were reacted at 490°C and 0.2 MPa for 6 hours to obtain green cokes, which were then calcined at 1400°C to obtain needle cokes. The results indicated that when the refined pitch had a HS content of 20.6% and a TI-QS content of 12.2%, it resulted in the best green coke fiber structure with a corresponding needle coke linear expansion coefficient of only 0.7302×10^{-6} /°C, which was the lowest observed value.

Keywords: Coal tar pitch, Correlation, Composition, Softening point, Needle coke

Correlation between properties of coal-based mesophase pitch with sea-island texture prepared by co-carbonization and graphite fibers

Ruiyang Li[1], Xingwei Zhang[1,2] and Yan He[2]

Beijing university of chemical technology, Chaoyang District, Beijing, China

Utilizing a co-carbonization approach, we've developed a high mesophase content (>80%) coal-based mesophase pitch (CMP) from refined coal tar pitch (RCTP), bypassing the need for pre-hydrogenation. This method simplifies the production of CMP, which traditionally requires thermal sedimentation or centrifugation. The resulting CMP was successfully used to produce graphite fibers with an impressive thermal conductivity of 1191 W/m·K. The inclusion of 5 wt% tetrahydronaphthalene (THN) was key, enhancing pitch reactivity and facilitating the transition from non-mesophase to mesophase molecules, leading to a high anisotropy CMP. The fibers' high thermal conductivity is attributed to the orientation and condensation of mesophase and non-mesophase molecules during spinning, forming larger carbon layers. This study significantly advances the efficient use of CMP in the creation of high-value carbon materials.

References

[1] Zhang X, Ma Z, He Y, et al. Correlation between properties of coal-based mesophase pitch with sea-island texture prepared by hydrogenation and graphite fibers[J]. Carbon, 2024, 216: 118510.

Effect of iron chloride on the pre-oxidation of pitch-based carbon nanofibers

Zhenhai Cui [1,3], Tao Yang* [1], Yan Song*[1,2], Xiaodong Tian[1], Yangang Zang [3]

[1]Shanxi Institute of Coal Chemistry, Chinese Academy of Sciences, Key Laboratory of Carbon Materials, Taiyuan, China
[2]Center for Materials and Photovoltaics, University of Chinese Academy of Sciences, Beijing, China
[3]North Central University, School of Energy and Power Engineering, Taiyuan, China
*corresponding author: yansong1026@126.com

In this work, the effect of FeCl3 addition (0%, 0.5%, 1% and 2%) on the pre-oxidation of pitch-based carbon nanofibers prepared by using coal tar pitch (CTP) and polyacrylonitrile (PAN) as raw material were studied. The morphological characteristics of pitch-based nanofibers (C-PBCF-0, C-PBCF-0.5, C-PBCF-1 and C-PBCF-2) after primary spinning, pre-oxidation and carbonization were investigated. The results showed that FeCl3 had an important effect on the process of fibers stabilization. Severe fusion were observed after pre-oxidation, in the absence of FeCl3. With increasing of the amount of ferric chloride, the degree of adhesion gradually decreased and the fiber diameter distribution became more uniform, Meanwhile, carbon nanofibers displayed the same pattern. The results demonstrated that FeCl3 facilitated the oxidation of pitch molecules. In the process, oxygen was stored in various forms of oxygen-containing functional groups, which acted as a cross-linking bridge to promote the formation of stable macromolecules, and thus increased the thermal stability of nanofibers. Finally, the fiber morphology was well maintained.

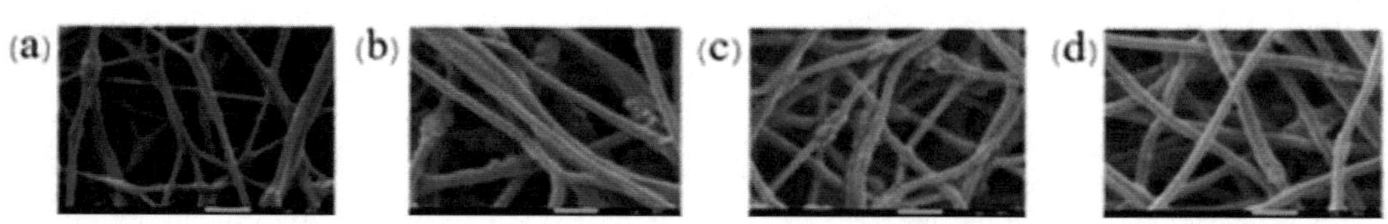

Figure 1. SEM images of (a) C-PBCF-0 (b) C-PBCF-0.5 (c) C-PBCF-1 (d) C-PBCF-2

Structure and conductive properties of polyimide-based graphite fibres under drawing

Yan He[1], Zhaokun Ma[1]*, Zihan Chen[1], Xingwei Zhang[1], Yuhong Liu[2], Huaihe Song[1]

[1]Beijing University of Chemical Technology, Beijing, 100029, PR China
[2]Grimat Engineering Institute Co,.Ltd, Beijing, 101400, PR China

Polyimide (PI) with a designable molecular structure and a simple carbonization process can be used as a potential precursor for high thermal conductivity carbon materials. In this paper, a partial chemical imidization method was used to synthesize PI fibers with planar molecular chains. Then, PI-based graphite fibers were prepared by undergoing successive thermal imidization, carbonization, and graphitization. Different drawing tension (from 0 to break) were applied during thermal imidization and carbonization, with a focus on examining the structure and conductive properties of PI-based graphite fibres undergoing different drawing. The experimental results show that the microstructures of PI-based graphite fibers prepared by certain drawing during thermal imidization and carbonization are denser, more compact and smoother than the fibers without any drawing. Under ultimate drawing tension during thermal imidization (77.51 MPa) and carbonization (7.88 MPa), the tensile strength and thermal conductivity of the graphite fiber are 1331 MPa and 373 W·m-1·k-1, which increase by 73% and 131%, respectively. These results suggest that the drawing in heat treatment makes it possible for PI fibers to become a new and promising precursor material for high thermal conductivity carbon fiber..

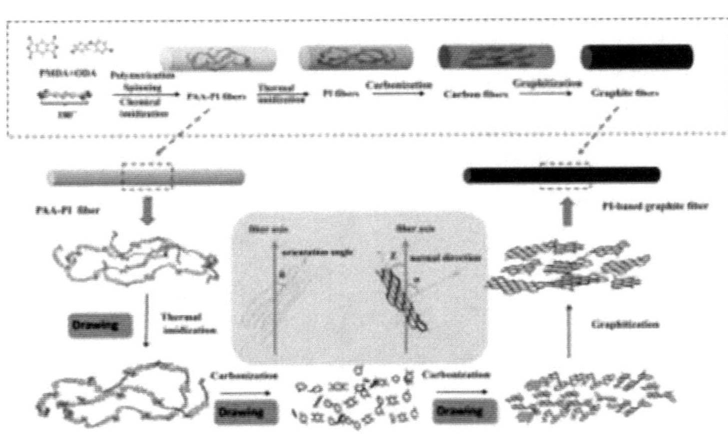

Figure 1 The graphical abstract of this research

Synthesis and structural evolution of coal-based mesophase pitch by hydrogenation-vacuum thermal polycondensation

Yan Yang [1,2], Wenzhong Shen [1,*]

[1] State Key Laboratory of Coal Conversion, Institute of Coal Chemistry, Chinese Academy of Sciences, 27 South Taoyuan Road, Taiyuan, PR China

[2] Center of Materials Science and Optoelectronics Engineering, University of Chinese Academy of Sciences, No. 19 (A) Yuquan Road, Shijingshan District, Beijing, PR China

The properties of high-performance pitch carbon fibers are conclusively determined by the structure and properties of mesophase spinnable pitches, and the controlled synthesis of mesophase pitches has received widespread attention. This work used coal pitches as raw materials to explore the pathway of synthesizing mesophase pitches through hydrogenation and vacuum polycondensation. Firstly, the effects of reaction temperature, reaction time, and THN addition on the structure and properties of hydrogenated pitches were investigated. Secondly, the mesophase pitches were prepared by light component removal of hydrogenated pitches under 0.001 MPa at 230 oC. The results revealed that the introduction of THN could effectively increase the H content in hydrogenated pitches, and the transformation of HN to Har was observed with the extension of the reaction time. The softening point of the resultant mesophase pitches was 230-245 oC, the Mn ranged from 1400 to 2000 Da, the anisotropic content was up to 96 %, and the contents of the four subfractions were THFI > HI-TS > HS > TI-THFS in descending order. This work provided an efficient and convenient approach for the synthesis of mesophase spinnable pitches, with a view to providing a theoretical and practical foundation for spinning-grade mesophase pitch preparation.

The Influence of Graphene Addition on the Formation of Mesophase Pitch and the Structure and Properties of Its Carbon Fibers

Xingwei Zhang [1], Ruiyang Li [1], Yan He [1], Zhaokun Ma [1] and Huaihe Song [1]

[1]School of Materials Science and Engineering, Beijing University of Chemical Technology, Beijing 100029, People' s Republic of China

Mesophase pitch (MP) possesses advantages such as easy orientation, easy graphitization, and high carbonization yield, making it an excellent precursor for many high-end carbon materials. Nucleation is the initial stage of mesophase formation, and the system undergoes a long period of metastability before nucleation occurs. Therefore, nucleation is the controlling step for the overall rate of mesophase generation and development, and studying this process is of great significance for the controllable preparation of high-quality spinnable MP. In this study, the influence of adding different amounts of graphene to refined coal tar pitch on mesophase nucleation and development was investigated. The correlation between graphene content and the structure and properties of MP-based carbon fibers was also explored. The results indicate that the addition of 0.01 wt% graphene promotes the aromatization and polymerization reactions of pitch molecules, shortening the time for mesophase formation and obtaining wide-area streamlined MP [1]. Moreover, graphene in the mesophase can disturb the radial arrangement of polycyclic aromatic hydrocarbons to some extent during the spinning process, and high-temperature heat treatment can induce the formation and growth of graphite microcrystals, resulting in high thermal conductivity carbon fibers with folded radial cross section morphology [2].

References

[1] X. W. Zhang. Z. K. Ma, H. H. Song, et al. J. Anal. Appl. Pyrol. 140 (2019) 274-280.
[2] X. W. Zhang. S. L. Ning, Z. K. Ma, H. H. Song, et al. Carbon 156 (2020) 499-505.

Study on the ablation performance of SiC-coated high thermal conductivity three-dimensional C/C composites

Ruoxi Zhang[1], Qiang Song[1] and Hejun Li[1]

[1] State Key Laboratory of Solidification Processing,
Shaanxi Key Laboratory of Fiber Reinforced Light-Weight Composites,
Northwestern Polytechnical University, Xi' an, Shaanxi, 710072, PR China

Ceramic coating ablation performance is critical for carbon/carbon (C/C) composites in hypersonic applications. A novel strategy that the improved heat transfer capability of the C/C substrates would significantly reduce the ablation temperature and thus the ablation rate of the coating is proposed. Carbon/carbon composites were prepared by densifying a three-dimensional preform made of mesophase pitch-based carbon fiber woven fabric lamination and PAN-based carbon fiber z-pin. Afterwards, SiC coating was prepared on them by pack cementation, and their ablation performances were investigated. Results show that the stacked of pitch-based carbon fiber woven fabrics in Z direction induces a high thermal conductivity of composites in in-plane direction and a sawtooth-structured surface of SiC coating, which helps to decrease the surface temperature (from 1778 to 1614 ℃) and the thermal stress of coating during ablation. Thus, a greatly improved ablation resistance is obtained for the coating in the X-Z plane, showing a mass ablation rate of -0.01 mg/s and a linear ablation rate of -0.12 μm/s, respectively, 90.9 % and 105.2 % lower than that of the X-Y plane. Our findings can provide a meaningful way for the development of high-performance ablation-resistant C/C composites.

Effect of spinnable pitch composition on spinning and performance of pitch-based carbon fibers

Qian Li[1,2], Wenzhong Shen[1,*]

[1]State Key Laboratory of Coal Conversion, Institute of Coal Chemistry, Chinese Academy of Sciences, 27 South Taoyuan Road, Taiyuan, PR China

[2]Center of Materials Science and Optoelectronics Engineering, University of Chinese Academy of Sciences, No. 19 (A) Yuquan Road, Shijingshan District, Beijing, PR China

Understanding the chemical components, molecular structure, melting behavior and rheological property of spinnable pitch is essential to improve the spinning stability and pitch-based carbon fibers performance. In here, isotropic spinnable pitches with different softening points were prepared from coal tar pitch by air-blowing thermal polycondensation. The spinnable pitches with softening point of 210 oC and 264 oC were synthesized by air-blowing at 320 oC for 4 or 5 h, respectively; both pitches showed good fiber-forming behavior with viscosity of 200 pa·s at spinning temperatures. The strengths and modulus of the resultant general-purpose grade pitch-based carbon fibers were 0.62±0.15 GPa and 0.64±0.18 GP, 30.84±7.26 GPa and 32.56±2.48 GP, respectively.

Preparation and characteristics of high strength activated carbon fiber derived from polyimide by dry-jet wet spinning

Jeong-Seo Park, Yong-Sik Chung

Jeonbuk National University, 567, Baekje-daero, Deokjin-gu, Jeonju-si, Jeonbuk-do, Republic of Korea1

The correlation between the imidization degree of polyimide fibers and characteristics of activated carbon fibers was studied. Poly(amic acid) solution was spun by dry-jet wet spinning method and poly(amic acid) fibers were transferred into polyimide fibers by heat treatment of IR heater at 300℃, 400℃, 500℃, 600℃. Polyimide fibers were carbonized and activated at 900℃. The imidization degree of polyimide fibers increased up to 500℃ but decreased at 600℃. The thermal and mechanical properties of polyimide fiber exhibited a dependency on the imidization degree. The mechanical properties and compact morphologies of activated carbon fiber strongly depended on the imidization degree of polyimide fiber. Activated carbon fiber was showed the maximum tensile strength value of 898 MPa and specific surface area composed of micropores value of 920 m2/g at IR heater temperature of 500℃ and the imidization degree of 96%.

Preparation and properties of graphene fiber carbon-based composites

Guanyu Wang[1], Ruoxi Zhang[1] and Qiang Song[1]

State Key Laboratory of Solidification Processing, Shaanxi Key Laboratory of Fiber Reinforced Light-Weight Composites, Northwestern Polytechnical University, Xi' an, Shaanxi, 710072, PR China

Graphene fibers have excellent comprehensive properties with high electrical conductivity, high thermal conductivity, high strength, excellent mechanical and electromagnetic shielding properties, which is a promising carbon fiber. Therefore, using graphene fibers instead of traditional carbon fibers and utilizing ICVI technology to densify graphene fiber preforms, the preparation of graphene fiber carbon based composites was achieved. The electrical, thermal, mechanical and electromagnetic shielding properties were investigated and compared with the performance of conventional carbon fiber reinforced carbon-based composites. This work provides a meaningful approach for the development of high-performance graphene fiber carbon based composites.

Fabrication and Properties of Polyimide-Based Graphite Fibers

Byoung-Min Lee, Yong-Sik Chung

Jeonbuk National University, 567, Baekje-daero, Deokjin-gu, Jeonju-si, Jeonbuk-do, Republic of Korea1

Some polyimide(PI)-based carbon fibers and films have been reported even with high precursor cost. PI as precursor showed high carbon yield and extremely high thermal and electrical conductivity when carbonized and graphitized. PI fibers were prepared by wet spinning of polyamic acid and subsequent thermal imidization. The fibers were carbonized in the presence of polydimethylsiloxane(PDMS) and finally graphitized at 2,200°C. The addition of PDMS altered the pyrolysis behavior and residue of PI fibers. PDMS pretreatment improved tensile strength and electrical conductivity of the resultant carbon fibers. Another important effect of PDMS was that the carbon fibers with PDMS were more flexible and much softer than PI fibers alone.

Co-carbonization of coal tar pitch and low-cost ethylene tar hydrocracking oil for preparation of spinnable mesophase pitches and their carbon fibers with high thermal conductivity

Disheng Lu[a,b], Jianguang Guo[a,b], Ge Zhang[a], Hui Zhu[a,b], Ye Cong[a,b], Baoliu Li[a,b], Zhijun Dong[a,b], Xuanke Li[a,b]

[a]Hubei Province Key Laboratory of Coal Conversion and New Carbon Materials, Wuhan University of Science and Technology, Wuhan 430081, China

[b]State Key Laboratory of Coking Coal Exploitation and Comprehensive Utilization, Pingdingshan 467000, China

Hydrogenation modification is usually a necessary step in preparing spinnable mesophase pitches from coal tar pitch (CTP). However, the high price of common hydrogenation agent such as tetrahydronaphthalene severely limits its wide application of this method in high performance carbon materials. Herein, a low-cost ethylene tar hydrocracking oil (ETHO) was used to co-carbonize with CTP, and then spinnable mesophase pitches were prepared by thermal polymerization process from the co-carbonized CTPs. The effects of different proportions of ETHO co-carbonization agents on the molecular structure of mesophase pitch and the crystal structure of graphitized carbon fiber were investigated. The molecular structure analysis results show that some aliphatic groups and naphthenic structures were introduced into the co-carbonized CTPs, further improve the molecular fluidity of the mesophase pitches and reduce their softening point, promote the molecular stacking of aromatic hydrocarbons and their orientation along the fiber axis during spinning. Benefit from this, the average thermal conductivity of carbon fibers prepared from mesophase pitch obtained by adding 50% ETHO in co-carbonization reaches 1125.8 W/(m · K). This work will provide some insight views for preparation of the low-cost CTP based spinnable mesophase pitches and their carbon fibers.

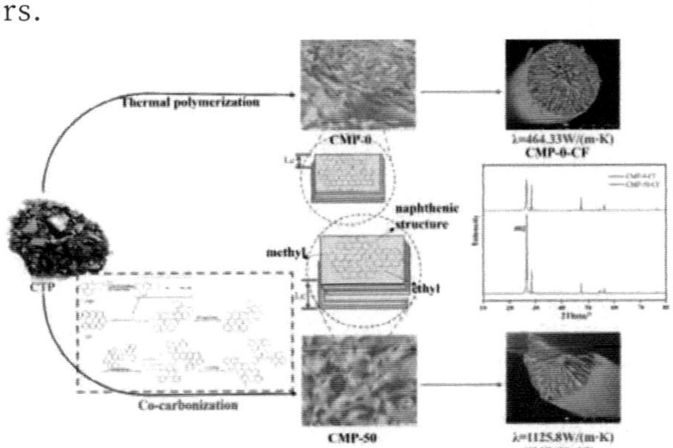

Preparation and oxidation ablation behavior of TaxHf1-xC (x=0.2, 0.25, 0.33, 0.5) single phase solid solution ceramic powders and their modified C/C composites

Zongyu Liu[1,2], Zhijun Dong[1,2], Hui Zhu[1,2], Guanming Yuan[1,2], Ye Cong[1,2], Baoliu Li[1,2], Jianguang Guo[1,2], Qin Zhang[1,2], Jiang Zhang[1,2], Xuanke Li[1,2]

[1]The State Key Laboratory of Refractories and Metallurgy, Wuhan University of Science and Technology, Wuhan, Hubei 430081, PR ChinaHubei Province Key

[2]Laboratory of Coal Conversion & New Carbon Materials, Wuhan University of Science and Technology, Wuhan, Hubei 430081, PR China

This study achieved controllable preparation of TaxHf1-xC ceramic powders with ideal single-phase solid solution structure characteristics via liquid phase precursor method using TaCl5, HfCl4 and phenolic resin as metal and C sources. On this basis, TaxHf1-xC modified C/C composites were prepared by precursor impregnation cracking process (PIP). The conversion process from TaxHf1-xC precursor to single-phase solid solution ceramic powders, the oxidation behavior of ceramic powders and the ablation behavior of C/C-TaxHf1-xC composite materials were studied. The results indicate that the organic precursors of TaxHf1-xC occur successively carbon thermal reduction reactions of Ta2O5 and Hf6Ta2O17, and solid solution reactions of HfC and TaC during the pyrolysis process. As the solubility of HfC increases, the temperature required to prepare TaxHf1-xC ceramic powders also increases. After pyrolysis in the temperature range of 1800 ℃ to 2000 ℃, both HfC and TaC in TaxHf1-xC (x=0.2, 0.25, 0.33, 0.5) undergo sufficient solid solution, forming a single-phase solid solution with a particle size of 150-300 nm. It can be found that the oxidation resistance of TaxHf1-xC ceramic powders and the anti-ablation ability of C/C-TaxHf1-xC composite materials are closely related to the composition of its oxidation products. In contrast, TaxHf1-xC (x=0.25) ceramic powder has better oxidation resistance.

Ablation behavior and mechanism analysis of (Zr, Hf)B2-SiC coating on carbon-carbon composites above 2300 °C

Qi Zhou[1,2], Feng Chen[1,2], Hu Yuan[3], Hui Zhu[1,2], Baoliu Li[1,2], Jianguang Guo[1,2], Ye Cong[1,2], Xuanke Li[1,2], Zhijun Dong[1,2,*]

[1]The State Key Laboratory of Refractories and Metallurgy, Wuhan University of Science and Technology, Wuhan, Hubei 430081, P. R. China

[2]Hubei Province Key Laboratory of Coal Conversion & New Carbon Materials, Wuhan University of Science and Technology, Wuhan, Hubei 430081, P. R. China

[3]College of Chemistry and Chemical Engineering, Chongqing University of Technology, Chongqing 400054, P.R. China

ZrB2 and HfB2, whose oxidation products exhibit low vapor pressures with great potential for application in ablation-resistant coatings for C/C composites. However, the crystal phase change during the cooling process of ZrB2 and HfB2, which causes volume expansion and leads to cracking or detachment, is a big barrier for the development of ablation-resistant coatings for C/C composites. This will diminish the protective ability of the coatings, reducing their ablative resistance. Therefore, it is essential to inhibit the phase transition and form a dense oxide layer to improve the ablation resistance of the coating. Herein, a novel integrating coating, (Zr,Hf)B2 solid solution, and SiC with embedded (C/C) composites was prepared by reactive melt infiltration (RMI) and packet cementation (PC) method. These unique (Zr,Hf)B2-SiC solid solution coatings with synergistic antioxidant effects (ZrO2, HfO2, SiO2, and B2O3) can be subjected to high-temperature (up to 2300 °C) ablation for 180 s by plasma flames and exhibit low mass and linear ablation rates (1.25 mgs-1 and 2.86 μms-1, respectively). The additional components, such as silicon carbide (SiC), result in the formation of a highly viscous silicate glass during the high-temperature ablation process, which in turn enhances the ablation resistance of the coating.

Preparation and thermal conductivity of 2.5D Carbon/Carbon composites

He Huang[1,2], Zhijun Dong [1,2], Baoliu Li[1,2]

[1]The State Key Laboratory of Refractories and Metallurgy,
Wuhan University of Science and Technology, Wuhan, Hubei 430081, P. R. China
[2]Hubei Province Key Laboratory of Coal Conversion & New Carbon Materials,
Wuhan University of Science and Technology, Wuhan, Hubei 430081, P. R. China

In the preparation of 2.5D Carbon/Carbon composite, three kinds of reinforcement with different T700 fiber content were densified in different densification modes((Mesophase asphalt impregnation/Resin impregnation/Chemical vapor impregnation/CVI+MPI/CVI+RI)+1000℃ HT). Test results show that Fiber content plays a leading role in heat conduction. The mesophase carbon has the best performance which XY direction thermal conductivity is 251.2 W·m·K-1, Z direction thermal conductivity is 121.4 W·m·K-1. Resin carbon is second. This is because most of the components of pyrolytic carbon are smooth layers or even isotropic, and thermal stress occurs between resin carbon and fiber, resulting in additional graphitization after heat treatment at 3000℃, e.g. [1].

References

[1] Luz A P, et al. Ceramics International 43 (2017) 8171-8182

Scalable Compliant Graphene Fiber Based Thermal Interface Material with Metal-Level Thermal Conductivity via Dual-Field Synergistic Alignment Engineering

Jiahao Lu, Yingjun Liu*, Zhen Xu*, Chao Gao*

Department of Polymer Science and Engineering, Zhejiang University,
866 Yuhangtang Road, Hangzhou 310027, China

High-performance thermal interface materials (TIMs) are highly desired for high-power electronic devices to accelerate heat dissipation. However, the inherent trade-off conflict between achieving high thermal conductivity and excellent compliance of filler-enhanced TIMs results in unsatisfied interfacial heat transfer efficiency of existing TIM solutions. Here we report the first graphene fiber (GF) based elastic TIM with metal-level thermal conductivity via mechanical-electric dual-field synergistic alignment engineering[1]. Compared with the state-of-the-art carbon fiber (CF), GF features both record high-thermal conductivity of ~1200 W/mK and outstanding flexibility[2]. Under dual-field synergistic alignment regulation, GFs are vertically aligned with a high orientation (0.88) and superior array density (33.5 mg/cm2), forming continuous thermally conductive pathways. Even at a low filler content of ~17 wt%, GF-based TIM demonstrates extraordinarily high through-plane thermal conductivity up to 82.4 W/mK, exceeding most CF-based TIMs, and even comparable to commonly used soft indium foil. Benefited from the low stiffness of GF, GF-based TIM shows lower compressive modulus down to 0.57 MPa, excellent resilience rate of 95% after compressive cycles, and diminished contact thermal resistance as low as 7.4 Kmm2/W. Our results provide a new paradigm for directed assembly of thermally conductive and flexible GFs to achieve scalable and high-performance TIMs, overcoming the long-standing bottleneck of mechanical-thermal mismatch in TIM design.

References

[1] Jiahao Lu, Yingjun Liu, Zhen Xu, Chao Gao, et al. ACS Nano (2024, Accepted).
[2] Hang Shi, Jiahao Lu, Chao Gao et al. Carbon, 221(2024), 118947.

Thermal Management with Graphite Fiber Backboned Graphitic Foam Composites: Enhanced Multidimensional Heat Transfer and Anti-Leakage Properties for Phase Change Systems

Yu Qin[1,2], Junhao Liu[1,2], Yafang Zhang, Ting Ouyang [1,2*]

[1]College of Materials Science and Engineering, Hunan University,
Changsha 410082, People's Republic of China.

[2]College of Materials Science and Engineering, Hunan Province Key Laboratory for Advanced Carbon Materials and Applied Technology, Hunan University, Changsha 410082, People's Republic of China.

*oyt@hnu.edu.cn

Phase-change materials (PCMs) are promising thermal storage medium for thermal management due to their efficient thermal energy harvesting capabilities. However, the low thermal conductivity (TC) and poor shape stability of PCMs have hindered practical applications. Carbon materials can effectively address these problems. While traditional methods have relied on highly oriented carbon fillers to enhance thermal conductivity in a single direction, resulting in pronounced anisotropy, the challenges associated with aligning these fillers have led to a pivot toward the exploration of three-dimensional (3D) materials with isotropically enhanced TC. Herein, we developed a novel 3D graphite fiber backboned graphite foam composite (GFBGF) to address the challenges. Our approach involves the infiltration of a graphite fiber matrix, exhibiting a preferred xy-plane orientation, with mesophase pitch. This is followed by a confined foaming process that induces a vertically alignment of graphite crystal structure, significantly enhancing the through-plane TC in a single, scalable operation. The resulting GFBGF-PCM composite demonstrates impressive TC of 33.05 W·m−1·K−1 at in-plane and 18.76 W·m−1·K−1 at cross-plane directions, with a remarkably low anisotropy ratio of just 1.75. Moreover, the unique interconnected architecture of the composite effectively encapsulates PCMs, preventing leakage during phase transitions. This work presents a scalable and straightforward fabrication method, offering significant advancements for the application of PCMs in thermal energy storage, solar energy harvesting, and other renewable technologies.

Correlation between the crack formation and mechanical properties of carbon fibers and the anisotropic content of mesophase pitch

Jingdong Yang, Yiting Zhao, Gang Li, Yongqi Li, Haiyong Zhang, Yonggang Wang

School of Chemical & Environmental Engineering, China University of Mining and Technology (Beijing), D11 Xueyuan Road, Haidian District, Beijing, 100083 PR China

The anisotropic content of mesophase pitch (MP) is closely related to the properties of mesophase pitch-based carbon fibers (MPCFs). To elucidate the effect of anisotropic content on the properties of MPCFs, MP with varying anisotropic content of 62%-99% were prepared by self-pressurization/N2-blowing two-stage thermal condensation without additives using refined coal tar pitch as raw material, and MPCFs were obtained by melt spinning, pre-oxidation, carbonization processes. Results showed that the aromaticity of MP was enhanced as the anisotropic content of MP increased, and the carbon layer stacking orderliness of MPCFs was significantly enhanced. The radial transverse texture in the cross-section of MPCFs became more pronounced, with closer stacking of carbon layers, leading to significant improvements in tensile strength and modulus. MPCFs have a maximum tensile strength of 1515 MPa and a maximum tensile modulus of 185 GPa. However, when the anisotropic content of MP was increased from 92% to 99%, the structure of the MPCFs was more prone to cracking during circumferential shrinkage. The probability of cracking of MPCFs increased from 24.15% to 42.57%, and the formation of defects led to a significant decrease in the tensile strength of the MPCFs.

Keywords: Mesophase pitch, Carbon fibers, Anisotropic content, Crack

Fig. 1. SEM images of (a) MPCF-1L, (b) MPCF-2L, (c) MPCF-3L, (d) MPCF-4L and (d, f) MPCF-5L

N, P-coordinated Fe atomic Sites with Fe3C nanoparticles on porous carbon nanosheets for promoting adsorption-catalysis functions in lithium-sulfur batteries

Zhou Zhiqiang[1], Wang Jitong [1*], Ling Licheng [1,*]

[1]State Key Laboratory of Chemical Engineering, East China University of Science and Technology, Shanghai, 200237, China

Lithium-sulfur (Li-S) batteries are considered as one of the most promising next-generation electrochemical energy-storage systems due to its superb theoretical energy density, low production cost and environmental friendliness. However, the poor conductivity of S and Li_2S, severe shuttle effect and sluggish reaction kinetics of phase transformation greatly hinder the large-scale commercialization of Li-S batteries. Carbon materials have achieved widespread attention due to their high specific surface area, excellent electrical conductivity, and structural diversity. Nevertheless, non-polar carbon materials are difficult to interact closely with highly polar polysulfides, resulting in a low sulfur utilization and serious shuttle effect. Due to the advantages of rapid redox kinetics and rich adsorption sites of single atom/nanoparticle catalysts, integrating them with carbon-based materials is essential to enhance chemical adsorption and electrochemical reaction activity for lithium polysulfides (LiPSs).

Vanadium Oxide/Graphene Oxide/SWCNTs Composite for Flexible Zinc-Ion Batteris

Zhouyang Qin[1], Ruitao Lv[1], Feiyu Kang[1,2] and Zheng-Hong Huang[2]

[1]Key Laboratory of Advanced Materials of Ministry of Education,
School of Materials Science and Engineering, Tsinghua University, Beijing 100084, China
[2]Shenzhen Geim Graphene Center, Institute of Materials Research,
Shenzhen International Graduate School, Tsinghua University, Shenzhen 518055, China

Aqueous zinc-ion batteries (AZIBs) are identified as a serious contender for flexible energy storage devices on account of their safety and low cost. Nevertheless, the rapid capacity decay attributed to the dissolution and structural deterioration of active materials has hindered their further development, which is more evident at low current density. Herein, we prepared a flexible self-supported cathode material composed of V2O5·1.6H2O (VOH), graphene oxide (GO), and single-walled carbon nanotubes (SWCNTs) through vacuum filtration, which delivers extremely high specific capacity of 453.8 mAh g-1 at 0.1 A g-1 and exhibits remarkable cycle stability with a capacity retention of 86.7% after 5000 cycles at 0.5 A g-1. In this composite, GO restrains the direct contact between the VOH and the electrolyte, thereby effectively avoiding its dissolution. The existence of SWCNTs could improve the kinetics of electrochemical reaction due to its good conductivity. For the active material VOH, Zn2+ is embedded into the interlayers to form Zn-O bonds through in-situ electrochemical reaction in the first cycle, which act as pillars between adjacent [VO] layers to maintain structural stability. This work provides insights on the combination of active materials with carbon materials and in-situ electrochemical conversion.

Carbon-supported Tri-metal-atom Catalyst towards Bifunctional Oxygen Catalysis

Zhiping Zeng [1]*, Guowei Yang [1]

[1]State Key Laboratory of Optoelectronic Materials and Technologies, School of Materials Science and Engineering, Sun Yat-sen University, Guangzhou, 510275, China

It is challenging to reveal the configuration-activity relationship and synergistic enhancement mechanism (e.g., electron spin state, d-orbital energy level) and the intrinsic bifunctional properties of carbon-based metal atomic catalysts.[1-3] We propose a carbon-supported tri-metal-atom catalyst with a tri-nuclear active structure, which exhibits excellent and stable bifunctional oxygen catalytic activities (oxygen reduction reactions and oxygen evolution reaction), which is far superior to carbon-supported single-atom catalysts. The electronic state analysis further shows that the synergistic interaction of carbon-supported metallic atoms forms asymmetric metal sites, optimizes the position of d orbital energy level, shifts the center of the d band and weakens the binding strength with oxygen intermediates. Metal-air batteries equipped with a tri-metal-atomic catalyst cathode offer ultra-high rechargeable stability (3940 cycles/1970 hours at 2 mA cm-2 at room temperature) and superior resistance to freezing test cycles (over 2040 cycles/340 hours at 1 mA cm-2 at -60°C). The coordination design and in-depth mechanistic analysis have further promoted the promotion of carbon-supported atomic catalysts in various electrocatalytic applications.

References

[1] L. Han, et al., Nat. Mater. 21 (2022) 681.

[2] Z. Zeng, et al., Nat. Commun. 12 (2021) 4088.

[3] J. Zhong, et al. ACS Nano 18 (2024) 5258-5269

Structural Design, Interface Optimization, and Performance Research of Metal Phosphides Catalyst for Advanced Lithium-Sulfur Batteries

Zhenxin Zhao[1], Xiaomin Wang

[1]College of Materials Science of Technology , Taiyuan University of Technology, Taiyuan, China

Regulating the behavior of polysulfides through reasonable methods is rather important for improving the overall electrochemical performance of lithium-sulfur batteries. Based on the reasonable design of phosphorus-based compounds, this dissertation prepared a series of iron phosphide/carbon-based composite materials, constructed electrochemical reaction interfaces with excellent adsorption and catalytic performance. The mechanism of materials and interface structure on polysulfides was studied through experiments and calculations. The relationship of materials for improving lithium-ion diffusion, inhibiting polysulfides shuttles and reducing activation energy has been clarified.

References

[1] Li HJ, Zhao ZX, Wang XM, et al. Journal of Materials Science & Technology 99 (2022) 184–192

[2] Cheng XQ, Li HJ, Zhao ZX, et al. Chemical Engineering Journal 442 (2022) 136222

[3] Zhao ZX, Yi ZL, Wang XM, et al. Nano Energy 81 (2021) 105621

[4] Zhao ZX, Yi ZL, Wang XM, et al. Chemical Engineering Journal 450 (2022) 138310

Space-Confined Molecular Catalysis toward Highly Efficient Electrocatalytic CO2 Reduction on Metal Phthalocyanine@Nitrogen-doped Carbon Nanosheet

Hongbing Zheng [1], Jitong Wang [1,*] and Licheng ling [1,*]

[1]State Key Laboratory of Chemical Engineering, East China University of Science and Technology, Shanghai 200237, China

Metal complexes composed of metal centers and macrocyclic chelating ligands, such as metalloporphyrins and phthalocyanines, have emerged as attractive materials for electrocatalytic carbon dioxide reduction. However, finding a support that is structurally stable and can better disperse phthalocyanine while maintaining the catalytic activity and catalyst loading remains elusive. Porous carbon materials are considered reliable substrates for supporting molecular catalysts. Herein, a defective nitrogen-rich porous carbon (NiPc(CoPc)/NMCN) was synthesized based on a dynamically competitive cascade coating methods and used for the stretcher-loaded phthalocyanine catalyst. The multilayer composite structure of NMCN guided the molecular-scale dispersion of phthalocyanine and played a domain-limiting role in its catalytic process. Therefore, the excellent CO2 electroreduction properties of phthalocyanine are in the foreground. At -0.73 V vs. RHE, NiPc/NMCN achieved the highest CO faradaic efficiency (FECO) of 96.0%, while CoPc/NMCN achieved the CO partial current density (jCO) of 16.10 mA cm-2 (-0.83V vs.RHE). Consequently, the strategy of constructing confined multilayer mesoporous carbon structures provides a new avenue for the design of efficient molecular catalysts for CO2 reduction.

Improvement of Li-ion Battery Performance of Hard Carbon by the Pyrolytic Carbon Coating

Xiazhe Zhai [1], Kenta Morinaga[1], Saki Nishida[1], Hyeonseok Yi[2], Yusuke Okumura[3], Hikaru Kobayashi[3], Ryuta Haga[4], Mikito Suto[3], Koji Nakabayashi[1,2], Seong-Ho Yoon[1,2], Jin Miyawaki[1,2]

[1]Interdisciplinary Graduate School of Engineering Sciences, Kyushu University, 6-1 Kasuga-koen, Kasuga, Fukuoka 816-8580, Japan

[2]Institute for Materials Chemistry and Engineering, Kyushu University, 6-1 Kasuga-koen, Kasuga Fukuoka, 816-8580, Japan

[3]Steel Research Laboratory, JFE Steel Corporation,1 Kawasaki-cho, Chuo-ku, Chiba 240-0835, Japan

[4]Chemical Research Laboratory, JFE Chemical Corporation, 1 Kawasaki-cho, Chuo-ku, Chiba 240-0835, Japan

Hard carbon (HC) has emerged as a promising anode material for lithium-ion batteries (LIBs), because the unique storage site for lithium ions, called closed pore, of HC is expected to provide potentially high discharge capacity and rate performance. However, the practical application of HC has been hindered by the insufficient discharge capacity and low initial Coulombic efficiency, and structure and role of closed pore have not been fully elucidated.

In this study, we aimed to improve the LIB performance of HC by coating with pyrolytic carbon layers. Chemical vapor deposition (CVD) using a hydrocarbon gas gave rise to the pyrolytic carbon coating of the HC surface, and significant improvement in the first-cycle discharge capacity was observed particularly at lower CVD temperatures. Notably, the initial Coulombic efficiency was enhanced irrespective of the CVD temperature, whilst a simple mixing of HC with pyrolytic carbon rather resulted in worsening of the initial Coulombic efficiency, indicating the crucial role of pyrolytic carbon coating on HC surfaces.

Our findings suggest that the quantity and quality of the pyrolytic carbon coating, influenced by CVD parameters, play a pivotal role in enhancing the LIB performance of HC.

Instructions for Preparation of "CARBON 2024" Short Abstract
Design of High-entropy layered hydroxide and its solid-state electrolyte performance

Liu Yuting [1,2], Wang Jianlong [1] and Guan Taotao[1]

[1]Institute of Coal Chemistry, Chinese Academy of Sciences, Shanxi, Taiyuan, China
[2]North University of China, Shanxi, Taiyuan, China

In recent years, the research on solid-state electrolytes has gradually increased, but neither inorganic solid-state electrolytes nor polymer solid-state electrolytes can meet the requirements of high ionic conductivity, good mechanical properties, good contact, and wide electrochemical window at the same time, and cannot meet the practical use of all-solid-state lithium metal batteries. The combination of organic polymers and inorganic fillers to form composite solid electrolytes can solve this problem well, and has been widely studied and concerned. Among them, polyvinylidene fluoride-hexafluoropropylene copolymer (PVDF-HFP) based gel polymer electrolytes are considered promising solid electrolytes, while High-entropy layered hydroxides have excellent electrochemical properties. The solid electrolyte prepared by combining the two has a high ionic conductivity of 8.2×10-4S/cm at room temperature. At 60°C, the ionic conductivity reaches 1.2×10-3S/cm. The assembled Li symmetrical battery can be stably cycled for 1000h at room temperature, and the lithium metal battery with lithium iron phosphate as the positive electrode can reach 160mAh/g at the charge and discharge rate of 0.1C, which has excellent performance and can meet the use of solid-state batteries, which proves the good application prospect of composite solid-state electrolyte.

Instructions for Preparation of "CARBON 2024" Short Abstract

Yiming Li[1], Xiaohong Chen *

Beijing University of Chemical Technology, Beijing, 100029, P. R. China

S doping is an effective strategy to improve the potassium-ion storage performance of carbon-based materials. However, due to the large atomic radius of S and poor thermal stability, it is challenging to synthesize carbon materials with high sulfur content by solid-phase transformation. In this work, we designed a multi-cavity structure that can confine the molten S during heat treatment and make it fully react, then achieving high S doping (7.6 at. %). As we known, S doping can also effectively increase the active sites of carbon materials to obtain higher capacity. In addition, through different ex/in-situ characterizations and DFT calculations, we confirmed that the S atoms can effectively expand the interlayer spacing of carbon, which facilitates the intercalation/deintercalation reaction of K+, thereby significantly improving the rate performance. Therefore, benefiting from the effect of S-doping, the sample exhibits high reversible specific capacity (401.0 mAh g-1 at 0.1 A/g) and rate performance (167.2 mAh g-1 at 5 A/g). The as-assembled K+ hybrid capacitor delivers both high energy density and power density (138.5 W h kg-1 and 7692.5 W kg-1 respectively). This work provides a new approach to design S content carbon-based materials for high performance K+ storage.

Instructions for Preparation of "CARBON 2024" Short Abstract

Yiming Li[1], Xiaohong Chen *

Beijing University of Chemical Technology, Beijing, 100029, P. R. China

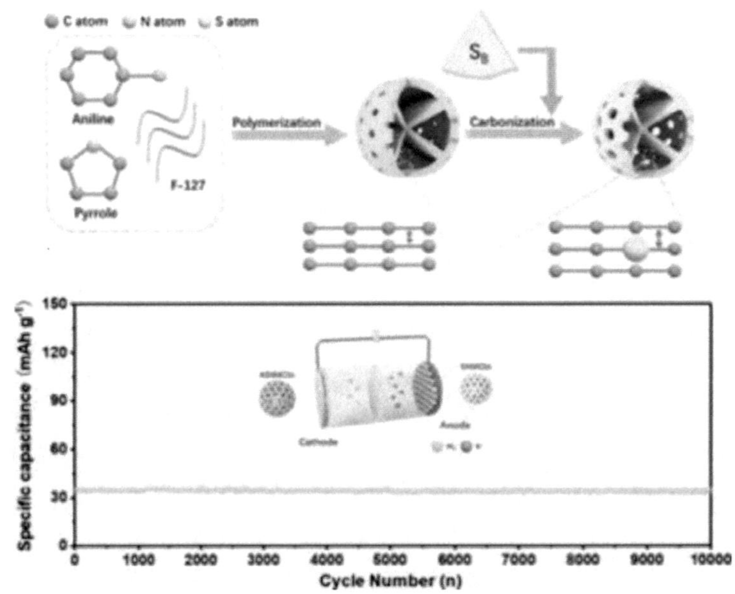

S doping is an effective strategy to improve the potassium-ion storage performance of carbon-based materials. However, due to the large atomic radius of S and poor thermal stability, it is challenging to synthesize carbon materials with high sulfur content by solid-phase transformation. In this work, we designed a multi-cavity structure that can confine the molten S during heat treatment and make it fully react, then achieving high S doping (7.6 at. %). As we known, S doping can also effectively increase the active sites of carbon materials to obtain higher capacity. In addition, through different ex/in-situ characterizations and DFT calculations, we confirmed that the S atoms can effectively expand the interlayer spacing of carbon, which facilitates the intercalation/deintercalation reaction of K+, thereby significantly improving the rate performance. Therefore, benefiting from the effect of S-doping, the sample exhibits high reversible specific capacity (401.0 mAh g-1 at 0.1 A/g) and rate performance (167.2 mAh g-1 at 5 A/g). The as-assembled K+ hybrid capacitor delivers both high energy density and power density (138.5 W h kg-1 and 7692.5 W kg-1 respectively). This work provides a new approach to design S content carbon-based materials for high performance K+ storage.

Instructions for Preparation of "CARBON 2024" Short Abstract
TMSs/C composite sodium-ion battery electrodes with a moderately elastic shell are synthesized through confined space CVD

Yu Xingbo[1,2], Guan Taotao[1] and Li Kaixi [1]

[1]CAS Key Laboratory of Carbon Materials, Institute of Coal Chemistry, Chinese Academy of Sciences, Taiyuan 030001, PR China;
[2]State Key Laboratory of Heavy Oil Processing, China University of Petroleum, Beijing 102249, PR China

Carbon-coated transition metal sulfides (TMSs) are considered as effective anode materials for sodium-ion batteries in terms of their superior electrochemical performance and stability. However, the application potential of these nanoparticles is limited because of the single composition of their carbon shells and a weak core-shell binding force. Herein, we report a confined space chemical vapor deposition method for the preparation of MoS2 coated with both soft carbon and hard carbon (MoS2@SC-HC). In this process, the formation of strong C-S bonds between the MoS2 core and elastic carbon shell significantly enhances the mechanical and chemical stability of MoS2. Benefiting from this unique carbon layer composition, MoS2@SC-HC exhibits excellent cycle stability and rate performance. After 100 cycles at 0.1 A g-1, its capacity can be increased by 20 times compared to pure MoS2; at 5 A g-1, its capacity can be increased by more than 200 times. Density functional theory (DFT) calculations demonstrate that the carbon-coated structure promotes enhanced Na storage capacity and conductivity in TMSs. Furthermore, employing this general synthesis strategy has enabled us to develop a series of TMSs@SC-HC with excellent performance, such as FeS@SC-HC and NiS2@SC-HC.

Deciphering the Triple Relationship: Precursor-Closed Pore-Plateau Capacity for Sodium-ion Batteries

Xiaotian Li, Jiapeng Zhang, Renlu Yuan, Boyang Zhao, Chuang Qiu, Dengke Wang, Man Yuan, Ang Li, Xiaohong Chen*, Huaihe Song*

State Key Laboratory of Chemical Resource Engineering, Beijing Key Laboratory of Electrochemical Process and Technology, Beijing University of Chemical Technology, Beijing 100029, China

Hard carbon anodes are pivotal for the future advancement of sodium-ion batteries. However, significant challenges persist in achieving hard carbon with high plateau capacity (< 0.1 V) and clearly defining the closed-pore structure corresponding to its plateau capacity. Here, we manipulate the closed pore structure of hard carbon at molecular dimension by altering the precursor molecular structure of phenolic resin. Our findings indicate that the chain length and curvature of the precursor molecules affect the closed pore structure. Furthermore, we establish a reliable conformational relationship among precursor molecular structure, hard carbon structure, and performance. The high plateau capacity depends on the closed pore volume at the appropriate sodium storage size rather than on the increase in the total closed pore volume or the average pore diameter. What' s more, this work also reveals the sodium storage behavior of closed pores with different sizes, with the optimal radius of the closed pore for sodium storage is approximately 1.2 nm. As a proof, the capacity of the optimized hard carbon at 50 mA g-1 was improved from 194.78 to 351.11 mA h g-1, and the plateau capacity was improved from 129.99 to 256.28 mA h g-1.

Hierarchical carbon nanotube hybrid films modified gas diffusion layer for high-performance proton exchange membrane fuel cells

Xiao-Fang Zhang [1], Yu-Ting Liu [1,2], Ting-Ting Yao [1], and Gang-Ping Wu [1,2]

[1]CAS Key Laboratory of Carbon Materials, Institute of Coal Chemistry, Chinese Academy of Sciences, 27 Taoyuan Road (South), Taiyuan 030001, China

[2]Center of Materials Science and Optoelectronics Engineering, University of Chinese Academy of Sciences, No.1 Yanqihu East Rd, Huairou District, Beijing 100049, China

The functional gas diffusion layer (GDL), serving as a water-gas transfer control medium, significantly influences the performance of proton exchange membrane fuel cells (PEMFCs). To improve simultaneously the water retention and gas transport rate, we present a straightforward and effective strategy to prepare carbon nanotube (CNT) hybrid films as the functional interlayer in GDLs. With single-walled CNTs and multi-walled CNTs in the weight ratio of 1:1, the as-prepared GDL exhibits the optimized electrochemical performance under all the experimental case. Especially, the current density is increased to 3.36 A cm-2 at 0.6 V under 26% relative humidity (RH) and 4.87 A cm-2 at 0.2 V under 100% RH. It is due to the CNT hybrid films with hierarchical pore structures in GDLs is valid to enhance both self-humidifying and mass transfer performance. The micropores with reasonable wettability serve as capillary vessel for absorbing water to promote the retention of moisture in the proton exchange membrane, while the mesopores provide preferential transport pathways for reaction gas. Our results demonstrate that the CNT-based GDL may illuminate a potential application for PEMFCs.

High-entropy materials based on deep eutectic solvent for boosting oxygen evolution reaction

Shengqin Guan[1], Jianlong Wang[1,2] and TaoTao Guan[2]

[1]CAS Key Laboratory of Carbon Materials, Institute of Coal Chemistry, Chinese Academy of Sciences, Taiyuan 030001, PR China

[2]Center of Materials Science and Optoelectronics Engineering, University of Chinese Academy of Sciences, 19A Yuquan Road, Beijing 100049, PR China

High-entropy materials (HEM) are promising electrocatalysts for oxygen evolution reaction (OER) because of their instinctive physicochemical properties and highly tailorable electrochemical properties. However, their synthesis and morphology control remains challenging due to the thermodynamic immiscibility of multimetallic elements under mild conditions. Deep eutectic solvents (DESs) as solvents and temples can affect the morphology and electrochemical performance of OER. Herein, a series of HEM electrocatalysts based on different DESs were synthesized to investigate the effect of DES on the morphology and OER performance of catalysts. The influence of metals on OER performance was further explored. Among them, poly-(ethylene glycol)/thiourea-based HEM catalyst (PEG/TU-NiFeCoMnAl) shows excellent OER performance with an overpotential of only 220 mV and 317 mV at10 mA cm-2 and 100 mA cm-2, respectively. The influence of different metal active sites on OER performance was analyzed by DFT calculations, and scanning electron microscopy (SEM), fourier transform infrared (FT-IR) spectroscopy, inductively coupled plasma-optical emission spectroscopy (ICP-OES), X-ray diffraction (XRD), X-ray photoelectron spectroscopy (XPS) and transmission electron microscopes (TEM) characterization testing technologies. This work provides a new direction for the construction of advanced high-entropy materials under mild conditions for energy conversion and storage.

Topologically Defective Carbon Anodes toward High-Rate and Large-Areal-Capacity Potassium-Ion Batteries

Shaohong Liu[1], Junlong Huang[1], Haozhen Yang[1], Ruowen Fu[1] and Dingcai Wu[1]

[1]School of Chemistry, Sun Yat-sen University, Guangzhou, China

Potassium-ion batteries (PIBs) have been regarded as one of the highly promising alternatives of lithium-ion batteries in terms of the abundant and widely distributed source of potassium (K) as well as the low redox potential of K/K+ couple. Carbonaceous materials, including graphite, hard carbon, and soft carbon, have been widely investigated as one of the most promising anodes for PIBs. Exploration of advanced preparation strategies, enabling facilitated kinetics of K+ de-solvation and diffusion across the SEI film as well as fast solid-phase K-diffusion, through simple structure optimization of carbon framework, holds significant importance for the development of high-rate PIBs. Herein, a simple temperature-programmed co-pyrolysis strategy is proposed for the efficient synthesis of topologically defective carbon anodes. We have demonstrated that the topological defects can effectively regulate the electronic structure of the resulted carbon anodes, which not only enables efficient potassium adsorption energy yet low potassium migration barrier, but also induce preferential decomposition of anions to produce high potassium ion conductive solid electrolyte interphase film with decreased potassium ion desolvation and transfer barrier. These synergistic structural advantages contribute to high potassium storage capacity, excellent rate capability, and long-term cycling stability of carbon anodes.[1-2]

References

[1] Junlong Huang, Yongqi Chen, Zongheng Cen, Tan Yi, Min Liang, Youlong Zhu, Ruliang Liu, Ruowen Fu, Shaohong Liu, Dingcai Wu. Advanced Materials, 2024, DOI: 10.1002/adma.202403033

[2] Haozhen Yang, Junlong Huang, Shaohong Liu, Yongqi Chen, Zongheng Cen, Chenguang Shi, Yuheng Lu, Ruowen Fu. Small, 2023, 19 (39), 2302537.

Electrochemical properties of activated carbon aerogel with flouorine functional groups introduced by carbon tetraflouride plasmas for supercapacitors

Seoyeong Cheon[1], Seongmin Ha[1], Seongjae Myeong [1], In Woo Lee [1], and Young-Seak Lee [1,2]

[1] Department of Chemical Engineering and Applied Chemistry, Chungnam National University, Daejeon, 34134, Republic of Korea

[2] Institute of Carbon Fusion Technology (InCFT), Chungnam National University, Daejeon, 34134, Republic of Korea

Fluorine heteroatoms were introduced by CF4 plasma treatment to overcome the limited specific capacitances of electric double-layer capacitors (EDLCs), and the effects of the fluorine atoms were analyzed. The ratios of semi-ionic to covalent C–F bonds among all C–F bonds were 62.68% for CA-F3, 74.00% for CA-F6, and 36.30% for CA-F9, with CA-F6 exhibiting the highest proportion of semi-ionic C–F bonds. These semi-ionic C–F bonds induced rapid charge transfer and redox reaction. As a result, we achieved a specific capacitance of 325.68 F/g for the CA-F6 sample at 0.5 A/g. By analyzing the contributions of the electric double-layer capacitance and the pseudocapacitance, we confirmed that the contribution from the pseudocapacitance was 37.57%. A specific capacitance retention rate of 95.87% was obtained over 1000 charge/discharge cycles with a high current density of 3 A/g. In addition, the reduced charge transfer resistance of 36.8% is due to the increase in semi-ionic C-F bonds. Therefore, the specific capacity was improved by fluorine heteroatoms, and semi-ionic C-F bonds played a decisive role in this improved performance.

Pyrolysis fuel oil-derived hollow-carbon nanocages as high-stability anode materials for potassium ion storage

Seongmin Ha[1], Chaehun Lim[1], Chung Gi Min[1], Seongjae Myeong[1], and Young-Seak Lee[1,2,*]

[1] Department of Chemical Engineering and Applied Chemistry, Chungnam National University, Daejeon, 34134, Republic of Korea
[2] Institute of Carbon Fusion Technology (InCFT), Chungnam National University, Daejeon, 34134, Republic of Korea

The utilization of graphite as an anode material in potassium ion batteries (KIBs) results in diminished electrochemical performance, largely due to significant interlayer expansion in potassation/depotassiation processes. Addressing this limitation, this study introduces a novel hollow-carbon nanocage (HCNC), innovatively designed using pyrolysis fuel oil (PFO), a cost-effective by-product. Heat-treated HCNC at 1500 °C exhibits a graphitic structure with high porosity and numerous closed pores, achieving a notable potassium ion storage capacity of 307.2 mAh g-1 at a current density of 50 mA g-1. Moreover, a structurally stable HCNC effectively withstands substantial volume expansion associated with potassium ion intercalation. The stability is attributed to an adsorption/insertion hybrid mechanism, a product of the synergistic interplay between the unique morphology of HCNC and its developed carbon layers. This configuration delivers a specific capacity of 106.1 mAh g-1 and maintains approximately 83 % capacity retention over 1000 cycles at a high current density of 2000 mA g-1. Consequently, the proposed HCNC, leveraging a PFO-based MgO template, marks a significant stride in developing efficient carbon materials for KIBs, addressing the critical challenges of graphite, such as unstable cycle stability.

“Disruption” vs. “repairing” of carbon materials by ball-milling

Renlu Yuan, Yue Dong, Ruoyang Hou, Lei Shang, Xiaoyu Xu, Ang Li, Xiaohong Chen, Su Zhang, Huaihe Song*

State Key Laboratory of Chemical Resources Engineering,
Beijing Key Laboratory of Electrochemical Process and Technology for Materials,
Beijing University of Chemical Technology, Beijing 100029, P. R. China

Mechanical ball-milling is a commonly used method to shattering crystals or introducing defects in carbon materials. In our works, we revealed that the ball-milling of highly crystalline expanded graphene is a “destructive” defect engineering to prepare defect-enriched graphene block (DGB) with controllable defect density and high packing density. The introduced “self-doping” defects could act as active sites for aqueous supercapacitors, and sodium-ion capacitors and Zn-ion capacitors, possessing robust capacitive behavior and greatly improved capacitance. What’ s more, the efficient utilization of the active sites in DGB can be realized by rational functionalization or pore structure design. While in our other works, we further observed that during ball-milling, the disordered and porous structure of activated carbon (AC) is gradually “repaired” to order carbon layers along with the introduction of oxygen in the bulk phase. The well-developed pores are gradually eliminated, leading to obviously decreased surface area from 1364 to 156 m2 g-1 and improved compaction density from 0.573 to 1.423 g cm-3. Due to the continuous particle cleavage, re-assembly, and cold-welding, the as-formed carbon blocks show 2 times larger in size and 4 times higher bulk oxygen content. This simultaneous crystallinity enhancement and oxygen functionalization is barely observed in other methods, leading to significantly improved volumetric capacitances and high-rate performance for aqueous supercapacitors. Furthermore, the ball milling-induced order rearrangement of carbon layers promotes the development of the large graphenic crystallites toward graphite structure during the subsequent 2800 ℃ treatment. A similar structure repair effect also be observed for various hard carbons derived from polymers and biomass. Therefore, rational ball-milling may be a universal pre-treatment strategy for “destroying” high crystalline carbons while “repairing” the disordered structure of hard carbons.

Interlayer Sodium Plating/Stripping in Van der Waals-Layered Quantum Dot Superstructure

Peng Liu and Jisheng Zhou*

State Key Laboratory of Chemical Resource Engineering,
Beijing Key Laboratory of Electrochemical Process and Technology for Materials,
Beijing University of Chemical Technology, Beijing 100029, P. R. China.
E-mail: zhoujs@mail.buct.edu.cn

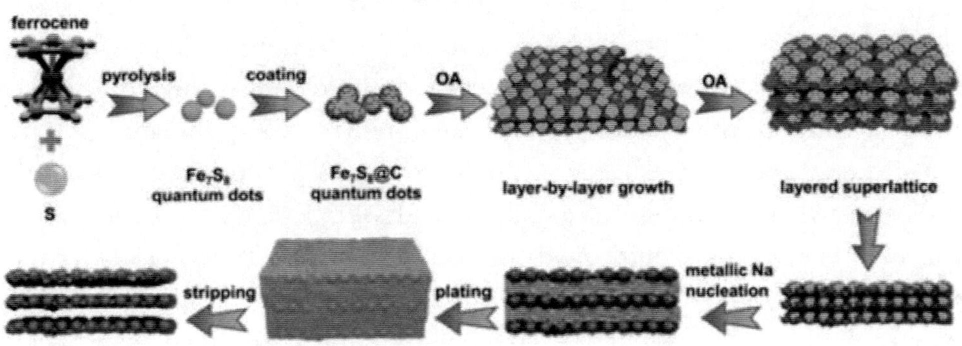

Scheme 1. Schematic Illustration of the OA Growth Process and Interlayer Metallic Na Plating/Stripping Behavior of Fe7S8@CNS.

Assembling quantum dots (QDs) into van der Waals (vdW)-layered superstructure holds great promise for development of high-energy-density metal anode. However, designing such superstructure remains to be challenging. Here, we propose a chemical-vapor Oriented Attachment (OA) growth strategy to achieve the synthesis of vdW-layered carbon/QDs hybrid superlattice nanosheets (Fe7S8@CNS) with a large vdW gap of 3 nm. The Fe7S8@CNS superstructure is assembled by carbon-coated Fe7S8 (Fe7S8@C) QDs as building blocks. Interestingly, the Fe7S8@CNS exhibits two kinds of edge dislocations similar to traditional atom-layered materials, suggesting that Fe7S8@C QDs exhibit quasi-atomic growth behavior during the OA process. More interestingly, when used as host materials for sodium metal anodes, the Fe7S8@CNS shows the interlayer sodium plating/stripping behavior, which well suppresses Na dendrite growth. As a result, the cell with Fe7S8@CNS anode can keep stable cycling for 1000 h with a high Coulombic efficiency (CE) of ~99.5% at 3.0 mA/cm2 and 3.0 mAh/cm2. Noticeably, the Na@Fe7S8@CNS || Na3V2(PO4)3 full cells can attain a capacity of 88.8 mAh/g with a retention of 97% after 1000 cycles at 1.0 A/g (~8C), showing excellent cycle stability for practical applications. This work enriches the vdW-layered QDs superstructure family and their application toward energy storage.

Synergetic effect of Metals in Carbon Materials to Enhance the Photothermal Performance

Mengxue Shen, Guanyu Chen, Jiapeng Zhang and Ang Li *

State Key Laboratory of Chemical Resources Engineering,
Beijing Key Laboratory of Electrochemical Process and Technology for Materials,
Beijing University of Chemical Technology, Beijing 100029, P. R. China.

With the increasing demand for the application of renewable energy, solar-driven photothermal materials has been widely explored due to their high energy conversion capability and eco-friendliness. Herein, a metal-carbon fiber composites were prepared to investigate the synergistic enhancement of photothermal performance. The composite materials were subjected to solar-driven photothermal tests alongside pure metals and pyrolytic phenolic resin materials (Phf). The results showed that the addition of metals effectively improved the photothermal efficiency of the materials, carbonized phenolic resin fibers containing nanometal Cu (Phf-Cu) exhibited the most optimal performance which reached up to 54.6 %, higher than those of Phf (49.9%) and Cu metal (27.1%). Under sunlight, the water evaporation rate of 2 wt% Phf-Cu evaporator reached 2.04 kg/(m2h). Furthermore, the analysis based on the Kubelka-Munk theory and Fermi level calculation demonstrates that the addition of metal reduces the bandgap width of carbon. Specifically, the bandgap width of Phf-Cu is the narrowest at 1.70 eV, while Phf has the widest bandgap width at 2.15 eV. This facilitates easier electron transition to excited states and a greater proportion returns to the ground state through non-radiative transitions. These findings indicate a strong synergistic enhancement effect between metal and carbon materials.

Carbon dioxide conversion of ethanol on copper-containing catalysts

M. Mambetova[1,2], M. Anissova[1,2], B. Serkebayev[2], G. Yergaziyeva[1,2], K. Dossumov

[1]Center of Physico-Chemical Methods of Research and Analysis, 96A Tole bi str., Almaty, Kazakhstan
[2]Institute of Combustion Problems, 172 Bogenbay batyr str., Almaty, Kazakhstan

In recent decades, the problem of climate change and energy security has become increasingly pressing, stimulating the search for environmentally friendly and efficient methods of energy conversion. In this regard, the process of carbon dioxide conversion of ethanol (CDCE), which is a promising technology that can play a key role in the production of clean energy and chemical products, is of particular importance. Syngas production by carbon dioxide conversion of ethanol (CDCE) is a promising method because it uses both renewable bio-derived ethanol and waste CO2 as feedstock. One of the most promising approaches to CDCE is the use of copper-based catalysts. Copper oxide, due to its unique chemical activity, shows high activity in the carbon dioxide conversion of ethanol at relatively low temperatures and pressures. In this work, low-percentage catalysts based on copper oxide were studied in the process of CDCE. Copper oxide containing 3 wt.% was applied to supports 5A, HZSM-5, Al2O3, SiO2 by capillary impregnation. All prepared catalysts were calcined at 300 °C for 2 hours, then at 500 °C for 3 hours. Before studying the carbon dioxide conversion of ethanol, the catalysts were reduced in a flow of H2:N2=1:1 at a temperature of 500 °C for 1 h. The synthesized catalysts were studied in the process of CDCE at CO2:C2H5OH = 1:1 in the temperature range 500-850 °C on a laboratory flow-type installation. The results obtained showed that on low-percentage copper-containing catalysts in the temperature range 500-850 °C and the ratio CO2:C2H5OH = 1:1, regardless of the nature of the carrier, ethylene, methane, and synthesis gas are formed in the reaction products. High concentrations of ethylene are observed at lower temperatures (500, 550 °C); a further increase in temperature to 800 °C leads to a decrease in the yield of ethylene. The highest concentrations of synthesis gas and methane are formed in the range of 650-80 0°C. The yield of reaction products is influenced by the nature of the copper catalyst support, so the highest concentration of ethylene, 41 vol.%, is formed on a 3 wt.% CuO/SiO2 catalyst at a reaction temperature of 550 °C. The highest concentration of CO 37.5 vol.% and H2 16.6 is formed on the catalyst 3 wt.% CuO/ γ-Al2O3 at a reaction temperature of 800 °C.

This research has been funded by the Science Committee of the Ministry of Science and Higher Education of the Republic of Kazakhstan (Grant No. APAP19678248).

Controllable Synthesis of Atomic Sn-anchored Carbon Host for Excellent Long-Cycle-Life Zinc Metal Batteries

Lantao Liu[1], Weiwei Pang*

[1] Petrochemical Research Institute, PetroChina Company Limited,102206 Beijing, P. R. China

Aqueous Zinc (Zn) metal batteries are considered to be promising next-generation safe energy storage devices, but severe hydrogen evolution reaction (HER) and uncontrollable dendrite growth hinder their commercial deployment. In this work, atomic Sn-anchored carbon sphere with multicavity (Sn@HMCS) electrode material is developed, to guide/control Zn metal deposition and inhibit HER and dendrites formation. XANES analysis, DFT simulations and in-situ/ex-situ characterization of the material reveal that single-atom Sn can enhance the zincophilicity of host materials, lower the nucleation barrier for Zn2+ adsorption, redistribute the charge density of the interface between Zn and carbon matrix, and accelerate the deposition kinetics of Zn2+, thereby suppressing the formation of dendrites and hydrogen gas during the plating/stripping process. As a result, in Zn-Zn symmetric cells, Sn@HMCS can cycle stably for more than 2300 h at 1 mA cm-2 for 1 mAh cm-2 with ultralow voltage hysteresis of ≈134 mV. In aqueous full cells, Sn@HMCS endows the devices with high energy density (up to 190 Wh kg-1) and stable cycling performance (900 cycles with a capacity retention of 89 % at 1 A g-1).

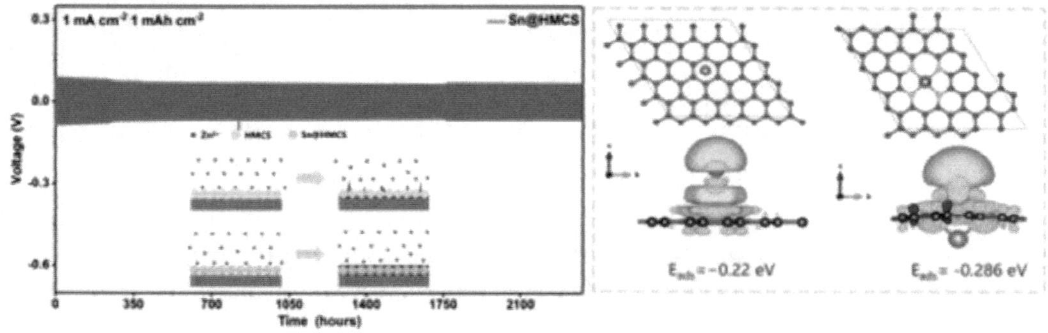

Effect of heat treatment on performance of setaria viridis-based anode materials for lithium ion batteries

Dong Ki Kim[1], Chaehun Lim[1], Seongjae Myeong[1], Naeun Ha[1], Chung Gi Min[1], and Young-Seak Lee[1,2]

[1]Department of Chemical Engineering and Applied Chemistry,
Chungnam National University, Daejeon, 34134, Republic of Korea
[2]Institute of Carbon Fusion Technology (InCFT),
Chungnam National University, Daejeon, 34134, Republic of Korea

Setaria viridis are distributed all around the world, and due to the high production volume, a large amount is discarded. As oxygen exist on the heat-treated surface of setaria viridis, it has the advantage of being able to omit a separate process of introducing heterogeneous elements into the anode material for lithium-ion secondary batteries. In this study, anode materials (SV-Cs) for lithium-ion secondary batteries were manufactured by heat-treating setaria viridis, a weed known to hinder plant growth, and their electrochemical performances were examined. In SV-C-750, the specific capacity (1003.3 mAh/g, at 0.1 C) is high due to the negative charge of oxygen present on the surface attracting lithium ion, but the capacity retention rate (61.0% at 500 cycles) and 1 C) was low. In SV-C-1150, the capacity retention rate (89.7% at 500 cycles and 1 C) was high, but the specific capacity (471.7 mAh/g) was low. Within the scope of this study, it was confirmed that the setaria viridis-based carbon anode materials manufactured according to the heat treatment temperature have a high specific capacity, and it was found that the surface oxygen content and crystallinity have high reliability in the electrochemical properties of the anode material.

References

[1] Su F.Y., You C.H., He Y.B., Lv W., Yang Q.H., et al. Flexible and planar graphene conductive additives for lithium-ion batteries. Journal of Materials Chemistry, 20, 9644-9650 (2010).

Boosting the Molecular Crosslinking in a Novel Phenolic Resin for Spherical Hard Carbon with Enriched Closed Pores towards Enhanced Sodium Storage Ability

Liewen Guo[1], Huaihe Song [1]

[1]Beijing University of Chemical Technology, Chaoyang District, Beijing, China.

ABSTRACT: Due to the shortage of lithium resources, sodium-ion batteries (SIBs) are being investigated as viable alternatives to lithium-ion batteries (LIBs) owing to the abundant sodium resources, long life and high safety performance. Graphite regrettably cannot be used in SIBs because of difficulty in forming a stable interlayer compound. Fortunately, hard carbon (HC) is regarded as the most promising anodes material for SIBs. Phenolic resin (PF) is considered as a promising precursor of HC because of its facile designability and high residual carbon yield. However, understanding how the structure of PF precursors influences sodium storage in their derived HC remains a significant challenge. Herein, the microstructure of HC is controlled by the cross-linking degree of resorcinol-benzaldehyde (RB) resin. We reveal that the robust molecular crosslinking in RB resin induced by hydrothermal treatment promotes the closed-pore formation in the derived HC. Consequently, the optimized hydrothermal treatment HC anode shows a higher specific capacity of 372.7 mAh g-1 and better rate performance than the HC anode without hydrothermal treatment (276.0 mAh g-1). This strategy can provide feasible molecular crosslinking engineering for the development of closed pores in PF-based HC towards enhanced sodium storage.

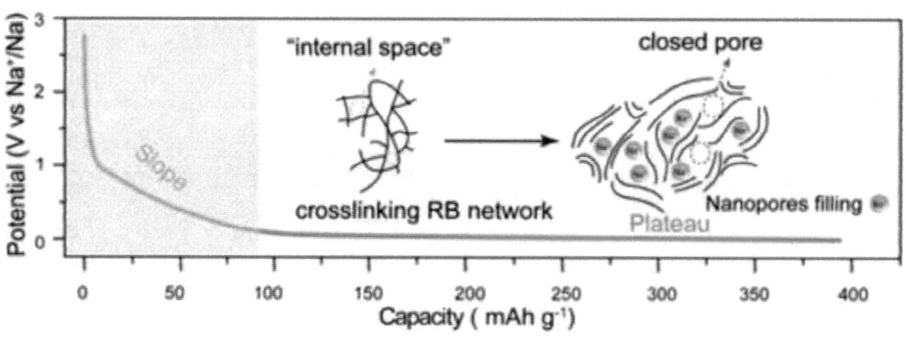

Design of the particle size and morphology of hard carbon anode materials for sodium-ion batteries through hydrothermal carbonization

Grigorii Lakenko[1], Zoya Bobyleva[2], Oleg Drozhzhin[2] and Evgeny Antipov[2,3]

[1] Skoltech Center for Energy Science and Technology,
Skolkovo Institute of Science and Technology, Nobel str.3, 143026, Moscow, Russia
[2] Department of Chemistry, Lomonosov Moscow State University,
Leninskie gory GSP-1, 119991 Moscow, Russia

With sodium-ion batteries (SIBs) finding widespread application, the demand grows for hard carbon, the most popular anode material for SIBs. Hydrothermal carbonization facilitates the production of hard carbon with desired characteristics from various sources. Despite the considerable volume of literature addressing this subject, there exists a notable absence of investigations elucidating the relationship between synthesis conditions and the electrochemical characteristics of the product. Here we study systematically the influence of hydrothermal carbonization parameters on hard carbon characteristics and emphasize the potential of hard carbon as an anode material for SIBs. The initial Coulombic efficiency (ICE) is significantly affected by the particle size of the glucose-derived hard carbon, which, in turn, depends on glucose concentration in the initial solution, pH, and stirring regime. By optimizing the hydrothermal carbonization parameters, the ICE up to 91% and a good reversible capacity of ~300 mAh g−1 in a half cell are achieved. Full cells with Na3(VO)2(PO4)2F cathode material demonstrate ICE of about 80% and reversible capacity of up to 100 mAh g−1cath. Considering the effective performance of pouch-cell SIB prototypes based on Na3(VO)2(PO4)2F and hard carbon, hydrothermal carbonization of glucose yields hard carbon with the necessary characteristics required for its successful application in SIBs.

This work was supported by the Russian Science Foundation (Project No. 24-23-20052)

Enhanced electrochemical performance of lithium-ion batteries utilizing O_2 plasma-treated needle coke-based anode materials

Chung Gi Min[1], Chaehun Lim[1], Seongjae Myeong[1], Naeun Ha[1], and Young-Seak Lee[1,2]

[1] Department of Chemical Engineering and Applied Chemistry,
Chungnam National University, Daejeon, 34134, Republic of Korea
[2] Institute of Carbon Fusion Technology (InCFT),
Chungnam National University, Daejeon, 34134, Republic of Korea

High-rate capacities and specific capacities are important indicators in evaluating lithium-ion batteries (LIBs) anodes. To improve the electrochemical properties of needle coke-based anode materials, oxygen plasma treatment was conducted. Various analyses were conducted to assess the impact of oxygen plasma treatments on the chemical, structural, and morphological properties of the needle coke samples. The oxygen plasma treatment effectively introduced the oxygen functional groups (hydroxyl, ether, carbonyl, and carboxyl groups) without significant structural changes in the needle cokes, and the O/C ratio increased up to 0.23 after the treatment. The synergistic effect of increased oxygen contents and defects formed on the surface by plasma treatment caused the improvement of the electrochemical properties of the needle coke-based anodes. The discharge capacity of the oxygen plasma-treated needle coke-based LIBs increased to a maximum of 20.48%, and the retention rate was 74.51% at 5 C compared with 0.1 C. Oxygen plasma-treated needle cokes are materials that can play an important role in improving the rate capacities and specific capacities of LIBs in a simple method.

One-Step Construction of Closed Pores Enabling High Plateau-Capacity Hard Carbon Anodes for Sodium-Ion Battery: Closed-Pore Formation and Energy Storage Mechanisms

Chuang Qiu, Huaihe Song*

State Key Laboratory of Chemical Resources Engineering,
Beijing Key Laboratory of Electrochemical Process and Technology for Materials,
Beijing University of Chemical Technology, Beijing 100029, P. R. China.

Closed pores play a crucial role in improving the low-voltage (< 0.1 V) plateau capacity of hard carbon anodes for sodium-ion batteries (SIBs). However, the lack of simple and effective closed-pore construction strategies, as well as the unclear closed-pore formation mechanism, have severely hindered the development of high plateau-capacity hard carbon anodes. Herein, we present an effective closed-pore construction strategies by one-step pyrolysis of zinc gluconate and elucidate the corresponding mechanism of closed-pore formation. The closed-pore formation mechanism during the pyrolysis of ZG mainly involves i) the precipitation of ZnO nanoparticles and the ZnO etching on carbon under 1100 °C to generate open pores of 0.45-4 nm, and ii) the development of graphitic domains and the shrinkage of the partial open pores at 1100-1500 °C to convert the open pores to closed pores. Benefiting from the considerable closed-pore content and suitable microstructure, the optimized hard carbon achieves an ultra-high reversible specific capacity of 481.5 mAh g-1 and an extraordinary plateau capacity of 389 mAh g-1 for use as the anode of SIBs. Additionally, some in-situ and ex-situ characterizations demonstrate that the high-voltage slope capacity and the low-voltage plateau capacity stem from the adsorption of Na+ at the defect sites and Na-cluster formation in closed pores, respectively.

Composite hydrogen adsorbent based on nanoporous carbon and metal nanoparticles

B. Lesbayev[1,2], M. Nazhipkyzy[1,2] G. Ustayeva[1,2], N. Rakhymzhan, A. Maltay [1,2], S. Taupikhova[1,2]

[1]Al-Farabi Kazakh National University, Almaty, Kazakhstan
[2]Institute of Combustion Problems, Almaty, Kazakhstan

Energy production is an important factor in socio-economic progress, but the extraction and use of fossil fuels, leading to environmental pollution, increased greenhouse gas emissions and irreversible deterioration of the planet' s climate, dictate the need to search for alternative environmentally friendly energy sources. As part of the global strategy for sustainable energy development, the use of hydrogen as a secondary energy carrier expands the possibilities for long-term storage of excess electricity generated by alternative energy sources (solar, wind, geothermal and nuclear) by producing environmentally friendly hydrogen through the electrolysis of water. On the other hand, one of the key factors slowing the widespread use of hydrogen as an alternative fuel is the lack of safe, economical, and efficient hydrogen storage systems. In connection with the above, in the proposed work, research was carried out to study the possibilities of creating a composite hydrogen adsorbent based on nanoporous carbon and metal nanoparticles operating at moderate temperatures and pressures.

Instructions for Preparation of "CARBON 2024" Short Abstract

Microstructure regulation of coal-derived hard carbon for high performance sodium ion battery

Beilei Yuan*, Rui Li, Jian Chen, Wei Jiao

Sichuan University of Science & Engineering, Zigong, China, 643000

*Email: ybl129921@163.com

As a new type of anode material, hard carbon material has good electrical conductivity, low temperature performance and excellent electrochemical performance, and has shown a good application prospect in the field of sodium ion battery (SIB). However, the difficulty of pre-treatment and low yield limit the commercial development of hard carbon materials[1,2]. Based on this, we present a strategy to inhibit graphitized microcrystals of anthracite-derived carbon, which reveals the correlations between the initial Coulombic efficiency (CE) and heteroatom/defects. By introducing NH3 during the carbonization process of anthracite, it is possible to effectively remove the heteroatoms (O, S, etc.) in anthracite, while also expanding the spacing of carbon layers derived from anthracite. This results in obtaining high capacity and high initial CE. Using Nitrogen-doped anthracite hard carbon (N-AC-1200) as an anode material for SIBs, a high reversible capacity of 220 mAh·g-1 at 0.1 A·g-1 is achieved. Furthermore, a full cell assembled by Na2/3Ni1/3Mn2/3O2 cathode and N-AC-1200 anode delivers a high capacity of 76 mAh·g-1 at 5 C. The above work not only provides new ideas for the microstructure regulation of hard carbon, but also helps to promote the development of sodium ion batteries.

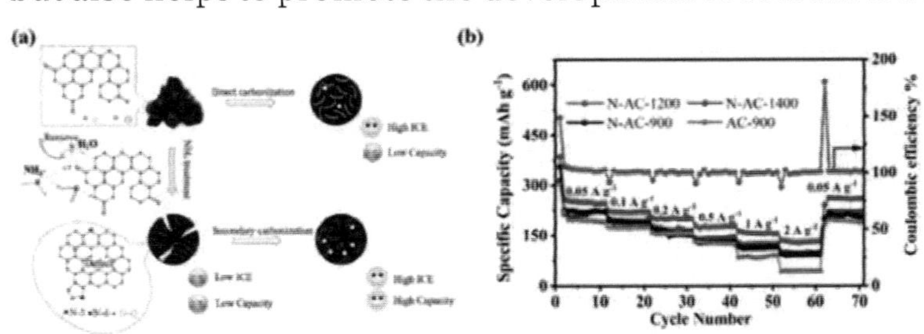

Fig1. (a) Schematic illustration of the synthesis process of AC and N-AC.; (b) Rate performance.

Keywords: Anthracite; Hard carbon; Microstructure regulation; Sodium-ion batteries

References

[1] F. Sun, H. Wang, Z. Qu, K. Wang, et al., Adv. Energy Mater. 2021, 11, 2002981.

[2] K.F. Wang, F. Sun, H. Wang, et al., Adv. Funct. Mater. 2022, 16, 2203725.

Understanding the effect of surface area to the performance of hard carbons as negative electrodes of Na-ion batteries

Berke Karaman [1], Helene Tonnoir [1], Jimena Castro Gutiérrez [2], Zoé Deckers [1], Bryan Carré [1], Alexandre F. Léonard [3], Vanessa Fierro [2], Alain Celzard [2,4], Nathalie Job* [1]

[1]Department of Chemical Engineering - NCE (Nanomaterials, Catalysis, Electrochemistry), University of Liège, 4000 Liège, Belgium

[2]Université de Lorraine, CNRS, Institut Jean Lamour - IJL, 88000 Épinal, France

[3]Department of Chemical Engineering - CARPOR, University of Liège, 4000 Liège, Belgium

[4]Institut Universitaire de France - IUF, 75231 Paris, France

(*corresponding author: berke.karaman@uliege.be)

The increase in energy storage requires the development of better materials and manufacturing processes. In that respect, Na-ion batteries are seen as a key technology. At the negative electrode, carbonaceous materials are widely studied given their highly tunable character. Especially, hard carbons are materials that can be improved to be used efficiently in Na-ion batteries. This study focuses on understanding the effect of specific surface area (SBET) and surface area accessible to the electrolyte to the performance of negative electrodes of Na-ion batteries that are prepared using a water-based process.

Hard carbons with large nodule size (~2 μm) were synthesized by pyrolysis of phenolic resins. The obtained materials were microporous, with large SBET (~600 m^2/g). Those samples were modified by Chemical Vapor Deposition and/or CO2 activation in order to tune the final SBET: a wide value range, from 1100 m2/g to 3 m2/g was obtained while keeping the same nodule size. Those various surface areas resulted in significantly different electrode capacities, from 62 mAh/g to 300 mAh/g, with a wide range of Initial Coulombic Efficiencies (18% to 84%). Those efficiencies were however not related to the SBET of the powder carbon, but strongly depended on the real electrode-electrolyte contact surface.

Targeted Manner for Achieving High-Capacity Hard Carbon Anodes in Li-ion and Na/K-ion Batteries

Wang Zhou[1], Rui Tang[1], and Jilei Liu[1]

[1]College of Materials Science and Engineering, Hunan University, Changsha 410082, China

The intricate microstructures of hard carbons together with the diverse physicochemical properties of alkali metal ions contribute to the enigmatic nature of Li/Na/K ions storage mechanisms[1-2]. The current state of research still lacks comprehensive understanding of the similarities and differences in Li-ion and Na/K-ion storage behaviors in hard carbon, making it challenging to optimize the performance of hard carbons for Li/Na/K-ion batteries in a targeted manner. Herein, we quantitatively investigated the extrinsic defects, crystallite parameters, and nanopores of resin-derived hard carbons and correlated these microstructural parameters with the electrochemical performance. Our findings demonstrate that adsorption, intercalation and nanoclustering, are all available mechanisms for Li/Na/K-ion, when the d002 >3.7 Å. As the d002 decreases, only adsorption and intercalation remain available for Li ions, whereas the adsorption and nanoclustering become predominant for Na/K ions. Leveraging these findings, we have successful synthesized hard carbons with a sufficient large d002 to facilitate Li migration into the closed pores for clustering, resulting in satisfactory reversible capacity (497 mAh/g). Furthermore, by fabricating abundant closed pores with a suitable graphene interlayer spacing ($\approx$ 3.7 Å), we achieved remarkable reversible capacity in Na/K-ion batteries, with values of 530 mAh/g for Na and 520 mAh/g for K.

References

[1] Wang Zhou, et al. Adv. Energy Mater. 13 (2023) 2202874.
[2] Wang Zhou, et al. Adv. Funct. Mater. 34 (2024) 2312994.

Both Resilience and Adhesivity Define Solid Electrolyte Interphases for a High Performance Anode

Yue Zhai[1,2,3], Shichao Wu[1,2,3] and Quan-Hong Yang[1,2,3,4]

[1]Nanoyang Group, Tianjin Key Laboratory of Advanced Carbon and Electrochemical Energy Storage, School of Chemical Engineering and Technology, and Collaborative Innovation Center of Chemical Science and Engineering (Tianjin), Tianjin University, Tianjin, 300072, China.
[2]National Industry-Education Integration Platform of Energy Storage, Tianjin University, Tianjin, 300072, China.
[3]Haihe Laboratory of Sustainable Chemical Transformations, Tianjin, 300192, China.
[4]Joint School of National University of Singapore and Tianjin University, International Campus of Tianjin University, Fuzhou, 350207, China.

Solid-electrolyte interphases (SEIs) are sought to protect high-capacity anodes, which suffer from severe volume changes and fast degradations. The previously proposed effective SEIs were high-strength yet abhesive inducing a yolk-shell structure to decouple the rigid SEI from anode for accommodating the volume change. Ambivalently, the interfacial void-evolved electro-chemo-mechanical vulnerabilities become inherent defects. Here, we establish a new rationale for SEIs that resilience and adhesivity are both requirements and pioneer a design of a resilient yet adhesive SEI (re-ad-SEI), integrated into a conjugated surface bilayer structure. The re-ad-SEI and its protected particles exhibit excellent stability almost free from the thickening of SEI and the particle pulverization during cycling. More promisingly, the dynamically bonded intact SEI-anode interfaces enable a high-efficiency ion transport and provide a unique mechanical confinement effect for structural integrity of anodes. The high Coulombic efficiency (>99.8%), excellent cycling stability (500 cycles) and superior rate performance have been demonstrated in micro-sized Si-based anodes.

References

[1] Yue Zhai, et al. Journal of the American Chemical Society 146(22) (2024) 15209-15218

High-Stability CO2-Philic & Hydrophobic Interface Directed by Interconnected sp2/sp3 Carbon Boosts Electrochemical CO_2-to-Formate

Wenbin Li[1], Chang Yu[1] and Jieshan Qiu[2]

[1]Dalian University of Technology, Dalian, China
[2]Beijing University of Chemical Technology, Beijing, China

Carbon materials have attracted much attention due to their high electronic conductivity, excellent stability and cost advantages, especially in the emerging electrocatalytic CO2 conversion field. In addition to improving intrinsic catalytic activity directly as electrocatalysts, functionalized carbon materials can be used as an important component of catalysts to construct interconnected molecular diffusion channels and conductive networks, which can promote the diffusion of reactants and rapid electron conduction within the catalysts, and enhance the structural stability of the catalysts under severe electrocatalytic conditions. Formic acid are the most common products and have the highest commercialization potential in the CO2 catalytic conversion. Herein, a cross-linked carbon coupled with active bismuth catalyst with a stable spatial structure and unique molecular sieving effects is designed, achieving the formate production at an industrial large current. [1] Moreover, based on the highly active and stable carbon-based materials, we have made attempts and efforts in the optimization and scaling-up of electrochemical devices to enhance device and system stability. [1, 2]

References

[1] Wenbin Li, et al. ACS Catalysis, 14 (2024) 8056-8061.
[2] Wenbin Li, et al. Chemistry of Materials, 36 (2024) 1602-1611.

Ni/N-modulated biomass pyrolysis to make carbon-based catalysts for electrochemical CO_2-to-CO conversion

Yafang Zhang[1], Chang Yu[1] and Jieshan Qiu[2]
[1]Dalian University of Technology, Dalian, China
[2] Beijing University of Chemical Technology, Beijing, China

There are tremendous requirements in CO2 electrolysis to produce value-added chemicals, where the functional carbon-based catalysts possess the unique superiorities for CO2 conversion. Pyrolysis kinetics as a theoretical guidance is pivotal to construct the high-activity carbon-based catalysts precisely, which is decoupled urgently. Herein, the pyrolysis behaviors of ipomea batatas leaves were well optimized by Ni/N modulators and unveiled systematically, where the residence time of biomass was prolonged during the pyrolysis process and the emerged out-of-plane orientation sp3 N–C species were increased. Additionally, the presented first-order kinetics revealed that with Ni/N modulators, the higher apparent energy barrier was determined than single biomass, corresponding to an enhanced thermostability. As a guidance, the derivative catalyst with high N content (17.22 at%) can drive electrochemical CO2-to-CO reaction efficiently. The current density was up to 200 mA cm−2 and the CO Faradaic efficiency kept over 95% at a wide work window (–0.66 to –1.46 V versus reversible hydrogen electrode) in flow cell. [1]

References

[1] Yafang Zhang, et al. Nano Energy, 126 (2024) 109658.

One-Step Construction of Closed Pores Enabling High Plateau-Capacity Hard Carbon Anodes for Sodium-Ion Battery: Closed-Pore Formation and Energy Storage Mechanisms

Chuang Qiu, Huaihe Song*

State Key Laboratory of Chemical Resources Engineering,
Beijing Key Laboratory of Electrochemical Process and Technology for Materials,
Beijing University of Chemical Technology, Beijing 100029, P. R. China.

Closed pores play a crucial role in improving the low-voltage (< 0.1 V) plateau capacity of hard carbon anodes for sodium-ion batteries (SIBs). However, the lack of simple and effective closed-pore construction strategies, as well as the unclear closed-pore formation mechanism, have severely hindered the development of high plateau-capacity hard carbon anodes. Herein, we present an effective closed-pore construction strategies by one-step pyrolysis of zinc gluconate and elucidate the corresponding mechanism of closed-pore formation. The closed-pore formation mechanism during the pyrolysis of ZG mainly involves i) the precipitation of ZnO nanoparticles and the ZnO etching on carbon under 1100 °C to generate open pores of 0.45-4 nm, and ii) the development of graphitic domains and the shrinkage of the partial open pores at 1100-1500 °C to convert the open pores to closed pores. Benefiting from the considerable closed-pore content and suitable microstructure, the optimized hard carbon achieves an ultra-high reversible specific capacity of 481.5 mAh g-1 and an extraordinary plateau capacity of 389 mAh g-1 for use as the anode of SIBs. Additionally, some in-situ and ex-situ characterizations demonstrate that the high-voltage slope capacity and the low-voltage plateau capacity stem from the adsorption of Na+ at the defect sites and Na-cluster formation in closed pores, respectively.

Efficient Pseudocapacitive Energy Storage of Conductive PANI Composites in Alkaline Electrolytes

Yiran Jia[1], Hongyuan Yu[1], Ying Tao[1], Yang Quan-Hong[1,2,3]

[1]Nanoyang Group, State Key Laboratory of Chemical Engineering,
School of Chemical Engineering and Technology, Tianjin University, Tianjin 300072, China.
[2]Haihe Laboratory of Sustainable Chemical Transformations, Tianjin 300192, China.
[3]Joint School of National University of Singapore and Tianjin University International Campus of Tianjin University Binhai New City, Fuzhou 350207, China.

Protonation has been considered essential for the pseudocapacitive energy storage of polyaniline (PANI) for years, as proton doping in PANI chains not only activates electron transport pathways, but also promotes the proceeding of redox reactions. Rarely has the ability for PANI of storing energy without protonation been investigated, and it remains uncertain whether PANI has pseudocapacitive charge storage properties in an alkaline electrolyte. Here, this work first demonstrates the pseudocapacitive energy storage for PANI without protonation using a PANI/graphene composite as a model material in an alkaline electrolyte. Using in situ Raman spectroscopy coupled with electrochemical quartz crystal microbalance (EQCM) measurements, this work determines the formation of −N= group over potential on a PANI chain and demonstrates the direct contribution of OH− in the nonprotonation type of oxidation reactions. This work finds that the PANI/graphene composite in an alkaline electrolyte has excellent cycling stability with a wider operation voltage of 1 V as well as a slightly higher specific capacitance than that in an acidic electrolyte. The findings provide a new perspective on pseudocapacitive energy storage of PANI-based composites, which will influence the selection of electrolytes for PANI materials and expand their application in energy storage fields.

Ionic Modulation Layer Triggered by Polyaniline Enabling High-Efficiency CO_2 Electroreduction in Strongly Acidic Electrolytes

Zhenpeng Hu[1], Chang Yu[1] and Jieshan Qiu[2]

[1]Dalian University of Technology, Dalian, China

[2]Beijing University of Chemical Technology, Beijing, China

Electrocatalytic reduction of CO2 offers a viable pathway to produce high-value fuels and chemicals[1]. Acidic electrolytes not only exhibit low ohmic resistance and efficiency carbon efficiency, but also effectively reduce carbonate clogging in gas diffusion electrodes (GDE). Carbon materials exhibit excellent conductivity, stability, and abundant resource availability, showing significant potential in acidic electrolytes[2]. However, in strongly acidic electrolytes, H+ reduction is kinetically favorable and can corrode the structural stability of carbon materials. In addition, the -NH- groups in polyaniline are readily protonated by acids, allowing them to repel H+ ions through short-range forces, thereby preventing these ions from attacking the catalysts. In this study, we have modified the surface of cathode catalysts with doped polyaniline, constructing a conductive ion-regulating layer. The local pH at the catalyst-electrolyte interface has been increased by this redistribution of H+ in the electric double layer, thus stabilizing the local reaction environment. Furthermore, this ion regulatory layer strategy has been applied to various carbon-based materials, revealing the potential of the ion-regulating layer for high-efficiency CO2 reduction in strongly acidic electrolytes

References

[1] Xinyi,Tan, Chang Yu and Jieshan Qiu, et al. Energy Environ. Sci., 14(2021) 765-780.

[2] Wenbin Li, Yafang Zhang, Chang Yu and Jieshan Qiu, et al. ACS Catalysis, 14 (2024) 8056-8061.

Catalytic conversion of Li_2S for high-performance lithium-sulfur batteries

Chuannan Geng[1,2], Wei Lv[2] and Quan-Hong Yang[1]

[1]Nanoyang Group, State Key Laboratory of Chemical Engineering,
School of Chemical Engineering and Technology, Tianjin University, Tianjin 300072, China
[2] Shenzhen Geim Graphene Center, Engineering Laboratory for Functionalized Carbon Materials,
Tsinghua Shenzhen International Graduate School, Tsinghua University, Shenzhen, 518055, China
Email: gengcn@tju.edu.cn

The rechargeable lithium-sulfur (Li-S) batteries have become one of the most promising systems for next-generation high energy-density energy storage. However, the sluggish kinetics of the sulfur evolution reaction (SER) that occur because of the high oxidation barrier of Li2S causes low sulfur utilization and the poor rate performance of Li-S batteries. The design of the catalysts to solve this problem is still hard to achieve because it is difficult to precisely correlate the catalytic oxidation ability with the electronic structure. Here, a layer transition metal oxide, NaxTi0.5Co0.5O2, is used as a model catalyst to probe such a correlation because it has a tunable electronic structure and good stability in the working potential window of Li-S batteries. By removing Na+, a partial phase change gradually increases the concentration of Co active sites while decreasing the work function with an upshift of the Fermi level, accelerating charge transfer on the catalyst surface and therefore improving its catalytic oxidation activity of Li2S.

On this basis, we further developed the homogeneous catalyst trithiocyanuric acid trilithium salt (TTCA-Li) dissolved in the electrolyte to improve the catalytic efficiency. It can not only catalytically decrease the initial overpotential of Li2S, but also form an interfacial layer by in-situ polymerization during charging. The produced coating effectively confines the shuttling of lithium polysulfides, thus inhibiting the redistribution of sulfur species and active sulfur loss upon cycling. The prototype full cell where the coated Li2S cathode couples with the Li anode has an extremely high CE of over 99.5%.

References

[1] Qu W. J., Geng C. N., Lv W.*, Yang Q.H.*, et al., Advanced Energy Materials 2022, 2202232.

[2] Geng C. N., Lv W.*, Yang Q.H.*, et al., Advanced Energy Materials 2023, 2204246.

A long/short-range interconnected carbon with well-defined mesopore for high-energy-density supercapacitors

Jiani Chen[1], Chang Yu[1] and Jieshan Qiu[2]

[1]Dalian University of Technology, Dalian, China
[2]Beijing University of Chemical Technology, Beijing, China

As representative and widely adopted carbon-based electrode materials in supercapacitors, the biomass-derived porous carbon possesses the well adsorption capability towards the electrolyte ions due to large SSA. Nevertheless, its short-range ordered structure, generally produced at a low annealing temperature, leads to poor conductivity and the correspondingly sluggish electron transfer in charge/discharge process. Herein, in order to in-situ implant the long-range ordered structure into amorphous bio-based carbon, we presented a petroleum asphalt-mediated co-conversion technique to modify the solid-phase conversion/carbonization process of biomass for configuring the soft-hard carbon materials (SHCs) with coupled long- and short-range ordered structure in a smallscaled range by employing the intrinsic features of liquidphase carbonization for petroleum asphalt. Thanks to the well configured structure in small scale, the as-made co-converted carbon can be operated in high-viscosity EMIMBF4 electrolyte with a superior capacitance (315 F·g−1@1 A·g−1). Besides, the as-assembled symmetric supercapacitor can deliver a super-high specific energy of 174 Wh·kg−1@2.0 kW·kg−1. This work provides a new version for designing highly porous biomass-derived carbon with long/short-range alternating structure at molecular level. [1]

References

[1] Yuanyang Xie, et al. Nano Res. 2022, 15(2): 1399–1408.

Optimizing Pore Structure of Pitch-Based Carbon Materials for Efficient Na Ion Storage

Sheng Zhou, and Ye Cong*

Wuhan University of Science and Technology, Wuhan, China

Modulating pore structure is an effective strategy to enhance electrolyte infiltration, increase transport channels, and improve the diffusion kinetics and cycling stability of carbon materials in sodium ion batteries. However, the imbalance in electrochemical performance caused by porous structure presents significant challenges for commercialization. Herein, microporous/mesoporous carbon were synthesized using a template method with asphalt as the raw material. The hierarchical pores play crucial roles in sodium ions storage. Mesopores provide abundant surface adsorption sites, while micropores effectively confine Na^+ ions, preventing the entry of large solvent molecules and enhancing Na^+ storage performance. Further pore size modification was achieved by filling the pores with graphite-like carbon via chemical vapor deposition, resulting in abundant ultramicropores ($\approx$0.8 nm). Consequently, the hierarchical porous carbon anode materials exhibit a high reversible capacity (313 mAh g−1 at 0.02 A g−1), excellent rate capability (201 mAh g−1 at 10.0 A g−1), and outstanding cycling stability ($\approx$ 98% capacity retention after 1000 cycles), offering a promising avenue for the application of asphalt-derived porous carbon materials in energy storage.

Preparation of hydrophobic activated carbon fibers and their adsorption behavior for dichloromethane

LinHong Wen[1], Jianxiao Yang[1], Xin Zhang[2]

[1] Hunan University, Changsha 410082, China

[2] University of Chinese Academy of Sciences, Beijing, 101408, China

In recent years, it has become necessary to prepare hydrophobic materials for adsorbing the emitted dichloromethane (DCM) during the production process of lithium battery separators. In order to improve the adsorption capacity of DCM under humid conditions, hydrophobic activated carbon fibers (ACFs) were modified by grafting hydrophobic alkyl chains via amide reaction of three organic acids (formic acid, n-octanoic acid and palmitic acid), respectively. The results showed that the hydrophobic long alkyl chains were successfully grafted onto the ACFs surface due to the formation of -CONH bonds. Moreover, the hydrophobicity of ACFs increases gradually with the increase of the carbon chain length. As a result, the DCM adsorption capacity of the pristine ACFs decreased by 85.8% when the relative humidity was gradually increased from 0% to 90%, whereas that of the P-N-ACF prepared with palmitic acid decreased by only 36.1%. In addition, the static adsorption, heat of adsorption, and activation energy of the modified ACFs were also investigated, which further confirmed that the hydrophobic alkyl chains in the modified ACFs played a dominant role in preventing the H2O molecules from combining with the oxygen-containing functional groups in the competitive adsorption of the DCMs with H2O. This study provides a new feasible strategy to improve the performance of ACFs for adsorbing DCM under high humidity conditions and explains the competitive adsorption behavior of DCM with H2O.

Modified graphene film powder scraps for re-preparation of highly thermally conductive flexible graphite heat spreaders

Ke Xiong[1], Zhipeng Sun[1], Tingting Yang[1], Wenming Qiao[1], Licheng Ling[1]

[1]State Key Laboratory of Chemical Engineering,
East China University of Science and Technology, Shanghai, 200237, China

The graphene film (GF) is extensively utilized for heat dissipation in portable electronic devices due to its lightweight and high thermal conductivity (K). However, during the die cutting process of GF, a significant amount of scraps with high K is inevitably generated. Therefore, it is worthwhile to investigate the secondary

utilization of this waste material. This contribution first used polydopamine (PDA) to surface-modify the crushed large-sized graphite nanosheets (GNs) powder scraps to obtain GNs@PDA. Subsequently, by self-assembling GNs@PDA with graphene oxide (GO) sheets and undergoing graphitization, we achieved a covalent bond-based "welding" effect between graphitized PDA and GNs that facilitated long-range ordered defect-free lamellar stacking structure formation. This structural arrangement effectively enhanced phonon and electron transfer within the resulting

graphene/GNs@PDA (G/GP) composite films, leading to significantly improved the K and electrical conductivity values. Specifically, the in-plane (IP) K and electrical conductivity of G/GP films reached 1301.8 W/mK and 5939.0 S/cm respectively. Furthermore, G/PG films also exhibit superior mechanical properties, foldability, and electromagnetic shielding performance. This work provides ideas for the preparation of flexible heat spreaders and realizes the reuse of GF scraps.

Controlling surface chemistry of activated carbon fibers for high product sulfate selectivity of H2S at room temperature catalytic oxidation

Yue Liu[1], Jianxiao Yang [1], Xin Zhang [2]

[1]Hunan University, Changsha 410082, China

[2]University of Chinese Academy of Sciences, Beijing, 101408, China

In recent years, the issue of odor control has become increasingly significant due to the rise in civil complaints and concerns regarding environmental and human health. Wastewater treatment plants located in cities are a major source of malodorous gases such as H_2S, which can cause great nuisance to neighboring communities. Hence, this work investigated the catalytic oxidation properties of H_2S at room temperature on a series of modified activated carbon fibers (ACFs) prepared from carbon fibers through NH_3 activation and non-thermal plasma (NTP) treatment in various modification atmospheres (N_2, O_2, Air). Among them, NH_3-ACF-A10 achieved a high sulfur capacity (0.21 gH_2S/g), the highest products sulfate selectivity (85.74%), and excellent performance in water scrubbing regeneration (82.98 %). During catalytic oxidation, oxygen-containing functional groups (OFGs) could react with HS^- and regenerate by O_2. $O_2\cdot^-$required for the reaction was facilitated by N-6 and sp^2 ring carbon defects. H_2O was responsible for transporting catalytic oxidation products and continuously exposing active sites. Lewis acid sites (COOH) were proved that were the specific sites affected products sulfate selectivity, rather than the strength of the acid sites. This work presented a novel methodology for the design of high-regeneration catalysts and provided a comprehensive and detailed theoretical framework for the further understanding of the catalytic oxidation effect of ACFs.

Photodegradation of toluene on UV-irradiated TiO2: Influence of molecular density of adsorption on Nanoporous Carbons

Juan Matos[1], Po S. Poon[2]

[1] Unidad de Cambio Climático y Medio Ambiente (UCCMA),
Instituto Iberoamericano de Desarrollo Sostenible (IIDS), Facultad de Arquitectura,
Construcción y Medio Ambiente, Universidad Autónoma de Chile, Temuco, Chile.
[2] Unidad de Desarrollo Tecnológico (UDT), Universidad de Concepción,
Barrio Universitario s/n, Concepción, Chile.

Toluene is one of the most frequently found VOCs pollutant in our indoor air environments. Chronic exposure to toluene affects respiratory and the central nervous system [1]. In this context, the influence of the molecular adsorption density of toluene on nanoporous carbons has been revisited and correlated with the photoactivity of TiO_2 in toluene photooxidation. The efficiency of the composite TiO_2-C was verified. Results showed that surface pH of carbons play a minor role while the textural and porosimetry properties strongly influence the photoactivity of the semiconductor. Results suggest that toluene adsorbed on the nanoporous carbon diffuses to TiO_2 particles where photooxidation takes place. In conclusion, TiO_2-nanoporous carbon was able to perform the photooxidation of toluene for two consecutive runs more efficiently than TiO_2 alone. Thus, combination of "adsorption on nanoporous carbon" and "oxidative photocatalysis on TiO_2" is a promising direction for indoor air purification by removing hazardous gaseous pollutants. The present findings permit the use of the molecular density of adsorption as a parameter to predict the photocatalytic activity of photoactive semiconductors to gaseous NOx and SOx. Accordingly, present studies are being conducted to verify the photoactivity of UV- and UV-vis irradiated TiO_2 in the degradation of these molecules.

Acknowledgements. This work was financially supported by Chilean National Agency for Research and Development (ANID) projects: ANILLO ATE220014, FONDEF ID23I10085 and FONDECYT 1220228.

References

[1] Martijn Rooseboom, et al. Regulatory Toxicology and Pharmacology 141 (2023) 105387.

Tailoring of Water State of Sulfonated Corncob for Boosting Solar Interfacial Evaporation

Weijun Zhong, Qian Yin, Wenke Gu, Congqian Pei, Feng Gao*, Jiangying Qu*, Yunhao Zang

Dongguan Key Laboratory of Low-Carbon Recycling and Utilization,
School of Environment and Civil Engineering, Dongguan University of Technology,
E-mail: qujianggaofeng@163.com, fenggao2003@163.com

Previous researches on carbon-based solar interfacial evaporation (SIE) have mainly focused on improving the solar absorption, neglecting the effect of water state on the SIE performance. Modulation of the ratio of intermediate water to free water (IW/FW) on sulfonated corncobs (SCCs) to improve the performance of evaporation devices was demonstrated [1]. The IW/FW ratio of SCCs in the range of 0.53–0.96 were obtained by tailoring the amount of hydrophilic sulfonic acid groups. The optimized SCC sample with a high IW/FW ratio of 0.96 exhibits the low enthalpy of water evaporation of 1800 J g−1, achieving the evaporation rates of 3.16 kg m−2 h−1, evaporation efficiency of 159% and salt resistance value of 1.62 g. Above results show that the sulfonic acid groups play an important role in activating the water state and conferring salt resistance. This study provides an effective technique for regulating water state for carbon-based SIE devices on the molecular level.

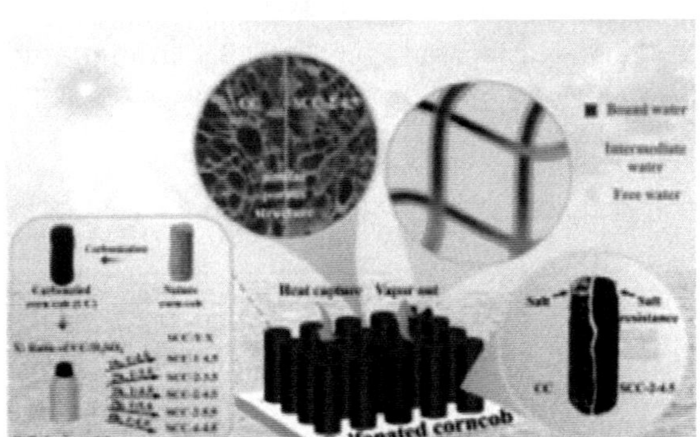

Fig.1 Schematic illustration of the fabrication of SCC-Y-X for solar-driven seawater desalination.

References

[1]Weijun Zhong, Feng Gao, Jiangying Qu, Yunhao Zang, Zhe Jiao, Carbon, 224, 119087

Oxygen functional group introduced activated carbon through ozone treatment for cesium ion adsorption

Eunseon Chae[1], Chung Gi Min[1], Chaehun Lim[1], Seoyeong Cheon[1], and Young-Seak Lee[1,2]

[1]Department of Chemical Engineering and Applied Chemistry, Chungnam National University, Daejeon, 34134, Republic of Korea
[2]Institute of Carbon Fusion Technology (InCFT), Chungnam National University, Daejeon, 34134, Republic of Korea

Cesium is harmful to the human body as it emits potent gamma radiation and shares a biological resemblance to potassium, potentially inducing cancer or DNA mutations. In this study, oxygen functional groups were introduced on activated carbons (ACs) through ozone treatment to augment the adsorption capacity of cesium. As the ozone treatment time increased, the oxygen contents on the ACs surface increased. Subsequently, the electrostatic interaction between ACs and cesium enhanced, resulting in higher cesium ion adsorption efficiency across all samples. In particular, the sample treated for 7 minutes had roughly 12% greater oxygen functional group content and achieved the highest cesium removal effectiveness (97.6%). Meanwhile, the sample treated for 5 minutes showed only a marginal 0.3% difference in cesium removal rate compared to that treated for 7 minutes. This slight difference was attributed to the surface chemical similarity between the two samples caused by the reactive characteristics of ozone gas. Nevertheless, the cesium adsorption performance of ozonated ACs seems to be primarily influenced by the amount of oxygen functional groups introduced to the surface, although the specific surface area and pore structure of the ACs also play crucial roles.

Synthesis of magnesium-based MOFs and their derived carbon materials for the catalytic oxidation of H2S at room temperature

Shengwei Chen[1], Jitong Wang [1,*] and Licheng ling [1*]

[1] State Key Laboratory of Chemical Engineering,
East China University of Science and Technology, Shanghai 200237, China

With the gradual increase of environmental protection awareness and requirements, the efficient removal and conversion of air pollutants has become the focus of environmental protection. Carbon materials have well-developed textural features and easily modified surface chemistry, making them effective catalysts for the catalytic oxidation of hydrogen sulfide (H2S) at room temperature. In this paper, magnesium-based MOFs are obtained as porous carbon materials with high MgO loading through the screening of synthesis conditions and precise morphology tuning, followed by one-pot solvothermal and one-step pyrolysis. The catalyst with 2D morphology shows the highest sulfur capacity and exhibits the best catalytic performance. This is mainly due to the huge external surface area of the 2D nanomaterial, which allows gas molecules to diffuse freely to be efficiently trapped by the catalyst. At the same time the stacking space between the layers provides storage space for the solid sulfur of the product and avoids the loss of reactive sites. Proposed mechanism elucidates that the carbon matrix acts as an activator of oxygen, and MgO in the aqueous film dissociates H2S into HS-, and the two react to produce sulfur.

Photo-responsive Surface Molecularly Imprinted Adsorbent for Efficient Phenol Removal and Recovery from Wastewater

Boxi Jing[1], Lei Qin[1,2], Yong Zhong[1], Min Zhang[1], Yongzhen Yang[1,2] and Xuguang Liu[3]

[1] Key Laboratory of Interface Science and Engineering in Advanced Materials, Ministry of Education, Taiyuan University of Technology, No.79 West Yingze Street, Taiyuan 030024, China
[2] Shanxi-Zheda Institute of Advanced Materials and Chemical Engineering, No. 67 Zhengyang Street, Taiyuan 030600, China
[3] College of Materials Science and Engineering, Taiyuan University of Technology, No.79 West Yingze Street, Taiyuan 030024, China

The introduction of photo-responsive materials facilitates the adsorption and recovery of molecularly imprinted polymer SMIP on the powder surface. In this work, carbon microspheres were prepared by hydrothermal method, and then TiO2 was loaded through magnetic stirring to prepare C-TiO2 carrier material. Then, C-TiO2 synthesized a photo responsive surface molecularly imprinted adsorbent using o-phenylenediamine as the functional monomer, trimethylolpropane trimethylacrylate as the crosslinking agent, and phenol as the template molecule through a series of functionalization, grafting, and elution processes (SMIP@C-TiO2) . During adsorption, SMIP@C-TiO2 Under UV excitation, hydrophilic, improving mass transfer rate; Stop UV after adsorption, under the action of hydrophobic crosslinking agent, SMIP@C-TiO2 Hydrophobic, improving solid-liquid separation efficiency. The morphology and structure of the product were characterized using methods such as field emission scanning electron microscopy, transmission electron microscopy, X-ray diffraction, and Fourier transform infrared spectroscopy. Using a UV visible spectrophotometer to test adsorption capacity, SMIP@C-TiO2 Showing a certain adsorption capacity (24.90 mg g-1). And use a contact angle measuring instrument to measure the range of UV/dark water contact angle changes from 60 to 130 °, and can achieve multiple wettability switching, which is helpful for the recovery and reuse of adsorbates and adsorbents.

Preparation and characteristics of high strength activated carbon fiber derived from polyimide by dry-jet wet spinning

Jeong-Seo Park, Yong-Sik Chung

Jeonbuk National University, 567, Baekje-daero, Deokjin-gu, Jeonju-si, Jeonbuk-do, Republic of Korea1

The correlation between the imidization degree of polyimide fibers and characteristics of activated carbon fibers was studied. Poly(amic acid) solution was spun by dry-jet wet spinning method and poly(amic acid) fibers were transferred into polyimide fibers by heat treatment of IR heater at 300℃, 400℃, 500℃, 600℃. Polyimide fibers were carbonized and activated at 900℃. The imidization degree of polyimide fibers increased up to 500℃ but decreased at 600℃. The thermal and mechanical properties of polyimide fiber exhibited a dependency on the imidization degree. The mechanical properties and compact morphologies of activated carbon fiber strongly depended on the imidization degree of polyimide fiber. Activated carbon fiber was showed the maximum tensile strength value of 898 MPa and specific surface area composed of micropores value of 920 m2/g at IR heater temperature of 500℃ and the imidization degree of 96%.

Fabrication and Properties of Polyimide-Based Graphite Fibers

Byoung-Min Lee, Yong-Sik Chung

Jeonbuk National University, 567, Baekje-daero, Deokjin-gu, Jeonju-si, Jeonbuk-do, Republic of Korea1

Some polyimide(PI)-based carbon fibers and films have been reported even with high precursor cost. PI as precursor showed high carbon yield and extremely high thermal and electrical conductivity when carbonized and graphitized. PI fibers were prepared by wet spinning of polyamic acid and subsequent thermal imidization. The fibers were carbonized in the presence of polydimethylsiloxane(PDMS) and finally graphitized at 2,200°C. The addition of PDMS altered the pyrolysis behavior and residue of PI fibers. PDMS pretreatment improved tensile strength and electrical conductivity of the resultant carbon fibers. Another important effect of PDMS was that the carbon fibers with PDMS were more flexible and much softer than PI fibers alone.

Construction of mesocarbon microbeads/polyimide ordered skeleton for the fabrication of thermal interface materials

Zhipeng Sun[1], Ke Xiong[1], Wenming Qiao[1], Jitong Wang[1], Licheng Ling[1]

[1] State Key Laboratory of Chemical Engineering,
East China University of Science and Technology, Shanghai, 200237, China

With the development of the 5G era, smart electronic products are constantly upgrading and iterating, followed by the increasing power density and heat production of the chip. To keep electronic components " cool " constantly, one of the most convenient and effective method is to utilize thermal interface materials (TIMs) assisting in heat dissipation. Mesocarbon microbeads (MCMB) have excellent properties such as high thermal conductivity and excellent thermal stability, which can be utilized as thermally conductive fillers for the fabrication of novel TIMs. In this work, the PM (PI/MCMB) foam was obtained by directional freezing and high temperature thermal annealing technique. The radial arrangement of crystals in MCMB making PI easy to assists the directional alignment of MCMB during the composite process through directional freezing, which is conducive to the transfer of phonons. This promotes an enhancement in the through-plane (TP) thermal conductivity (K) of PM foam. Consequently, anisotropic PDMS/PM composite was fabricated by backfilling polydimethylsiloxane (PDMS), which has a high TP K of 9.97 W/mK and meets the development requirements of smart electronic products.

Preparation of hydrophobic activated carbon fibers and their adsorption behavior for dichloromethane

LinHong Wen[1], Jianxiao Yang[1], Xin Zhang[2]

[1] Hunan University, Changsha 410082, China

[2]University of Chinese Academy of Sciences, Beijing, 101408, China

In recent years, it has become necessary to prepare hydrophobic materials for adsorbing the emitted dichloromethane (DCM) during the production process of lithium battery separators. In order to improve the adsorption capacity of DCM under humid conditions, hydrophobic activated carbon fibers (ACFs) were modified by grafting hydrophobic alkyl chains via amide reaction of three organic acids (formic acid, n-octanoic acid and palmitic acid), respectively. The results showed that the hydrophobic long alkyl chains were successfully grafted onto the ACFs surface due to the formation of -CONH bonds. Moreover, the hydrophobicity of ACFs increases gradually with the increase of the carbon chain length. As a result, the DCM adsorption capacity of the pristine ACFs decreased by 85.8% when the relative humidity was gradually increased from 0% to 90%, whereas that of the P-N-ACF prepared with palmitic acid decreased by only 36.1%. In addition, the static adsorption, heat of adsorption, and activation energy of the modified ACFs were also investigated, which further confirmed that the hydrophobic alkyl chains in the modified ACFs played a dominant role in preventing the H2O molecules from combining with the oxygen-containing functional groups in the competitive adsorption of the DCMs with H2O. This study provides a new feasible strategy to improve the performance of ACFs for adsorbing DCM under high humidity conditions and explains the competitive adsorption behavior of DCM with H2O.

Facile synthesis of ultrathin polymer-derived porous carbon nanostructures on zeolite template for high-performance capacitive energy storage and desalination

Keyang Li [1] and Qianming Gong [1,*]

[1]School of materials science and engineering, Tsinghua University, Beijing, China

Ultrathin porous carbon nanostructures are considered potent alternatives for electrode materials for electric double-layer capacitors (EDLCs) and capacitive deionization devices due to their high electrochemical stability, abundant pore structure, as well as excellent conductivity. Here in this work, two facile strategies were designed to obtain novel carbon nanostructures by using ZSM-5, one of the most widely applied zeolite materials as templates [1-2]. For strategy I, in-situ growth of ultrathin resorcinol-formaldehyde resin on ZSM-5 surface was realized and after the carbonization/alkali-etch procedures, intact nitrogen-doped hollow carbon nanoboxes (HCNBs) with thin shell and zeolite-like porous structure were obtained, which exhibited typical EDLCs performance and high deionization capacity of 32.3 mg g−1 as well as high salt removal retention ability up to 95.40% after 50 cycles [1]; For strategy II, by co-sintering of blended fluororesin/ZSM-5 composite, carbon nanosheets (CNSs) with excellent capacitance (382.4 F·g-1 at 0.5 A·g-1) and capacitance retention ability (98.38% after 10000 cycles at 10A·g-1) can be obtained. Moreover, due to the high porosity and Lewis-acidic-site-abundance of ZSM-5, no significant signals of hazardous fluorinated-hydrocarbon species were detected during the sintering, which enabled this method as a powerful wasted-fluoroplastics conversion strategy [2]. Possessed with high scalability and extendibility, zeolite-templated carbon nanostructures own the potential for promising applications in capacitive energy storage desalination in the future.

References

[1] Keyang Li, et al. Journal of Materials Chemistry A 12 (2024) 899-910

[2] Keyang Li, et al. Unpublished article.

Freeze-casting-based fabrication of zwitterion-decorated carbon nanotubes/ZIF-8 composite aerogel beads for effective protein-bound uremic toxins removal

Keyang Li[1], Chong Wang [2], Yuehong Li [2*] and Qianming Gong [1,*]

[1]School of materials science and engineering, Tsinghua University, Beijing, China
[2]Beijing Tsinghua Changgung Hospital, School of Clinical Medicine, Tsinghua University, Beijing, China

Protein-bound toxins are considered severe risk factors for chronic renal disease patients as they are closely related to severe cardiovascular complications and highly related to the risk of cardiovascular death in uremic patients. Blood purification therapy based on the highly selective adsorption characteristics of porous adsorption materials has become a novel and potent strategy for toxin removal for uremic patients [1]. Based on the characteristic that zwitterions can form a hydrated layer in the liquid phase to counteract albumin adsorption, a type of oxidized trimethylamine (TMAO) zwitterionic structure was covalently grafted onto carbon nanotubes through atomic transfer radical polymerization (ATRP), and composite with zeolite imidazole ester skeleton-8 (ZIF-8) nanoparticles with abundant porous structure [2]. Then composite aerogel beads with rich radial macropores were facilely prepared by using the freez-casting method, which exhibited excellent removal efficiency towards indole-3-acetic acid (IAA, 67.24 mg·g-1,), hippuric acid (HA, 147.63 mg·g-1), and indole-3-phenol sulfate (IS, 74.29 mg·g-1), along with low protein adsorption efficiency (~10 mg·g-1), desirable blood compatibility and low cytotoxicity. This adsorption material exhibits high potential as a novel alternative for blood purification devices and portable artificial kidneys for future renal disease therapy [2].

References

[1] Xiaohuan Zhao, et al. Chemical Engineering Journal. 454 (2023) 140213
[2] Keyang Li, et al. Unpublished work.

Spherical Carbon synthesized from bio-alcohol derived polymer for efficient chloroform removal from drinking water

Heon-Sik Byun[1], Jeong-Yeol Jo[1], Ki-Hoon Park[1], Jihn-Koo Lee[1], Chae-Ho Shin[1],

[1]PureSphere Co., Ltd., Nonsan, Chungnam 33020, Republic of Korea

In this study, we discuss the removal performance of chloroform (CHCl3) in water using furfuryl alcohol-based polymer spherical activated carbon (PureCarbon) extracted from bio-alcohol. PureCarbon properties analysis and chloroform removal experiments were compared based on commercialized carbon blocks and two types of polyacrylonitrile-based activated carbon fibers (PAN-ACF). To remove insoluble chloroform residual in water spherical type PureCarbon showed the most suitable physicochemical properties. In the chloroform removal test (based on 2000 L water), PureCarbon showed 97% removal performance.

Table 1. Physical characteristics of PureCarbon, Carbon block filter, PAN-ACF-1, 2.

Sample	Surface area (m^2 g^{-1})			Pore volume (cm^3 g^{-1})		Pore diameter [d]
	BET	Micropore [a]	External [b]	Total [c]	Micropore [a]	(nm)
PureCarbon	1081	1070	11	0.454	0.431	1.678
Carbon block	396	339	63	0.234	0.134	2.363
PAN-ACF-1	984	904	80	0.565	0.449	2.314
PAN-ACF-2	797	785	12	0.344	0.329	1.728

a Calculated from t-plot micropore method, b Calculated from BET surface area minus micropore area, c measured at p/p0 0.99, d a maximum pore diameter from BJH pore size distribution calculated from N2 adsorption isotherm branch.

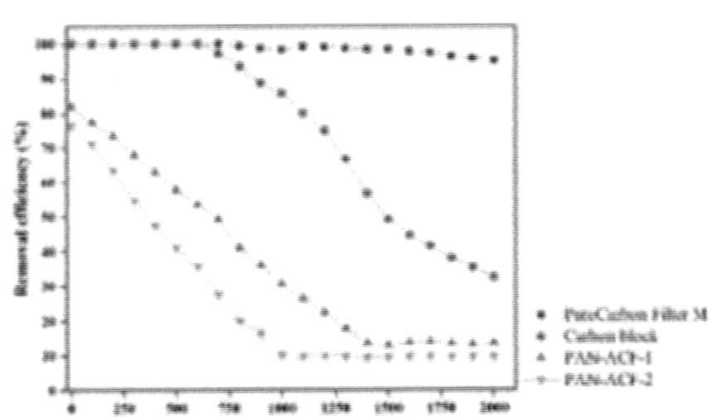

Figure 1. Result of chloroform removal performance of water purifier filter. The pH was adjusted for 7.0 and the initial concentration was 0.25 mg/L. The water flow pressure was 1.0 kgf·cm-2.

References

[1] Ivancev-Tumbas, I., et al., Water Research, 1999. 33(18): p. 3715-3722.
[2] Nakamura, T., et al., Toxicological & Environmental Chemistry, 1996. 57(1-4): p. 187-192.

Design of hydrogel with wound exudation inducing power generation and its tissue repair performance

Weiwen Liang[1], Qian Ma[2], Yuanbin Wang[1], Rongkang Huang[1], Hui Wang[1], Dingcai Wu[2*]

[1]The Sixth Affiliated Hospital of Sun Yat sen University

[2]School of Chemistry, Sun Yat sen University

The prolonged healing of chronic wounds is a major challenge in clinical treatment. The traditional treatment concept for wounds is mainly dry healing or wet healing. However, very few studies have focused on eliminating the inferior and retaining the good in wound exudate. In view of this, we developed a heterogeneous gel (HG) driven by percolation to generate electricity. Graphene oxide grafted polyacrylic acid molecular brush was compounded with amine rich gel network to build an external gel electrode with ion exchange function and high capacitance; Glucuronolactone was compounded with hydrophilic gel network to construct a hygroscopic inner gel electrode; The inner and outer gel are tightly combined to form an integrated HG through the imine bond. The HG can instantly generate an electric field by absorbing seepage, and 100 μ L 0.01M PBS can drive the HG to generate an electric field of 180 mV mm-1, with ultra fast moisture absorption and power generation performance. In the full-thickness skin defect model of diabetes rats, it showed obvious healing promoting effect.

A cotton-like 3D graphene fiber aerogel prepared by wet-electrospinning for highly sensitive pressure sensor

Tingting Yang[1], Wenming Qiao[1]

[1] State Key Laboratory of Chemical Engineering,
East China University of Science and Technology, Shanghai, China

The demand for elastomeric graphene aerogels with high sensitivity and reliable mechanical properties has grown rapidly in conjunction with the accelerated development of wearable electronic devices. This research presents a novel fibrous graphene aerogel that displays remarkable flexibility, elasticity, and fatigue resistance.

The preparation process involves wet-electrospinning to obtain graphene oxide/polyacrylic acid sodium (GO/PAAS) fiber assembles, followed by molding, freeze-drying, and thermal annealing. The obtained product is a cotton-like three-dimensional graphene fiber aerogel (GFA) with a soft exterior and a firm inner structure. Macroscopically, GFA is an aerogel made of randomly stacked 1D graphene fibers. Microscopically, these fibers become stronger after high-temperature thermal annealing, which makes the aerogel more resistant to fatigue and more elastic. The fabricated pressure sensor has a high sensitivity of 18.55 kPa m-1 and a low detection limit of 2 Pa, a fast response time of 2 ms, and good stability even after 3000 cycles. The extraordinary properties of GFA indicate great potential for applications in medical monitoring and motion detection. Furthermore, the wet-electrospinning preparation of GFA represents a novel approach to the fabrication of graphene aerogels.

A novel MnO@C aerogels for electromagnetic wave absorption and infrared stealth

Shi Jin[1], Xianlong Zhang[1,2], Xueping Wu[1,2,3]

[1]School of Chemistry and Chemical Engineering, Hefei University of Technology, Hefei 230009, China
Anhui Province Key Laboratory of Value-Added Catalytic Conversion and Reaction Engineering,
[2]School of Chemistry and Chemical Engineering, Hefei University of Technology, Hefei 230009, China
[3]Engineering Research Center of Advanced Composite Materials Design& Application of Anhui Province, Hefei 230009, China

Given the growth of military detection chnologies, it has been increasingly essential for developing the materials that present efficient electromagnetic wave absorption (EMWA) and provide thermal stealth for radar-infrared compatibility [1-3]. In this study, the honeycomb-like MnO@C aerogels were fabricated by sol-gel method. The graded porous structure enhances impedance matching and efficiently dissipates electromagnetic waves through multiple loss mechanisms, including reflection and scattering [4,5]. At optimal matching thickness of 4.9 mm, the MnO@C-12 wt.% (Mn content is 12 wt.%) achieves a minimum reflection loss (RLmin) of -48.27 dB. Moreover, maximum effective absorption bandwidth (EAB) at 3.5 mm is 4.6 GHz. After being heated on a plate at 90 oC for 1 hour, the surface color of MnO@C-12wt.% is consistent with the surrounding. The synergistic interaction between manganese oxides (MnO) and carbon materials presents an innovative paradigm for the development of multifunctional materials with integrated radar and infrared stealth functionalities.

References

[1] X. Chen, et al. Carbon 213 (2023) 118313
[2] M. Zhou, et al. Nano-Micro Lett. 15 (2023) 176
[3] M. Zhang, et al. Appl. Surf. Sci. 604 (2022) 154626
[4] J. Lu, et al. Int. J. Biol. Macromol. 254 (2024) 127851
[5] S. Li, et al. J. Mater. Sci. Technol. 133 (2023) 238-248

Novel visible-NIR sensitive organic photodetectors for high resolution imaging and high-fidelity optical communications

Zilong Ye[1], Haoyu Huang[1,2], Wei He[1], Yuan Li[1], Kefei Shi[1,2], Kai-Ning Tong[1], Zhuhua Xu[1,2], Huan Li[1], Xin Wang[1], Zonglong Zhu[3], Yuze Lin[4], Hongyan Fu[1,2,*], and Guodan Wei[1,2,*]

[1]Institute of Materials Research, Tsinghua Shenzhen International Graduate School, Tsinghua University, Shenzhen, China
[2]Tsinghua-Berkeley Shenzhen Institute, Tsinghua University, Shenzhen, China
[3]Department of Chemistry, City University of Hong Kong, Hong Kong, China.
[4]Beijing National Laboratory for Molecular Sciences, CAS Key Laboratory of Organic Solids, Institute of Chemistry, Chinese Academy of Sciences, Beijing, China.

Sensing near-infrared (NIR) photons using organic semiconductors utilizes its intriguing properties including structure tunability, bio-compatibility and cost-effectiveness, paving pathway for large scale integration and diverse applications. In this study, high-performance narrowband organic photodetectors (OPDs) were constructed using resonant cavity, achieving FWHM as low as 25 nm at 910 nm in NIR, specific detectivity close to 1×1012 Jones, LDR of more than 140 dB, and a low response time as low as 440 ns. Combining FDTD simulation and experimental validation, an arbitrary tuning of visible-NIR peaks is achieved by adjusting the thickness of the cavity. The device was successfully applied to a single-pixel near-infrared image sensing system and achieved high-quality NIR imaging of 20×300 pixels. This research further constructed an organic short-wave infrared (SWIR) photodetector with wavelength response exceeding 1600 nm, which achieved optical communication in the 1550 nm of OPDs for the first time. Using the PAM4 modulation, the device successfully demonstrated high-fidelity transmission of 600×600 colored image and text, with a transmission rate of more than 1 MHz, indicating the excellent performance and broad prospects for novel OPDs in NIR and SWIR to achieve high-quality imaging and optical communications.

A Graphene Charge Transfer Transistor for High-speed Operation

Xinzhe Wang [1,2], Xiaoyue Wang [1,2], Yue Kong [1,3], Dongming Sun [1,2] and Chi Liu [1,2]

[1]Shenyang National Laboratory for Materials Science, Institute of Metal Research, Chinese Academy of Sciences, 72 Wenhua Road, Shenyang, 110016, China.

[2]School of Materials Science and Engineering, University of Science and Technology of China, 72 Wenhua Road, Shenyang, 110016, China.

[3]School of Information Science and Engineering, Northeastern University, Shenyang 110819, China.

The communication technology has driven the information age. While 5G technology has profoundly changed people's lives, future 6G technology demands transistors with cut-off frequencies over 1 THz, which is challenging for traditional bulk material technologies because of the limitation of channel length or base thickness [1]. Graphene, with its atomic thickness and ultra-high carrier mobility, can combine with bulk material to form various barriers for carrier modulation, potentially providing ultra-high-speed devices with novel mechanisms [2]. Here, we report a graphene charge transfer transistor composing of two graphene-germanium Schottky junctions. By accelerating carries in the graphene to their high energy state, hot carriers can be emitted from the modulated potential barrier into the collector. Thanks to the atomic thickness of graphene, the transistor shows a potential to work in the THz domain. In experiments, the device shows a common-emitter current gain of more than 10 and a power gain approaching 1000. A cut-off frequency (fT) and a maximum oscillation frequency (fMAX) up to several GHz has also been achieved. Optimization are needed in the future including using an insulator substrate in the future. Our charge transfer transistor offers a new option to fabricate ultra-high-speed transistors.

References

[1] International Roadmap for Devices and Systems. IEEE (2022).

[2] Liu Chi et al. Nat. Commun. 10 (2019) 1-7.

A high-sensitivity ammonia sensor based on carbon nanotubes and conductive protein nanowires

Yihan Guo[1,3], Mengjie Zhuang[1,2], Yun Sun[1,2], Huiming Cheng[1,2,3], Dongming Sun[1,2]

[1]Shenyang National Laboratory for Materials Science, Institute of Metal Research, Chinese Academy of Sciences, 72 Wenhua Road, Shenyang, 110016, China.

[2]School of Materials Science and Engineering, University of Science and Technology of China, 72 Wenhua Road, Shenyang, 110016, China.

[3]School of Physical Science and Technology, ShanghaiTech University, 393 Huaxia Middle Road, Shanghai, 200120, China.

In the field of biomedicine, the concentration of ammonia in the human breath can reflect the health level of the body, so high-sensitivity and selectivity ammonia sensors are necessary for health monitoring. Carbon nanotubes (CNTs) and conductive protein nanowires (Pili) show great potential in the field of gas sensors due to their unique one dimensional structure and electrical properties. In this study, an ammonia sensor based on CNT-Pili composite was proposed with high sensitivity and high selectivity for ammonia detection taking advantage of the high conductivity of CNT and the special biological function of Pili, which has a good application prospect in the fields of medical health, environmental monitoring and industrial safety.

Research on the Rapid Temperature Change Characteristics of Carbon Nanotube Transistors Based on Thermoelectric Coolers

Tong-Jian Liu[1,3], Wei-YuNing[1,2], Xin Yan[1,3], Kai-Ping Tai[1,2], Dong-Ming Sun[1,2]*

[1]Shenyang National Laboratory for Materials Science, Institute of Metal Research, Chinese Academy of Sciences, 72 Wenhua Road, Shenyang, 110016, P. R. China

[2]School of Material Science and Engineering, University of Science and Technology of China, 72 Wenhua Road, Shenyang, 110016, P. R. China.

[3]School of Information Science and Engineering, Northeastern University, Shenyang 110819, China

As electronic device integration increases and device sizes shrink, power density and operating temperatures rise, significantly impacting device performance due to variations in ambient temperature. Carbon nanotube field-effect transistors (CNTFETs), with their high speed, low power consumption, and thermal stability, are poised to become the mainstream technology in next-generation integrated circuits. To enhance device stability, broaden application scenarios, and optimize energy efficiency, it is crucial to deeply study the temperature effects on CNTFET performance. Recently, thermoelectric coolers based on the Peltier effect, which convert electrical energy into thermal energy, have been used for precise temperature control by adjusting the working voltage and current. These coolers offer fast response, simple structure, and low maintenance, enabling optimized integration with electronic devices.

In this study, a method for regulating the performance of carbon nanotube field-effect transistors compatible with thermoelectric coolers was proposed for the first time. This method enables rapid and precise temperature control through thermoelectric devices. A systematic investigation of the electrical performance changes in CNTFETs across various temperature ranges was conducted. Reversible polarity transitions were observed at different temperatures. Additionally, electrical performance parameters such as threshold voltage, subthreshold swing, current on/off ratio, and transconductance of CNTFETs under different temperature conditions were obtained. These findings offer new insights for the integration and application expansion of thermoelectric devices while providing novel methods for the performance regulation of carbon nanotube transistors.